Benchmark Papers
in Geology

Series Editor: Rhodes W. Fairbridge
Columbia University

A selection from the published volumes in this series

Volume

31 PALEOBIOGEOGRAPHY / *Charles A. Ross*

46 PALYNOLOGY, PART I: Spores and Pollen / *Marjorie D. Muir and William A. S. Sarjeant*

47 PALYNOLOGY, PART II: Dinoflagellates, Acritarchs, and Other Microfossils / *Marjorie D. Muir and William A. S. Sarjeant*

76 TERRESTRIAL TRACE FOSSILS / *William A. S. Sarjeant*

78 NANNOFOSSIL BIOSTRATIGRAPHY / *Bilal U. Haq*

79 CALCAREOUS NANNOPLANKTON / *Bilal U. Haq*

Related Titles

BIOSTRATIGRAPHY OF FOSSIL PLANTS: Successional and Paleoecological Analyses / *David L. Dilcher and Thomas N. Taylor*

CONCEPTS AND METHODS OF BIOSTRATIGRAPHY / *Erle G. Kauffman and Joseph E. Hazel*

ECOLOGIC ATLAS OF BENTHIC FORAMINIFERA OF THE GULF OF MEXICO / *C. Wylie Poag*

THE ENCYCLOPEDIA OF PALEONTOLOGY / *Rhodes W. Fairbridge and David Jablonski*

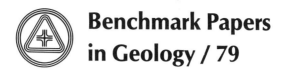

**Benchmark Papers
in Geology / 79**

A BENCHMARK® Books Series

CALCAREOUS NANNOPLANKTON

Edited by
BILAL U. HAQ
Exxon Production Research Company

Hutchinson Ross Publishing Company

Stroudsburg, Pennsylvania

This volume is dedicated to the memory of M. N. Bramlette, Georges Deflandre and Ervin Kamptner, investigators par excellence, *and founders of modern nannopaleontology.*

LIBRARY OF CONGRESS CATALOGING IN PUBLICATION DATA
Main entry under title:
Calcareous nannoplankton.
   (Benchmark papers in geology; 79)
   Includes bibliographical references and indexes.
   1. Coccoliths, Fossil—Addresses, essays lectures.
2. Plankton, Fossil—Addresses, essays, lectures.
I. Haq, Bilal U. II. Series.
QE955.C34   1983   561'.93   83-4366
ISBN 0-87933-090-2

Distributed worldwide by Van Nostrand Reinhold Company Inc., 135 W. 50th Street, New York, NY 10020.

CONTENTS

Series Editor's Foreword ix
Preface xi
Contents by Author xiii

Introduction: Historical Review of Calcareous Nannoplankton Research 1

PART I: EARLY DEBATE ABOUT THE NATURE OF COCCOLITHS

Editor's Comments on Papers 1 Through 4 12

1   WALLICH, G. C.:   Remarks on Some Novel Phases of Organic Life, and
          on the Boring Powers of Minute Annelids, at Great Depths
          in the Sea 14
              Ann. Mag. Nat. Hist., ser. 3, **8:**52-58 (1861)

2   SORBY, H. C.:   On the Organic Origin of the So-called 'Crystalloids' of
          the Chalk 21
              Ann. Mag. Nat. Hist., ser. 3, **8:**193-200 (1861)

3   HUXLEY, T. H.:   On Some Organisms Living at Great Depths in the
          North Atlantic Ocean 29
              Quart. Jour. Microsc. Sci. **8:**203-212 (1868)

4   WALLICH, G. C.:   Observations on the *Coccosphere* 41
              Ann. Mag. Nat. Hist., ser. 4, **19:**342-348, 349-350 (1877)

PART II: STRUCTURAL, BIOLOGICAL, BIOGEOGRAPHIC, AND PRESERVATIONAL ASPECTS OF NANNOPLANKTON

Editor's Comments on Papers 5 Through 16 52

5   BLACK, M.:   The Fine Structure of the Mineral Parts of
          Coccolithophoridae 58
              Linnean Soc. London Proc. **174:**41-46 (1963)

6   PAASCHE, E.:   Biology and Physiology of Coccolithophorids 65
              Ann. Rev. Microbiol. **22:**71-86 (1968)

7   WATABE, N., and K. M. WILBUR: Effects of Temperature on Growth,
          Calcification, and Coccolith Form in *Coccolithus Huxleyi*
          (Coccolithineae) 81
              Limnology and Oceanography **11:**567-575 (1966)

8   McINTYRE, A., A. W. H. BÉ, and M. B. ROCHE: Modern Pacific
          Coccolithophorida: A Paleontological Thermometer 90
              New York Acad. Sci. Trans., ser. 2, **32:**720-731 (1970)

Contents

9 OKADA, H., and S. HONJO: The Distribution of Oceanic
Coccolithophorids in the Pacific   102
  Deep-Sea Research **20:**355-374 (1973)

10 BRAMLETTE, M. N.: Significance of Coccolithophorids in Calcium-
Carbonate Deposition   124
  Geol. Soc. America Bull. **69:**121-126 (1958)

11 BERGER, W. H.: Deep-Sea Carbonates: Evidence for a Coccolith
Lysocline   129
  Deep-Sea Research **20:**917-921 (1973)

12 ROTH, P. H., M. M. MULLIN, and W. H. BERGER: Coccolith
Sedimentation by Fecal Pellets: Laboratory Experiments and
Field Observations   134
  Geol. Soc. America Bull. **86:**1079-1084 (1975)

13 HONJO, S.: Coccoliths: Production, Transportation and
Sedimentation   140
  Marine Micropaleontology **1:**65-79 (1976)

14 THIERSTEIN, H. R.: Selective Dissolution of Late Cretaceous and
Earliest Tertiary Calcareous Nannofossils: Experimental
Evidence   155
  Cretaceous Research **2:**165-176 (1980)

15 WISE, S. W., Jr., and K. R. KELTS: Inferred Diagenetic History of a
Weakly Silicified Deep Sea Chalk (Abstract)   167
  Gulf Coast Assoc. Geol. Socs. Trans. **22:**177 (1972)

16 NOËL, D., and H. MANIVIT: Nannofaciès de "black shales" aptiennes
et albiennes d'Atlantique sud (legs 36 et 40). Intérêt
sédimentologique   168
  Soc. Géol. France Bull. **20:**491-502 (1978)

**PART III: PALEOBIOGEOGRAPHY AND PALEOENVIRONMENTAL
APPLICATIONS**

Editor's Comments on Papers 17 Through 28   182

17 McINTYRE, A.: Coccoliths as Paleoclimatic Indicators of Pleistocene
Glaciation   186
  Science **158:**1314-1317 (1967)

18 GEITZENAUER, K. R.: Coccoliths as Late Quaternary Palaeoclimatic
Indicators in the Subantarctic Pacific Ocean   190
  Nature **223:**170-172 (1969)

19 McINTYRE, A., W. F. RUDDIMAN, and R. JANTZEN: Southward
Penetrations of the North Atlantic Polar Front: Faunal and
Floral Evidence of Large-scale Surface Water Mass
Movements over the Last 225,000 Years (Abstract)   193
  Deep-Sea Research **19:**61 (1972)

20 RUDDIMAN, W. F., and A. McINTYRE: Northeast Atlantic Paleoclimatic
Changes over the Past 600,000 Years (Abstract)   194
  Geol. Soc. America Mem. 145, 1976, pp. 111-112

21 HAQ, B. U., and G. P. LOHMANN: Early Cenozoic Calcareous
Nannoplankton Biogeography of the Atlantic Ocean
(Abstract)   195
Marine Micropaleontology **1**:119-120 (1976)

22 HAQ, B. U., I. PREMOLI-SILVA, and G. P. LOHMANN: Calcareous
Plankton Paleobiogeographic Evidence for Major Climatic
Fluctuations in the Early Cenozoic Atlantic Ocean   197
Jour. Geophys. Research **82**:3861-3865, 3866-3867, 3869,
3871, 3873-3876 (1977)

23 HAQ, B. U.:   Biogeographic History of Miocene Calcareous
Nannoplankton and Paleoceanography of the Atlantic Ocean   207
Micropaleontology **26**:414-432, 442-443 (1980)

24 ROTH, P. H., and J. L. BOWDLER: Middle Cretaceous Calcareous
Nannoplankton Biogeography and Oceanography of the
Atlantic Ocean (Abstract)   228
Soc. Econ. Paleontologists and Mineralogists Spec. Pub. 32, p. 517 (1981)

25 BUKRY, D.:   Coccoliths as Paleosalinity Indicators—Evidence from
Black Sea   229
Am. Assoc. Petroleum Geologists Mem. 20, 1974, pp. 353-354, 356,
358, 360, 362-363

26 MARGOLIS, S. V., P. M. KROOPNICK, D. E. GOODNEY, W. C. DUDLEY, and
M. E. MAHONEY: Oxygen and Carbon Isotopes from
Calcareous Nannofossils as Paleoceanographic Indicators   236
Science **189**:555-557 (1975)

27 GOODNEY, D. E., S. V. MARGOLIS, W. C. DUDLEY, P. KROOPNICK, and
D. F. WILLIAMS: Oxygen and Carbon Isotopes of Recent
Calcareous Nannofossils as Paleoceanographic Indicators
(Abstract)   239
Marine Micropaleontology **5**:31 (1980)

28 DUDLEY, W. C., J. C. DUPLESSY, P. L. BLACKWELDER, L. E. BRAND, and
R. R. L. GUILLARD: Coccoliths in Pleistocene-Holocene
Nannofossil Assemblages   240
Nature **285**:222-223 (1980)

PART IV: ORIGIN AND EVOLUTION OF NANNOPLANKTON

Editor's Comments on Papers 29 Through 33   244

29 DEFLANDRE, G.:   Présence de nannofossiles calcaires (coccolithes et
Incertae sedis) dans le Siluro-dévonien d'Afrique du Nord   248
Acad. Sci. Comptes Rendus, ser. D, **270**:2916-2921 (1970)

30 BUKRY, D.:   *Discoaster* Evolutionary Trends   258
Micropaleontology **17**:43-52 (1971)

31 HAQ, B. U.:   Evolutionary Trends in the Cenozoic Coccolithophore
Genus *Helicopontosphusaera*   268
Micropaleontology **19**:32-52 (1973)

32 GARTNER, S., and D. BUKRY: Morphology and Phylogeny of the
Coccolithophycean Family Ceratolithaceae   289
U.S. Geol. Survey Jour. Research **3**:451-454, 455, 460 (1975)

Contents

33 **HAQ, B. U.:** Transgressions, Climatic Change and the Diversity of Calcareous Nannoplankton 295
Marine Geology **15:**M25–M30 (1973)

PART V: NANNOPLANKTON AND THE CRETACEOUS/TERTIARY BOUNDARY

Editor's Comments on Papers 34 Through 37 302

34 **BRAMLETTE, M. N.:** Massive Extinctions in Biota at the End of Mesozoic Time 305
Science **148:**1696–1699 (1965)

35 **WORSLEY, T. R.:** The Terminal Cretaceous Event 309
Nature **230:**318–320 (1971)

36 **THIERSTEIN, H. R., and H. OKADA:** The Cretaceous/Tertiary Boundary Event in the North Atlantic (Abstract) 312
Initial Reports of the Deep Sea Drilling Project **43:**601–616 (1979)

37 **ROMEIN, A. J. T.:** Calcareous Nannofossils from the Cretaceous/Tertiary Boundary Interval in the Barranco Del Gredero (Caravaca, Prov. Murcia, S.E. Spain). I 313
Koninkl Nederlandse Akad. Wetensch. Proc., ser. B, **80:**256–268, 277–279 (1977)

Author Citation Index 331
Subject Index 335
About the Editor 338

SERIES EDITOR'S FOREWORD

The philosophy behind the Benchmark Papers in Geology is one of collection, sifting, and rediffusion. Scientific literature today is so vast, so dispersed, and, in the case of old papers, so inaccessible for readers not in the immediate neighborhood of major libraries that much valuable information has been ignored by default. It has become just so difficult, or so time consuming, to search out the key papers in any basic area of research that one can hardly blame a busy person for skimping on some of his or her "homework."

This series of volumes has been devised, therefore, as a practical solution to this critical problem. The geologist, perhaps even more than any other scientist, often suffers from twin difficulties—isolation from central library resources and immensely diffused sources of material. New colleges and industrial libraries simply cannot afford to purchase complete runs of all the world's earth science literature. Specialists simply cannot locate reprints or copies of all their principal reference materials. So it is that we are now making a concerted effort to gather into single volumes the critical materials needed to reconstruct the background of any and every major topic of our discipline.

We are interpreting "geology" in its broadest sense: the fundamental science of the planet Earth, its materials, its history, and its dynamics. Because of training in "earthy" materials, we also take in astrogeology, the corresponding aspect of the planetary sciences. Besides the classical core disciplines such as mineralogy, petrology, structure, geomorphology, paleontology, and stratigraphy, we embrace the newer fields of geophysics and geochemistry, applied also to oceanography, geochronology, and paleoecology. We recognize the work of the mining geologists, the petroleum geologists, the hydrologists, and the engineering and environmental geologists. Each specialist needs a working library. We are endeavoring to make the task of compiling such a library a little easier.

Each volume in the series contains an introduction prepared by a specialist (the volume editor)—a "state of the art" opening or a summary of the object and content of the volume. The articles, usually some twenty to fifty reproduced either in their entirety or in significant extracts, are selected in an attempt to cover the field, from the key papers of the last century to fairly recent work. Where the original works are in foreign languages, we have endeavored to locate or commission translations. Geologists, because of their global subject, are often acutely aware of the oneness of our world. The selections cannot therefore be restricted to any one country, and whenever possible an attempt is made to scan the world literature.

To each article, or group of kindred articles, some sort of "highlight commentary" is usually supplied by the volume editor. This commentary should serve to bring that article into historical perspective and to emphasize its particular role in the growth of the field. References, or citations, wherever possible, will be reproduced in their entirety—for by this means the observant reader can assess the background material available to that particular author, or, if desired, he or she too can double check the earlier sources.

A "benchmark," in surveyor's terminology, is an established point on the ground that is recorded on our maps. It is usually anything that is a vantage point, from a modest hill to a mountain peak. From the historical viewpoint, these benchmarks are the bricks of our scientific edifice.

RHODES W. FAIRBRIDGE

PREFACE

Calcareous nannoplankton [*nano* (Greek)=dwarf] is a collective term that includes the minute, unicellular marine algae belonging to Class Coccolithophyceae, and the cooccurring, but morphologically dissimilar, forms of uncertain affinity, that have left behind a fossil record at least since early Jurassic.

The past two decades have seen great strides in our knowledge of fossil calcareous nannoplankton, their biostratigraphic utility, and the beginning of applications in paleoenvironmental research. The rapid advancement and continued growth in these fields and the induction of calcareous nannoplankton in micropaleontological curricula, has stimulated the demand for related literature. This Benchmark volume (and a companion volume on purely biostratigraphic aspects of nannofossils, *Nannofossil Biostratigraphy*) is a response to these needs. It attempts to trace the development of nannoplankton research from their discovery in the mid nineteenth century to the most recent applications in paleoecology.

The space constraints in an anthology such as this has meant that many of the included papers had to be shortened considerably; for others, only the abstracts could be included, and still many more could not be included at all. I have endeavored, however, to reproduce as much of the classic literature as the space permits, so as to furnish a succinct survey of the developmental history of calcareous nannoplankton that can be easily adapted for undergraduate and graduate courses.

This volume contains a collection of papers on structural, biological, and biogeographic aspects, as well as examples of paleoenvironmental applications of nannoplankton. The biostratigraphic papers are included in the companion Benchmark volume, *Nannofossil Biostratigraphy*.

The anthology is organized into five parts. The first papers pick up the early debate about the nature of coccoliths from the time it was realized that coccoliths were not of inorganic origin, to the recognition of the true place of coccospheres amongst unicellular organisms. Part II is a mixed bag of papers on the ultrastructural, biological, distributional, and preservational aspects of nannoplankton. It includes papers on the mineralogy of coccoliths; biology and physiology and the extant distribution of coccolithophores; the fate of the coccospheres in the water column, on the sea floor and within the sediments; and the overall significance of nannoplankton for carbonate deposition in the oceans.

The third part includes selections from recent works on mapping the biogeography of nannoplankton in the Mesozoic and Cenozoic and their applications in paleoenvironmental/paleoclimatic reconstructions of the oceans. Papers in Part IV are selections on the origin and evolutionary aspects, including reports of their occurrence in the Paleozoic, examples of lineage studies, and the diversity fluctuations of nannofossils through time.

Part V includes selected papers that document the great extinction of calcareous nannoplankton at the end of the Cretaceous. The nannofossil evidence for this event is at the center of the recent controversy about the cause (terrestrial vs. extraterrestrial) of the mass mortality of marine biota recorded at the Cretaceous/Tertiary boundary.

I take this opportunity to extend my thanks to the colleagues who have given permission to reproduce their papers, many of whom also offered very helpful comments that influenced the organization and selections.

BILAL U. HAQ

CONTENTS BY AUTHOR

Bé, A. W. H., 90
Berger, W. H., 129, 134
Black, M., 58
Blackwelder, P. L., 240
Bowdler, J. L., 228
Bramlette, M. N., 124, 305
Brand, L. E., 240
Bukry, D., 229, 258, 289
Deflandre, G., 248
Dudley, W. C., 236, 239, 240
Duplessy, J. C., 240
Gartner, S., 289
Geitzenauer, K. R., 190
Goodney, D. E., 236, 239
Guillard, R. R. L., 240
Haq, B. U., 195, 197, 207, 268, 295
Honjo, S., 102, 140
Huxley, T. H., 29
Jantzen, R., 193
Kelts, K. R., 167
Kroopnick, P. M., 236, 239
Lohmann, G. P., 195, 197

McIntyre, A., 90, 186, 193, 194
Mahoney, M. E., 236
Manivit, H., 168
Margolis, S. V., 236, 239
Mullin, M. M., 134
Noël, D., 168
Okada, H., 102, 312
Paasche, E., 65
Premoli-Silva, I., 197
Roche, M. B., 90
Romein, A. J. T., 313
Roth, P. H., 134, 228
Ruddiman, W. F., 193, 194
Sorby, H. C., 21
Thierstein, H. R., 155, 312
Wallich, G. C., 14, 41
Watabe, N., 81
Wilbur, K. M., 81
Williams, D. F., 239
Wise, S. W., Jr., 167
Worsley, T. R., 309

INTRODUCTION:
HISTORICAL REVIEW OF CALCAREOUS
NANNOPLANKTON RESEARCH

The great nineteenth century microbiologist, C. G. Ehrenberg, who more than anyone else, deserves to be called the father of micropaleontology, is credited with the discovery of many of the phytoplankton groups, including silicoflagellates, ebridians, dinoflagellates, and of course, coccoliths and discoasters. In his talks to the Berlin Academy of Sciences in 1836 and later, Ehrenberg referred to small, flat, elliptical discs (coccoliths) and star-shaped crystalloids (discoasters) that he had observed in the chalk samples from the island of Rügen. He believed these objects to be of inorganic origin, formed by a process similar to the one that gave rise to concretions in limestones and clays. He even went so far as to attempt to produce (unsuccessfully!) these "chalk-morpholiths" in his laboratory. In 1854 in his monumental microbiologic opus *Mikrogeologie*, Ehrenberg illustrated coccoliths and discoasters (see Fig. 1), the smaller, matrix-bound components of the rich microfossil assemblages of the chalk (mostly Foraminifera, Radiolaria, and diatoms).

In the late 1850s, in preparation for the laying of the first trans-Atlantic telegraphic cable, deep sea oozes were recovered in the course of soundings taken by the H.M.S. *Cyclops*. T.H. Huxley, in 1858 reported the discovery of Ehrenberg's inorganic crystalloids in these ooze samples and called them "coccoliths" for the first time. Soon thereafter the concept of inorganic origin of coccoliths was challenged by H. C. Sorby and G. C. Wallich. In a short communication to the Annals of Natural History, in 1861, Wallich (Paper 1) made the observation that coccoliths were constituents of larger, spherical objects, which he termed "coccospheres." Wallich described and illustrated the coccosphere and compared it to the juvenile stages of Foraminifera. Similarly in 1861, Sorby (Paper 2) argued that coccoliths were not flat as described by Ehrenberg, but curved to fit a hollow sphere. Sorby regarded coccoliths as independent organisms, based on the differences in the optical character of coccospheres and

1

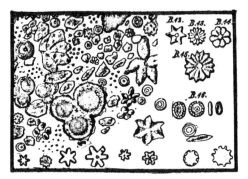

Fig. 1. The earliest illustrations of coccoliths and discoasters by C. G. Ehrenberg (1854), collected from plates in *Mikrogeologie.*

foraminiferal tests. Sorby also pointed out that coccoliths make up a quantitatively significant part of the chalk, and the chalk could therefore be thought to have an organic origin as well. It was Wallich, however, who, in 1865, reported the first free living coccospheres from the surface waters of the Indian Ocean and who eventually clarified their true nature (see Paper 4).

In 1868, T. H. Huxley published his famous essay "On Some Organisms Living at Great Depths in the North Atlantic Ocean" in which he not only clearly illustrated coccoliths and coccospheres (see Paper 3), but also presented the bold, but ill-advised thesis that the mass of gelatinous material in which the coccoliths and coccospheres were found embedded was comparable to the 'Urschleim' of Haeckel, and represented a new life form, to which Huxley assigned the name *Bathybius haeckelii.* He proposed that coccoliths and coccospheres were hard parts of the *Bathybius,* analogus to the spicules in the soft parts of sponges and Radiolaria.

Because of Huxley's considerable prestige within the scientific community, his ideas were not challenged for quite some time. On the contrary, such renowned scientists as Carpenter and Wyville Thomson (1868) jumped on the *Bathybius* bandwagon and claimed that they had confirmed Huxley's conclusions about the "new life form" in other deep sea dredgings. Huxley (1871) himself went so far as to envision *Bathybius* covering the vast expanse of the sea floor, girdling the entire surface of the world ocean.

In 1875, however, Wyville Thomson informed Huxley that in spite of the best efforts of the staff of H.M.S. *Challenger* they were unable to recover *Bathybius* in a fresh state. It had become clear to Wyville Thomson that this new "life form" was little more than the sulphate of lime precipitated in a flocculent state from the seawater by the alcohol used in the preservation of specimens. This gave Wallich (1875), who had been one of the few detractors of the *Bathybius* concept, the opportunity to take Messers Huxley, Carpenter and Wyville Thomson to task. He set the record straight by reminding them that in 1865 he had already reported free, living coccospheres in the surface waters of the open ocean, which should have forewarned Huxley and his followers about the untenability of the *Bathybius* concept.

Further reports of the occurrence of coccoliths and the controversy about their origin continued well into the later part of the nineteenth century. In 1891 when the reports of the expedition of H.M.S. *Challenger* were published, Murray and Renard reported a variety of nannoflora from the *Challenger* samples. They illustrated complete coccospheres of rhabdolithids, helicosphaerids, scyphosphaerids and ceratolithids and thoracosphaerids, and made observations about the geographic distribution of some of these forms.

During the early twentieth century, the study of living and fossil nannoplankton accelerated and entered an inventory-taking phase. The center of research shifted from England to the Continent, where, between 1902 and 1919, H. Lohmann was most prolific in describing living and fossil coccolithophores. Other noteworthy inventory-takers were C. H. Ostenfeld, who published between 1899 and 1913, and J. Schiller, who published between 1913 and 1937. In 1930, the latter collected what was known of living coccolithophores, for inclusion in Rabenhorst's *Kryptogamen-Flora*—this collection remains useful to the present day.

In the late 1920s and the 1930s, E. Kamptner in Austria, G. Deflandre and F. Barnard in France, and T. Braarud and his colleagues in Norway began describing the bulk of remaining extant coccolithophores and many fossil forms as well. Kamptner and Deflandre, in particular, described a large number of fossil genera and species. Much of the work of these and other authors is monographic in nature, making it difficult to include a meaningful sample in a volume like this. Kamptner published extensively between 1927 and 1967 and covered a wide spectrum of topics, including a description of nannoflora from the Adriatic and Mediterranean seas; the ultrastructure and optical character of coccoliths; fossil assemblages from the Pacific dredges and sections in the East Indies and from Vienna Basin. Deflandre's contributions to nannopaleontology were equally prodigious. Together with his coworkers, between 1934 and 1973, he published on practically every aspect of nannoplankton, including the earliest pictures of coccoliths seen under the electron microscope (see Deflandre and Fert, 1954).

Beginning in 1935, T. Braarud and his Norwegian colleagues

(P. Halldal, K. Gaarder, J. Markali, G. Hasle and E. Paasche) have contributed a large body of knowledge on the living coccolithophores, ranging from taxonomy to physiology, ecology and influence of environmental factors such as salinity, temperature, and nutrients on the growth of coccoliths. These workers were pioneers in the attempt to culture coccolithophores in the laboratory and in the routine use of the transmission electron microscope (TEM) for the study of the ultrastructure of coccoliths and their tissue (see e.g. Braarud, 1954).

The observations on coccolithophore growth and calcification by Mary Parke and her coworkers at Plymouth in England have added a new dimension to our knowledge of the life cycle of these microplankton algae. Parke and Adams (1960) found that during the two (motile and nonmotile) phases of their life cycle, the coccolithophore *Crystallolithus hyalinus* Gaarder and Markali produced two entirely different types of coccolith plates. During the motile phase, the coccoliths are composed of loosely-packed microrhombohedrals, whereas during the nonmotile phase the coccoliths typical of the taxon *Coccolithus pelagicus* (Wallich) result. The latter consist of modified microcrystallites arranged in two shields connected by a central tube. This discovery of holo- and heterococcoliths in the life cycle of a single species has obvious implications for the description of fossil forms and in quantitative studies of fossil assemblages.

During the 1950s, inventory taking of fossil nannoplankton started in earnest. With the publication of Bramlette and Riedel's (1954) paper pointing out the usefulness of nannofossils in biostratigraphy, a new descriptive phase of documenting the nannofloras of the Mesozoic and Tertiary began. M. N. Bramlette's subsequent publications with his coworkers (F. R. Sullivan, E. Martini, and J. Wilcoxon) broke new ground in the identification and description of a large majority of the biostratigraphically useful Paleogene and Miocene taxa. Other authors who contributed extensively to the inventory of nannofossils in the 1950s and 1960s were H. Stradner in Austria, E. Martini in West Germany, and W. W. Hay in the United States.

With the introduction of TEM to the study of microfossils, a new phase in the description of ultrastructure began. The earliest enthusiast in the use of TEM for the study of detailed mineralogy of nannoplankton was M. Black in England. Starting in the late 1950s Black and his coworker, B. Barnes, published electron micrographs that set the standard for shear good looks for future works. Because of the novelty of the TEM methodology and its often handsome product, Black often unknowingly redescribed taxa already described under the light microscope. Sometimes two different plan views of a single coccolith were confused as different genera. Black was a sedimentary petrologist by profession (and a coccolith enthusiast only by inclination), and thus his description of the mineral parts of coccoliths (see Paper 5) remains his best contribution to nannopaleontology. In spite of their taxonomic shortcomings, Black's electronmicrographs of coccoliths are still among the best available (see e.g. Black,1965).

With the help of TEM, utilitarian descriptions of nannofossils have been provided since the early 1960s by numerous workers in Europe and the United States. Among the early workers were: W. W. Hay and his coworkers (beginning in 1962). D. Noël (1964 and later), B. Haq (1966 and later), H. Stradner (1966 and later), K. Perch-Nielsen (1967 and later), S. Gartner (1967 and later), and D. Bukry (1969 and later). D. Noël collected a large number of transmission electronmicrographs from the Jurassic and scanning electronmicrographs from the Cretaceous of calcareous facies from Europe and North Africa in two monographs (see Noël, 1965 and 1970).

The application phase in nannopaleontology can be said to have begun with the publication of Bramlette and Riedel's (1954) groundbreaking paper on the biostratigraphic usefulness of nannofossils. In the 1960s, the contributions by Bramlette and his colleagues (Bramlette and Sullivan, 1961; Martini and Bramlette, 1963; Bramlette and Martini, 1964) laid the foundation for the later development of a Cenozoic nannofossil zonation. Stradner's (1963) classic synthesis of the Mesozoic assemblages provided a similar basis for the development of a Mesozoic nannofossil zonation. The greatest impetus to the development of a refined biochronologic scale was supplied by the availability of cores taken by R. V. *Glomar Challenger,* operated by the Deep Sea Drilling Project (DSDP), which began in 1968. The relatively continuous and better-preserved assemblages in the DSDP cores have in great part led to the establishment of a practical Cenozoic zonation (see Hay et al. 1967; Hay and Mohler, 1967; Bramlette and Wilcoxon, 1967; Martini, 1971; Bukry, 1973), and a Mesozoic zonation (Thierstein, 1976; Sissingh, 1977). (Calcareous nannofossil biostratigraphy is discussed in detail in the companion Benchmark volume where all relevant papers on the subject are included).

A second descriptive phase of the study of microfossils was entered with the introduction of the scanning electron microscope (SEM) in the late 1960s. The ability of the SEM to bounce off electrons directly from metal-coated objects, made the time-consuming process of carbon-replication which was essential for TEM work, quite unnecessary. The structural collapse of the carbon-replicas of the relatively thicker objects (such as discoasters) and the resulting distortion of the image could be avoided and the original objects could now be observed under the electron beam. This proved to be a breakthrough for the study of microfossils, particularly the detailed ultrastructure of nannoplankton. Soon after the first demonstration of its usefulness (see e.g., Hay and Sandberg, 1967), the SEM became the preferred instrument for the observation of coccoliths.

During the 1970s, concomitant to the development of a refined biostratigraphic zonation, coccolith research also entered an interpretive phase. The mapping of the biogeographic patterns of the modern oceans (see Papers 8 and 9) has aided in the interpretation of past environmental conditions. The work of McIntyre and his colleagues demonstrated the usefulness of nannofloral distribution data in the reconstruction of Quaternary climatic/oceanographic conditions (see Papers 17 through 20). Haq and his colleagues have mapped the Tertiary biogeographic distributions of nannoplankton and have attempted to reconstruct the paleoceanography on the basis of these data (see Papers 21 through 23). More recently, there have been attempts to use the coccolith component of the carbonate sediments for oxygen and carbon isotopic analyses as an aid in the reconstruction of the thermal and productivity history of the Cenozoic water masses (see Papers 26 through 28).

Work on the settling history of the coccolithophore particles (see Paper 13), their disolution in the water column (see Papers 11 and 12), and diagenesis on the sea floor and within the sediment (see Wise, 1977 and Papers 14 and 15) has increased our understanding of the dynamic role of coccolith sedimentation in the carbonate cycle and its response to changing environmental conditions.

One aspect of nannoplankton research that has received less attention than it deserves is the evolution and diversification of coccoliths through time. The evolutionary studies are limited to educated guesses about the ancestor-descendant relationships of coccolith morphotypes in some lineages (see e.g. Prins, 1971; Gartner, 1970; and Papers 30 through 32). The variations in the diversity of nannoplankton during the Mesozoic and Cenozoic have been documented by Haq (see Paper 33). These variations suggest that diversification is linked to environmental factors, such as changes in the climates and eustatic changes in the sea level (see Paper 33). A massive and sudden reduction in the late Maastrichtian diversity high of nannoplankton and other marine plankton occurred at the close of the Cretaceous. Several explanations for this biotic crisis have been offered recently (see Papers 34 through 37 and Gartner and Keany, 1978; and Thierstein and Berger, 1978).

During the 1980s, a concomitant refinement of the nannofossil biochronology and a wider use of nannoplankton in paleoenvironmental studies is to be expected. The understanding of their distribution patterns (paleobiogeography) and the response of major assemblages to changes in the environment will lead to valuable input in the mapping of areas of high productivity of the past. Such information is of prime importance for successful hydrocarbon exploration and thus acquires considerable practical significance.

REFERENCES

Black, M., 1965, Coccoliths, *Endeavour* **24:**131–137.

Braarud, T., 1954, Studiet av planktonalger i elektronmikroskop, [The Study of Plankton Algae in Electron Microscope], *Blyttia* **2:**102–108, 4 pl.

Bramlette, M. N., and E. Martini, 1964, The Great Change in Calcareous Nannoplankton Fossils Between Maastrichtian and Danian, *Micropaleontology* **10:**291–322.

Bramlette, M. N., and W. R. Riedel, 1954, Stratigraphic Value of Discoasters and Some Other Microfossils Related to Recent Coccolithophores, *Jour. Paleontology.* **28:**385–403.

Bramlette, M. N., and F. R. Sullivan, 1961, Coccolithophorids and Related Nannoplankton of the Early Tertiary in California, *Micropaleontology* **7:**129–188.

Bramlette, M. N., and J. A. Wilcoxon, 1967, Middle Tertiary Calcareous Nannoplankton of the Cipero Section, Trinidad, W. I., *Tulane Studies Geology* **5:**93–131.

Bukry, D., 1969, Upper Cretaceous Coccoliths from Texas and Europe, *Kansas Univ. Paleont. Contr., Article 51 (Protista 2),* pp. 1–79.

Bukry, D., 1973, Low-latitude Coccolith Biostratigraphic Zonation, in *Initial Reports of the Deep Sea Drilling Project,* **15:**685–703.

Carpenter, W. B., and C. Wyville Thomson, 1868, Preliminary Report of Deep Sea Dredgings, *Royal Society (London) Proc.* **16:**190–191.

Deflandre, G., and C. Fert, 1954, Observations sur les Coccolithophoridés actuels et fossiles en microscopie ordinaire et electronique, *Ann. Paléontol.* **40:**115–176.

Ehrenberg, C. G., 1854, *Mikrogeologie; Das Erden und Felsen schaffende Wirken des unsichtbar kleinen selbständigen Lebens auf der Erde,* Leopold Voss, Leipzig, 405p.

Gartner, S., 1967, Nannofossil Species Related to *Cyclococcolithus leptoporus* (Murray & Blackman), *Kansas Univ. Paleont. Contr. 28,* 7p.

Gartner, S., 1970, Phylogenetic Lineages in the Lower Tertiary Coccolith Genus *Chiasmolithus,* in *Proceedings of the North American Paleontological Convention, Part G, Ultra Microplankton,* pp. 930–957.

Gartner, S., and J. Keany, 1978, The Terminal Cretaceous Event: A Geological Puzzle with an Oceanographic Solution, *Geology* **6:**708–712.

Haq, B. U., 1966, Electron Microscope Studies on Some Upper Eocene Calcareous Nannoplankton from Syria, *Stockholm Contr. Geology.* **15:**23–37 6 pl.

Hay, W. W., and K. M. Towe, 1962, Electronmicroscope Studies of *Braarudosphaera bigelowi* and Some Related Coccolithophorids, *Science* **137:**426–428.

Hay, W. W., and H. P. Mohler, 1967, Calcareous Nannoplankton from Early

Tertiary Rocks at Pont Labau, France and Paleocene—Early Eocene Correlations, *Jour. Paleontology.* **41:**1505-1541.

Hay, W. W., and P. A. Sandberg, 1967, The Scanning Electron Microscope, A Major Breakthrough for Micropaleontology, *Micropaleontology* **13:**407-418.

Hay, W. W., H. P. Mohler, P. H. Roth, R. R. Schmidt, and J. E. Boudreaux, 1967, Calcareous Nannoplankton Zonation of the Cenozoic of the Gulf Coast and Caribbean-Antillean area and Transoceanic Correlation, *Gulf Coast Assoc. Geol. Socs. Trans.* **17:**428-480.

Huxley, T. H., 1858, Report in *Deep Sea Soundings in North Atlantic Ocean Between Ireland and Newfoundland Made by H. M. S. Cyclops,* (Comdr. J. Dayman), Admirality, London, p. 64.

Huxley, T. H., 1871, Discussion on Capt. S. Osborn's paper "On the Geography of the Bed of the Atlantic and Indian Oceans and the Mediterranean Sea". *Proceedings of the Royal Geographical Society London,* **15**(1):37-39.

Martini, E., 1971, Standard Tertiary and Quaternary Calcareous Nannoplankton Zonation, in *II Planktonic Conference, Rome, 1970, Proceedings of the* vol. 2, A. Farinacci, ed., Edizioni Tecnoscienza, Rome, pp. 739-785.

Martini, E., and M. N. Bramlette, 1963, Calcareous Nannoplankton from the Experimental Mohole Drilling, *Jour. Paleontology.* **37:**845-856.

Murray, J., and A. F. Renard, 1891, Report on deep-sea deposits based on the specimens collected during the voyage of H. M. S. *Challenger* in the years 1872 to 1876, in *Report on the scientific results of the voyage of H. M. S. Challenger during the years 1872-76,* part 3, *Deep-sea deposits,* H. M. Stationary Office, London, 548p., 29 pl., 35 text-figures.

Noël, D., 1964, Modalités d'utilisation du microscope électronique pour 1'étude des Coccolithes fossiles, *Acad. Sci. Comptes Rendus* **259:**3051-3054.

Noël, D., 1965, *Sur les coccolithes du Jurassique Européen et d'Afrique du Nord,* C. N. R. S., Paris, 209p. 29 pl.

Noël, D., 1970, *Coccolithes crétacés de la Craie companienne du Bassin de Paris.* C. N. R. S., Paris, 192p., 48 pl.

Parke, M., and I. Adams, 1960, The Motile *Crystallolthus hyalinus* (Garder and Markali) and the Non-motile Phases in the Life History of *Coccolithus pelagicus* (Wallich) Schiller, *Marine Biol. Assoc. United Kingdom Jour.* **39:**263-274.

Perch-Nielsen, K., 1967, Nannofossilien aus dem Eozän von Dänemark, *Eclogae geol. Helvetiae.* **60**(1):19-32.

Prins, B., 1971, Speculations on Relations, Evolution, and Stratigraphic Distribution of Discoasters, in Proceedings of the *II Planktonic Conference, Roma, 1970,* vol. 2, A. Farinacci, ed., Edizioni Tecnoscienza, Rome, pp. 1017-1037.

Schiller, J., 1930, Coccolithineae, in *L. Rabenhorst's Kryptogamen-Flora,* 10. Akadamie Verlagsgesellschaft, Leipzig, pp. 89-263.

Sissingh, W., 1977, Biostratigraphy of Cretaceous Calcareous Nannoplankton, *Geologie. en Mijnbouw* **56:**37-65.

Stradner, H., 1963, New Contributions to Mesozoic Stratigraphy by Means of Nannofossils, in *Proceedings of the 6th World Petroleum Congress, Frankfurt am Main, 1963,* Sect. 1, Paper 4, Verein zur Förderung des 6 Welt-Erdöl-Kongresses, Hamburg, 1964, 16p.

Stradner, H., and D. Adamiker, 1966, Nannofossilien aus Bohrkernen und ihre elektronemikroskopische Bearbeitung, *Erdol-Erdgas-Zeitsch.* **82:**(8):330–331.

Thierstein, H. R., 1976, Mesozoic Calcareous Nannofossil Biostratigraphy of Marine Sediments, *Marine Micropaleontol.* **1:**325–362.

Thierstein, H., and W. H. Berger, 1978, Injection Events in Earth History, *Nature* **276:**461–464.

Wallich, G. C., 1865, On the Structure and Affinities of the Polycystina, *Micr. Soc. London, Trans.* ser. 2, **13:**57–84.

Wallich, G. C., 1875, On the True Nature of the So-called *"Bathybius"*, and Its Alleged Function in the Nutrition of the Protozoa, *Ann. Mag. Nat. Hist.,* ser. 4, **16:**322–339.

Wise, S. W., Jr., 1977, Chalk Formation: Early Diagenesis, in *The Fate of Fossil Fuel* CO_2 *in the Oceans,* N. R. Anderson and A. Malahoff, eds., Plenum Publishing Corp., New York, pp. 717–739.

Part I

EARLY DEBATE ABOUT THE NATURE OF COCCOLITHS

Editor's Comments
on Papers 1 Through 4

1   WALLICH
Remarks on Some Novel Phases of Organic Life, and on the Boring Powers of Minute Annelids, at Great Depths in the Sea

2   SORBY
On the Organic Origin of the So-called 'Crystalloids' of the Chalk

3   HUXLEY
On Some Organisms Living at Great Depths in the North Atlantic Ocean

4   WALLICH
Observations on the Coccosphere

The four communications from the *Annals and Magazine of Natural History* and the *Quarterly Journal of Microscopical Science* reproduced here form the heart of the controversy about the origin of coccoliths and who deserves credit for unravelling their true nature. The debate grew rancorous at times, and continued well into the last part of the nineteenth century. It subsided only after the reports of the *Challenger* expedition were published. In one of the thirty-five volumes of this opus devoted to the deep-sea sediments, Murray and Renard (1891) recorded a variety of nannoflora from the deep-sea ooze samples and made some preliminary observations about the biogeography of the nannoplankton. They noted that both rhabdolithids and coccolithids were common in the tropical areas, but the former decreased sharply in the higher latitudes, rarely occurring in waters with temperatures below 18.5°C (65°F). Coccolithids, on the other hand, were encountered in waters with temperatures as low as 7.5°C (45°F).

The four selections included here are samples of the heated debate between Wallich and Huxley and his followers. This was, in a way, typical of many scientific controversies of the late nineteenth century in which the personalities of the authors were as significant as their ideas. Wallich's first communication on the subject of coccospheres

(Paper 1) clearly illustrates the coccosphere and tentatively suggests its affinity to Foraminifera. Sorby's communication (Paper 2), published a few months after Wallich's presents his own views on the subject. His essay contains observations of the coccoliths in chalk and puts forth the proposition that they are of organic origin and were not formed due to inorganic crystallization as suggested earlier by G. C. Ehrenberg. Sorby maintains that coccospheres were independent organisms, akin to, but not juvenile forms of, Foraminifera, as proposed by Wallich. Sorby also made the important observation that because coccoliths constitute a major part of the chalk, then chalk must have an organic origin as well.

Huxley's paper (Paper 3) has become a classic in microbiologic literature, not so much for its clear description of the coccosphere, but more so for the rash creation of the sea-floor hugging, globe engirdling, gelatinous genus, *Bathybius*. Huxley goes on to propose that coccoliths and coccospheres were not independent organisms, but constituted the microskeletal elements of *Bathybius,* in much the same way as spicules were related to sponges or Radiolaria. He also suggested that it was more probable that coccospheres resulted from the coalescence of coccoliths, rather than their dispersal, as suggested by Sorby.

In the fourth paper, Wallich summarizes the major elements of the debate on the coccoliths up until that time (1875) and presents a state of the art which contains both factual observations and reasonable conclusions about coccospheres.

REFERENCE

Murray, J. and A. F. Renard, 1891, Report on deep sea deposits based on the specimens collected during the voyage of H.M.S. Challenger during years 1872 to 1876, in *Report on the Scientific Results of the Voyage of H.M.S. Challenger During Years 1872-76, part 3, Deep-sea Deposits,* H. M. Stationary Office,London, pp. 257,258,pl.xi.

1

Reprinted from *Ann. Mag. Nat. Hist.*, ser. 3, **8**:52–58 (1861)

REMARKS ON SOME NOVEL PHASES OF ORGANIC LIFE, AND ON THE BORING POWERS OF MINUTE ANNELIDS, AT GREAT DEPTHS IN THE SEA

G. C. Wallich, M.D., F.L.S. & F.G.S.

In the notice of the material obtained by the soundings taken on board H.M.S. 'Cyclops' in 1857, appended to the official report of Captain Dayman*, Professor Huxley mentions having met with a number of small rounded bodies, which he describes as consisting of several concentric layers surrounding a minute clear centre, and looking, at first sight, somewhat like single cells of the plant " Protococcus." To these bodies Professor Huxley provisionally applied the designation of *Coccoliths*.

In the deepest soundings taken during the recent expedition

* " Deep-Sea Soundings in the North Atlantic Ocean, between Ireland and Newfoundland, made in H.M.S. Cyclops, Lieutenant-Commander Joseph Dayman, in June and July 1857, published by order of the Admiralty."

to the North Atlantic, I detected these very curious bodies in great numbers,—occurring not only in the free state, noticed by Professor Huxley, but as adjuncts to minute spherical cells, upon the outer surface of which they were adherent in such a manner as to leave no doubt of that being their normal position. Whilst alluding to their occurrence, in my published " Notes on the Existence of Animal Life at vast depths in the Ocean," I ventured a surmise as to their being a larval condition of some of the Foraminifera,—first, in consequence of their being invariably present in greatest quantity in such of the deep-sea deposits as were most prolific of these organisms ; secondly, because, in one or two instances, Coccoliths had been met with by me adherent to Foraminiferous shells in such a manner as to render it highly improbable that they could have attained their position by accident ; and lastly, because the spherical cells, to which reference has been made, when entirely freed from their adherent Coccoliths, presented no discernible points of difference, save as regards somewhat inferior dimensions, from the minute and nearly hyaline solitary cells of the earliest stage of the Globigerinæ.

On reference to the annexed woodcut it will be seen that the composite bodies to which I allude, and to which I propose to give the name of *Coccospheres,* are minute spherical cells (figs. 1 & 2) having a defined limitary wall, and that upon their outer surface the Coccoliths of Professor Huxley are arranged at nearly regular intervals. The cells, when crushed, are seen to contain a homogeneous, gelatinous, and almost colourless matter, exhibiting no visible trace of organization, and, in all probability, consisting of sarcode. The wall of the cell may be distinctly seen under a high power ; but from the minuteness of the entire structure, I have hitherto found it impossible to do more than attest its existence. Accordingly there is nothing visible to show whether the wall is formed of one or more than one layer. Cells are sometimes met with in a fractured condition ; but I have never observed a collapsed specimen, or flattened-out fragment, such as would frequently occur were the basis of the wall formed of anything more yielding than calcareous matter. In like manner, I have hitherto failed to detect markings or apertures in the limitary wall of the Coccosphere. The solitary cells vary in diameter from $\frac{1}{1600}$th to $\frac{1}{1250}$th of an inch, when seen separately. Forming part of a series, as in the specimen of *Textularia* presently to be described, some cells, however, attain a much larger size.

The Coccoliths, to which term I would restrict the minute bodies described by Professor Huxley, are of an oblong shape, concave on their internal aspect, namely that on which they are attached to the surface of the Coccosphere-cells, and convex externally; in short, they are spoon-shaped, only with a much less marked convexity and concavity. In some specimens, a single aperture, only, occurs at the central portion. In others the aperture appears to be double; or, rather, there are two perforations placed side by side, in the direction of the long axis of the body, and separated from each other by an extremely delicate transverse band; whilst the external marginal surface, which thus constitutes a quoit-like but oblong ring round the central perforated portion, is striated in a radiate manner. When the two perforations are present, the little mass closely resembles a miniature plate of *Synapta*. The Coccoliths, like the spheres upon which they rest, are transparent and devoid of colour. Their mode of attachment is undistinguishable, owing to their extreme minuteness. They appear, however, to be simply placed in contact with the surface of the Coccosphere-wall, and to be retained in position by the delicate gelatinous layer in which the entire organism is invested. We may thus account for the seeming facility with which the Coccoliths are detached, and the vast numbers of free Coccoliths which crowd many of the deposits.

It is necessary to state that a high magnifying power and very careful and brilliant illumination are requisite to enable us to see the structure of the Coccoliths to this extent. Their presence in the finer portion of the deposits may just be ascertained under a good $\frac{1}{2}$-inch lens; but in order to make out the apertures and striation, a $\frac{1}{6}$ or $\frac{1}{8}$, of first-rate construction, is indispensable,—the difficulty of obtaining clear definition being materially, and almost insurmountably, enhanced from the circumstance of its being necessary to mount the material in its normal state, inasmuch as subjection to acids at once annihilates all trace of the objects under notice.

The average length of each Coccolith is about $\frac{1}{2700}$th of an inch. Fig. 3, *a, b, c*, represents these bodies as seen from their external, lateral, and inferior aspects. Fig. 4 gives a still more enlarged view of one, as seen from its external or convex aspect.

In the adjoining woodcut, fig. 1 exhibits a specimen in which four Coccospheres, with their adherent Coccoliths, are united together in a linear series similar to that of the chambers of the *Nodosariæ*. The cells are all, however, of uniform size, and smaller than the majority of the separate single specimens. Fig. 2 represents an unmistakeable *Textularia* (probably *T. variabilis*, Will.), the chambers of which apparently consist of several Coccospheres, in this instance of different sizes, arranged according to the double alternating order typical of the genus referred to. The chambers, as will be seen, are seven in number, the smallest and oldest measuring $\frac{1}{1230}$th of an inch, whilst the largest and last-developed measures $\frac{1}{450}$th of an inch. The en-  tire specimen presents the transparency and delicacy of the normal Cocco- spheres. No septal apertures are visi- ble; but this may arise in a great measure from the position of the spe- cimen, which is preserved on a slide in balsam, and also from the imperfect manner in which it was necessary to clean the deposit before mounting it. On the exposed surface of each chamber the Coccoliths are di- stinctly visible. That their adherence in this fashion is not the result of accident is, I think, evident both from their dis- position and the circumstance of numerous Foraminifera, pre- sent throughout the whole of the same slide, and of equal delicacy and transparency, not exhibiting a single Coccolith on their surfaces, although great numbers occur around them on every side.

During my earlier examinations of these remarkable objects, I repeatedly detected Coccoliths adherent to Globigerina-shells; but in no other instance than that just cited have I found the whole, or indeed more than one chamber of any Foraminiferous shell, so studded, and in other respects presenting appearances so identical with those seen in the free Coccospheres of which I have spoken.

It is certainly strange that, during the examination of a large series of slides exhibiting the lighter particles of the material in which the Coccospheres and Coccoliths abound, only one good example of a Foraminiferous shell should have been observed in the condition alluded to,—the four Coccospheres spoken of as occurring united in a linear series, although closely re- sembling the Nodosarian type in point of arrangement, pre- senting no positive evidence of their Foraminiferous origin. But it must be borne in mind, in investigating the lower organic forms of the animal and vegetable kingdoms, that instances are far from rare in which early phases of development are so ephemeral as to render the chances of their taking place under the eye of the observer extremely scanty. And again, for reasons already assigned, it is far from improbable that, although actually present in the material under analysis, the appearances are con- stantly overlooked.

These minute bodies, however, possess a high degree of in- terest apart from that arising from their association with the deep-sea deposits actually taking place in our own day, and this renders it particularly desirable that the attention of observers should be directed towards them. I allude to the discovery in the Chalk, by Mr. H. C. Sorby of Sheffield, of objects either identical with them or so nearly identical as to leave no doubt of their close affinity, and to the important additional evidence herein furnished regarding the identity in origin of some of the recent and more ancient oceanic deposits. It is not my wish at present to do more than point out these facts. Whether it be eventually shown that the association between these bodies and the Foraminifera is purely fortuitous or otherwise, there cannot be a doubt that they have some important office to perform in the history of the deep-sea deposits, and that the investigation of this office will materially assist us in clearing up the mystery that surrounds the occurrence of similar objects in the Chalk.

I have also to direct attention to some curious facts which have presented themselves to my notice whilst investigating the structure of certain Foraminiferous shells, and which illustrate, in a remarkable manner, the soundness of the views first propounded by Professor Carpenter with reference to the transmutation of many of the reputed species of these organisms.

Having selected some well-developed *Biloculina*-shells, I bisected them in various planes, and found that the innermost chamber of each individual—for segment it cannot in this case properly be termed—was in reality a minute and perfect *Miliola*, —this innermost chamber being, of course, the primordial chamber of the group. As is well known, *Biloculina* is a symmetrically-developed Foraminifer, the segments of which are arranged in alternating series, but with their margins in the same plane. In *Miliola*, on the other hand, the arrangement of the segments is asymmetrical, the plane of growth being a revolving one. Wide as the interval between the typical *Miliolidæ* and *Biloculinæ*, at first sight appears, the examples to which I refer show that there is no true line of demarcation between them, and that the one is neither more nor less than the primordial chamber of the other; whilst the future development of the primordial portion, into what has heretofore been considered a typical adult *Miliola* or a perfect *Biloculina*, depends wholly therefore on the conditions under which its further growth is regulated.

In the specimens under notice, the minute *Miliolæ* were found situated between the two earliest segments formed on the *Biloculina* type; and within the boundary so constituted it was placed, apparently, in an unattached state, but so closely pressed on as to indicate that the growth of the new segments took place upon the external surface of the primordial *Miliola*-chamber, and, as it were, on a mould. The minute *Miliola*, disposed with its long axis towards the axis of the septal orifices of the *Biloculina*-segments, is of sufficient size to be visible by the naked eye, and presents but one perfect revolution, as shown in fig. 3 of the last woodcut, there being no trace of a fractured margin of attachment at any portion of its surface.

Before concluding, I would also mention having met with several examples of Foraminiferous shells, brought up from the greatest depths, perforated, in all probability, by the minute boring Annelids that construct and inhabit the tubes of which I have made mention in my " Notes." The extreme delicacy of the inhabitants of these tubes has, as yet, completely baffled me in all my attempts to extract them and determine their character. In addition, however, to the tubes, formed in so singular a manner, of innumerable carefully selected Globigerina-shells cemented together, there also occur other tubes, in which the internal layer is a cylinder of tough membranous material with a rich sienna tint, whilst its outer surface is strengthened and protected partly by numerous Globigerina-shells, as in the previous case, and partly by a layer of silicious spicules, probably derived from some minute sponge. The perforations in the shells are invariably of one character, and consist of an aperture bored through and through, but having the entire thickness of the shell-wall, from the inner surface to the outer one, as it were countersunk. Accordingly, in section, such a perforation presents a truncated cone, the apex of which is directed inwards.

It has repeatedly been observed that, in the manufacture of telegraphic cables, it is a matter of vital importance to guard against the attacks of Annelids capable of thus penetrating hard substances at the greatest depths. Owing to the difficulty of boring into gutta percha, unless under the action of chemical solvents, I cannot conceive much danger is to be apprehended from its employment as the coating medium. If Annelids are able to drive their boring apparatus through gutta percha, it can only be when it has assumed the brittle, macaroni-like structure peculiar to it after long-continued exposure to heat, or in the case of impure and adulterated gutta percha. Pure caoutchouc, I do not hesitate to say, is impervious to a boring apparatus like that of the *Teredines*,—unless, as before mentioned, under the action of a chemical solvent, when, of course, no boring tool is necessary to render the mischief complete and fatal. And, lastly, there is this consolation for the advocates of gutta percha and caoutchouc coverings for submarine cables : —If their little hidden enemies at the bottom of the sea are so far advanced in civilization as to be able to discriminate when they ought to use their augers, or fall back upon their chemical laboratories for aid, neither glass, nor iron, nor hemp, nor adamant itself, will suffice to defy them. If any material exists, the characters of which are so thoroughly dissimilar from those of any substance known to occur at the bottom of the sea as to render it in the highest degree improbable that such creatures as live there could improvise means to pierce it, whilst, at the same time, it would secure perfect insulation of the telegraphic wire, *caoutchouc is that material.*

2

Reprinted from *Ann. Mag. Nat. Hist.,* ser. 3, **8**:193-200 (1861)

ON THE ORGANIC ORIGIN OF THE SO-CALLED 'CRYSTALLOIDS' OF THE CHALK

H. C. Sorby, F.R.S. &c.

THE appearance of Dr. Wallich's interesting paper, published in this Magazine (vol. viii. p. 52), in which he alludes to my having found in chalk objects similar to Coccoliths, induces me to give an account of my researches on the subject. I do not claim the discovery of such bodies in the Chalk, but to have been the first to point out that they are not the result of crystalline action, that they are identical with the objects described as Coccoliths by Prof. Huxley*, and that these are not single separate individuals, but portions of larger cells.

So far as I am aware, the illustrious Ehrenberg was the first who pointed out the ovoid bodies occurring in chalk, in a paper read at the Berlin Academy, Aug. 18, 1836, on "New Microscopic Characters of earthy and compact Minerals†." After alluding to the various minute bodies constituting some kinds of kaolin and agaric-mineral, he says that the most remarkable of all are those found in chalk, which shows small, flat, elliptical disks, similar to each other, consisting of only a few concentric rings, usually only one, and an internal nucleus of irregular character, as shown in his figure, pl. i. 2 B, in Pogg. Ann. He again alludes to them in his Memoir on Chalk and Chalk-marl ‡, saying that in a former paper he had declared that the preponderating substance of chalk, which forms the cementing material, was minute, elliptical, flat, granular bodies and their fragments. He looked upon them then, as he still continued to do, as concretions of a crystalline character, whose

* Deep-Sea Soundings in the North Atlantic Ocean, made in H.M.S. Cyclops. London, 1858.
† Monatsberichte, 1836; Poggendorff's Annalen, 1836, xxxix. 101.
‡ Abhandlungen der k. Akad. der Wissen. zu Berlin, 1838, 67.

form is peculiar to the chalk. In a note at p. 68, he ascribes them to the same kind of action as gave rise to the larger concretions met with in limestone- and clay-deposits, and considered the force which produced them not simple crystallization, though in some respects analogous, and proposes for it the term 'Crystalloid-Bildung.' The same idea is followed out in his paper on Concretions, read at the Berlin Academy, June 29, 1840[*], in which he says he had endeavoured to make bodies like those in chalk by artificial chemical means, but had not succeeded, though he had made some to a certain extent similar. It must, however, be borne in mind that he looked upon them as *flat disks*, and not as *curved* in the manner shortly to be described. In his magnificent work, 'Microgeologie' (Leipzig, 1854), he also figures these ovoid bodies at pl. xxv. fig. B. 16, under the term 'Kreide-Morpholithe,' along with various minute radiating groups of crystals, evidently ascribing the whole to an inorganic action more or less closely connected with crystallization. In order to show to what extent such ovoid disks serve to make up some varieties of chalk, he gives (at pl. xxx. B) a highly magnified representation of the chalk of Rugen, and in various other plates shows that they constitute a very large proportion of the whole. It appears to me, however, a great exaggeration to affirm that chalk is *composed* of them, since a still larger part is made up of particles which we may attribute with confidence to the decomposed tissue of Foraminifera and other shells.

The inorganic nature of the ovoid bodies of the chalk has hitherto been almost universally adopted; for the only exception I am acquainted with is the supposition of the Rev. J. B. Reade[†], who appears to have ascribed them to Infusoria. But when, about ten years since, I commenced studying the microscopical structure of chalk, I soon became convinced that both these explanations were unsatisfactory. By examining the fine granular matter of loose, unconsolidated chalk in water, and causing the ovoid bodies to turn round, I found that they are not *flat disks*, as described and figured by Ehrenberg, but (as shown by the oblique side view, fig. 5, p. 197) *concave* on one side and *convex* on the other, and indeed of precisely such a form as would result from cutting out oval watch-glasses from a moderately thick hollow glass sphere whose diameter was a few times greater than their own. This is a shape so entirely unlike anything due to crystalline or any other force acting independently of organization—so different to that of such round bodies, formed of minute radiating crystals, as can be made artificially and do really occur

[*] Neues Jahrbuch für Mineralogie, &c. 1840, 680; Journal für prakt. Chemie, 1840, xxi. 95; Ed. New Phil. Journ. 1841, xxx. 353.
[†] Mantell's Wonders of Geology, 2nd ed. vol. ii. 953.

in some natural deposits—and pointed so clearly to their having been derived from small hollow spheres, that I felt persuaded that such was their origin. The small cells of Foraminifera occurring in the chalk being just the size and thickness that would agree with this supposition, I endeavoured for a long time to make out that the ovoid bodies were in some way or other derived from them. I thought that, when decomposition took place, perhaps the calcareous matter might have re-arranged itself into more or less circular concretions whilst still in the form of the cells of Foraminifera, and thus, on further decay, they might have broken up into ovoid bodies of the form described above. I sought diligently for proof of this, but in vain, though I convinced myself that a very considerable part of the minute particles of the chalk was certainly derived from the decomposed tissue of Foraminifera. Notwithstanding this, I still adhered to the supposition of their having originated from organic spheres, and endeavoured to clear up the difficulty by studying recent deposits. Some eight or nine years ago, when examining mud from our own shores, I found one single body which was obviously similar to those in the chalk, both in form and optical characters, but was unable to make out its true nature.

In 1858 appeared Prof. Huxley's Report on the Deep-Sea Soundings in the Atlantic, in which, at p. 64, he says that in all the specimens, from depths varying between 1700 and 2400 fathoms, he had found "a multitude of very curious rounded bodies, to all appearance consisting of several concentric layers surrounding a minute clear centre, and looking at first sight somewhat like single cells of the plant *Protococcus*; as these bodies, however, are rapidly and completely dissolved by dilute acids, they cannot be organic, and I will for convenience' sake simply call them *Coccoliths*."

Still nourishing the conviction that ovoid bodies like those in chalk would be found in deep-sea deposits, at my request I was kindly furnished by Prof. Huxley with some of the Atlantic mud from a depth of 2230 fathoms. I was at that time as ignorant of what he had written on the subject as he was of my object, and of the connexion between the bodies he had described and the chalk. Directly I examined it with the microscope, I perceived that my long-cherished belief was true, and that this deep-ocean mud would completely explain the peculiar characters of our Chalk formations. Nor was this all; for on the 27th of August of last year (1860) I found that, as I had predicted several years before, the ovoid bodies were really derived from small hollow spheres, on which they occur, separated from each other, at definite intervals. I therefore read a short paper on the subject at the meeting of the Sheffield Literary and Philosophical Society, on the 2nd of October, in which I showed that the so-called crystalloids of the chalk are not of crystalline or concretionary origin, but are similar to ovoid bodies forming part of spherical cells in some respects analogous to the cells of Foraminifera.

Nearly two months after this, I had the pleasure of making the acquaintance of Dr. Wallich, who had just returned from his voyage in H.M.S. 'Bull-dog,' and found that he also had discovered the true origin of the Coccoliths, as described at p. 13 of his "Notes on the presence of Animal Life at vast Depths in the Sea, &c.," published for private circulation in November 1860, without having been aware of their important relation to chalk. Mr. Roberts, however, in his paper on "High and Low Life*," when alluding to Dr. Wallich's interesting discoveries, says, "Their discovery in a living state in this ooze is of high geological importance; for microscopical investigation, undertaken by Mr. Sorby, proves their existence in chalk-rocks, associated there, as they are in this North Atlantic Ocean, with Globigerinæ. Indeed, chalk itself is seen to be little else than a compacted mass of Foraminifera-shells, whole and fragmentary, and may be best described by using the very words by which Dr. Wallich introduces to science this recent deposit."

Having thus given a history of the subject, I will proceed to describe some of the facts I have observed, but at the same time shall not attempt to give anything like a complete account of the microscopical structure of chalk, which could not be done without a number of illustrations. Moreover, there are some interesting questions requiring further investigation, which I hope to describe in detail when treating on the microscopical structure of rocks in general. The drawings of Coccospheres and Coccoliths which I made nearly a year ago agree very closely with the figures accompanying Dr. Wallich's paper (pp. 53 & 54). I must confess that, as he justly observes, one is tempted to conclude that there is some connexion between Coccospheres and Globigerinæ; but, at the same time, I feel inclined to think that they may be an independent kind of organism, related to, but not the mere rudimentary form of, Foraminifera. Their optical properties are entirely different. Each cell of Globigerinæ, when alone or attached, gives a splendid well-defined black cross and coloured rings when examined with polarized light, which is readily explained by the fact of the shell being made up of minute crystals of calcite, arranged with their principal axis perpendicular to the surface of the shell. No such cross is, however, seen in the case of Coccospheres; and the cell-wall between the Coccoliths has such a very weak depolar-

* 'Geologist,' 1861, iv. 1.

izing action, that I very much doubt its calcareous nature. The individual Coccoliths, when on the spheres, or, still better by far, when detached, each give an extremely well-defined black cross; and their depolarizing action is much too powerful to allow us to suppose that this is due to the same arrangement of the carbonate of lime as in the shell of Globigerinæ, and that the Coccoliths are the commencement of calcification. At the same time it is not impossible that they might come off from the cells before general calcification took place; and I have found some shells of Foraminifera which showed imperfectly-defined oval bodies, giving black crosses with polarized light, thus proving that such a radiate arrangement of the carbonate of lime as that in Coccoliths does occasionally, though rarely, occur in the shell of Foraminifera. With respect to the individual Coccoliths, their optical characters prove that they have an extremely fine radiating crystalline structure, as if they had grown by the deposition of carbonate of lime on an elongated central nucleus, in accordance with the oval ringed structure shown in fig. 1 (magnified 800 linear).

In order to obtain a satisfactory knowledge of chalk, we should commence with the study of thin sections of the harder varieties. I am not aware that any one but myself has employed this method of research, but I have by this means succeeded in proving most completely that entire Foraminifera are comparatively rare, and make up only quite a small proportion of the whole. More or less detached and broken cells are, however, very numerous, so much so that in some cases they are almost in contact throughout the whole mass, and it is only the spaces between them that are filled with fine granular matter, which in some other specimens constitutes nearly the whole rock. In general, however, the constitution of chalk is intermediate between these two extremes. The nature of the granular matter is best learned by an examination of those very soft specimens which have not been much altered since deposition. When seen in water, under a bit of thin glass, with a power of from 400 to 800 linear, it is easy to perceive that a considerable part is made up of the decomposed tissue of Foraminifera. There are often also small well-defined groups of radiating crystals, similar to those named by Ehrenberg 'Krystaldrusen,' and figured on pl. xxv. B. 12–15 of his 'Microgeologie;' the nucleus is sometimes a minute fragment of the decomposed tissue of Foraminifera; and there can be no doubt respecting their crystalline and inorganic origin. They, however, differ entirely from the well-defined oval bodies hitherto described as chalk-crystalloids. These, in form and optical properties, are exactly similar to the Coccoliths of the Atlantic mud. When made to turn round, they both are seen to be concave on one side and convex on the other, as shown by the oblique side view of an unusually large one from the chalk, fig. 5 (magnified 800 linear); and they give the same kind of well-defined black cross with polarized light. Hence we must abandon the idea of their being "peculiar to the chalk," and may possibly be rather led to conclude that they are *character-istic of deep-ocean deposits.* Many of those in the chalk have a decided granular character, as shown in fig. 2 (magnified 800 linear). The rings, instead of being simple, are, as it were, made up of separate beads, and the centre is also of a compound granular character, with various modifications. Judging from Ehrenberg's drawings, and from what he says at p. 136 of his paper on Chalk and Chalk-marl, he appears to look upon this granular structure as their universal character, and concludes that their minute constituent granules were derived from decom-posed Foraminifera, and were afterwards arranged into crystal-loids by means of some unknown crystalloidal force. However, as already stated, some show no such granular structure, but are precisely similar to those in the Atlantic mud; and the granular constitution of the others admits of a very simple ex-planation. As is well known, when shells become fossil, they often acquire a crystalline texture; and, in fact, this occurs in the recent dead shells found in the mud of the Mediterranean, described by Marcel de Serres and Figuier *. I have also suc-ceeded, beyond all expectation, in producing artificially the same change in recent shells by keeping them for a month or two in a dilute solution of caustic potash, at a temperature of about 145° C. (293° F.), which, by dissolving the organic matter, per-mits the carbonate of lime to crystallize according to a new arrangement; and not only do shells consisting of aragonite undergo this change, but also sometimes those made of calcite†, though, in the case of fossils, it has often only occurred in those composed of aragonite. If such a molecular re-arrangement were to take place in the Coccoliths of the Atlantic mud, they would become almost exactly like the granular specimens found in the chalk; and I shall be much surprised if I do not succeed in imitating them by such artificial means as I have just de-scribed.

* Annales des Sciences Nat. 3 sér. 1847, vii. 21; Comptes Rendus, 1846, xxii. 1050; Neues Jahrbuch für Mineralogie, 1848, 873; Edinburgh New Phil. Journ. 1847, xlii. 381.

† See Rose's second treatise on Carbonate of Lime, Abhandlungen d. k. Ak. d. Wiss. zu Berlin, 1858, 63, since confirmed and extended by my own experiments.

Though the facts I have already stated appear to me conclusive, yet it is of course satisfactory to find that, though rarely seen to advantage, compound Coccospheres do really occur in the chalk; and, indeed, I had seen and made a drawing of one nearly ten years ago, without having properly understood its nature. They, however, like the Foraminifera, appear to have undergone much more decomposition in the chalk than in the fresh mud of the Atlantic, which is only what might have been expected.

But, besides simple ovoid Coccoliths, and others modified by various marks and apertures, there occur in chalk minute bodies which are apparently somewhat related to them, but differ from anything hitherto found in the Atlantic mud. As an illustration of these, I refer to figs. 3 and 4 (magnified 800 linear). Those like fig. 4 are similar to Coccoliths in being oval and spoon-shaped, but show four marks, arranged in a cross, instead of two, or a single elongated nucleus. When bodies like fig. 3 are made to turn about, the under side of the broad end is seen to be like fig. 4, which is, in fact, so to speak, the ground-plan of fig. 3. There are various forms of these curious objects, which are obviously of organic origin, and may be described as Coccoliths with a sort of spine growing outwards from the centre. These spines are four-sided, are sometimes pointed, sometimes end in a small cross, and sometimes extend into four well-developed wings. When the ovoid base occurs alone, either owing to the spine having been broken off or never developed, it is difficult to distinguish them from some varieties of Coccoliths, or at all events to point out any essential and widely remote difference; and therefore, though I have not yet met with sufficient evidence to prove it, I cannot help thinking that at the Chalk period there was a form of Coccosphere in which the Coccoliths were to a greater or less extent developed into small spines.

It is not easy to determine the extent to which these various ovoid organic fragments serve to make up chalk; but, like the Coccoliths of the Atlantic mud, and to a very similar extent, they and their fragments do certainly constitute a very material proportion of the whole. If to them we add the more or less entire and broken Foraminifera, and such particles as can be shown to result from their decay and from the decomposition of the shells of *Inoceramus*, it appears to me that we are in a position to completely account for the origin of the deposit. The importance of the fact of thus being able to make out the true nature of the so-called 'crystalloids' is, that we can no longer doubt the almost entirely organic origin of chalk. Had they been due to a kind of crystalline action, we might indeed have had good reason for supposing, with Ehrenberg, that the carbonate of lime of which they are composed was derived from decayed Foraminifera; but at the same time a strict proof would have been wanting, and we might have adopted the opinion expressed by Haidinger in his paper on the Metamorphism of Rocks *, and concluded that, though, according to Ehrenberg, chalk does contain very many organic bodies, it does itself consist of rounded forms, which are a chemical deposit from water containing soluble salts of lime. Now, however, that their real origin appears to be established, it is no longer requisite to assume the existence of any unknown crystalloidal force differing from simple crystallization; and we can clearly perceive that, though presenting characteristic differences, chalk is in every respect analogous to what we should have, if the mud now being formed at great depths in the Atlantic, by the accumulation of various minute organic bodies, were to be subsequently more or less altered by molecular changes or chemical actions of a well-known character. There is, however, one striking difference; for the Atlantic mud contains many Diatomaceæ, spicula of Sponges, and other silicious organic bodies, which are very rare in, or absent from, the chalk: it contains, however, silicious concretions; and this contrast in the state and aggregation of the silicious matter in the two otherwise analogous deposits makes me very much inclined to conclude, with Ehrenberg†, that the silex of the flints was derived from disseminated silicious organic bodies, which has collected round various centres of segregational attraction,—though there are some difficulties to remove before that opinion can be finally adopted.

3

Reprinted from *Quart. Jour. Microsc. Sci.* **8**:203–212 (1868)

ON SOME ORGANISMS LIVING AT GREAT DEPTHS IN THE NORTH ATLANTIC OCEAN

Professor T. H. Huxley, F.R.S.

In the year 1857, H.M.S. "Cyclops," under the command of Captain Dayman, was despatched by the Admiralty to ascertain the depth of the sea and the nature of the bottom in that part of the North Atlantic in which it was proposed to lay the telegraph cable, and which is now commonly known as the "Telegraph plateau."

The specimens of mud brought up were sent to me for examination, and a brief account of the results of my observations is given in 'Appendix A' of Captain Dayman's Report, which was published in 1858 under the title of "Deep-Sea Soundings in the North Atlantic Ocean." In this Appendix (p. 64) the following passage occurs:

"But I find in almost all these deposits a multitude of very curious rounded bodies, to all appearance consisting of several concentric layers surrounding a minute clear centre, and looking, at first sight, somewhat like single cells of the plant *Protococcus;* as these bodies, however, are rapidly and completely dissolved by dilute acids, they cannot be organic, and I will, for convenience sake, simply call them coccoliths."

In 1860, Dr. Wallich accompanied Sir Leopold McClintock in H.M.S. "Bulldog," which was employed in taking a line of soundings between the Faröe Islands, Greenland, and Labrador; and, on his return, printed, for private circulation, some "Notes on the presence of Animal Life at vast depths in the Sea." In addition to the coccoliths noted by me, Dr. Wallich discovered peculiar spheroidal bodies, which he terms "coccospheres," in the ooze of the deep-sea mud, and he throws out the suggestion that the coccoliths proceed from the coccospheres. In 1861, the same writer published a paper in the 'Annals of Natural History,' entitled "Researches on some novel Phases of Organic Life, and on the Boring Powers of minute Annelids at great depths in the Sea." In this paper Dr. Wallich figures the coccoliths and the coccospheres, and suggests that the coccoliths are identical with certain bodies which had been observed by Mr. Sorby, F.R.S., in chalk.

The 'Annals' for September of the same year (1861) contains a very important paper by the last-named writer, " On the Organic Origin of the so-called 'Crystalloids' of the Chalk," from which I must quote several passages. Mr. Sorby thus commences his remarks :

" The appearance of Dr. Wallich's interesting paper published in this magazine (vol. viii, p. 52), in which he alludes to my having found in chalk objects similar to coccoliths, induces me to give an account of my researches on the subject. I do not claim the discovery of such bodies in the chalk, but to have been the first to point out (1) that they are not the result of crystalline action ; (2) that they are identical with the objects described as coccoliths by Professor Huxley ; and (3) that these are not single separate individuals, but portions of larger cells."

In respect of the statement which I have numbered (1), Mr. Sorby observes :

" By examining the fine granular matter of loose, unconsolidated chalk in water, and causing the ovoid bodies to turn round, I found that they are not flat discs, as described and figured by Ehrenberg, but, as shown in the oblique side view (fig. 5), *concave* on one side, and *convex* on the other, and indeed of precisely such a form as would result from cutting out oval watch-glasses from a moderately thick, hollow glass sphere, whose diameter was a few times greater than their own. This is a shape so entirely unlike anything due to crystalline, or any other force, acting independently of organization—so different to that of such round bodies, formed of minute radiating crystals, as can be made artificially, and do really occur in some natural deposits—and pointed so clearly to their having been derived from small hollow spheres, that I felt persuaded that such was their origin."

Mr. Sorby then states that, having received some specimens of Atlantic mud from me, he at once perceived the identity of the ovoid bodies of the chalk with the structures which I had called coccoliths, and found that, as he had predicted several years before, " the ovoid bodies were really derived from small hollow spheres, on which they occur, separated from each other at definite intervals."

The coccospheres themselves, Mr. Sorby thinks, may be

30

" an independent kind of organism, related to, but not the mere rudimentary form of, Foraminifera."

" With respect to the coccoliths, their optical character proves that they have an extremely fine, radiating, crystalline structure, as if they had grown by the deposition of carbonate of lime on an elongated central nucleus, in accordance with the oval-ringed structure shown in fig. 1 (magnified 800 linear)."

I am not aware that anything has been added to our knowledge of the " coccoliths" and " coccospheres" since the publication of Mr. Sorby's and Dr. Wallich's researches. Quite recently I have had occasion to re-examine specimens of Atlantic mud, which were placed in spirits in 1857, and have since remained in my possession. I have employed higher magnifying powers than I formerly worked with, or than subsequent observers seem to have used, my great help having been an excellent $\frac{1}{12}$th by Ross, which easily gives a magnifying power of 1200 diameters, and renders obvious many details hardly decipherable with the $\frac{1}{8}$th inch objective which I used in 1857.

The sticky or viscid character of the fresh mud from the bottom of the Atlantic is noted by Captain Dayman.* " Between the 15th and 45th degrees of west longitude lies the deepest part of the ocean, the bottom of which is almost wholly composed of the same kind of soft, mealy substance, which, for want of a better name, I have called ooze. This substance is remarkably sticky, having been found to adhere to the sounding rod and line (as has been stated above) through its passage from the bottom to the surface—in some instances from a depth of more than 2000 fathoms."

This stickiness of the deep-sea mud arises, I suppose, from the circumstance that, in addition to the *Globigerinæ* of all sizes which are its chief constituents, it contains innumerable lumps of a transparent, gelatinous substance. These lumps are of all sizes, from patches visible with the naked eye to excessively minute particles. When one of these is submitted to microscopical analysis it exhibits—imbedded in a transparent, colourless, and structureless matrix—granules, coccoliths, and foreign bodies.

The *granules* vary in size from $\frac{1}{10000}$th of an inch to $\frac{1}{8000}$th, and are aggregated together into heaps of various sizes and shapes (Pl. IV, fig. 1), some having the form of mere irregular streaks, but others possessing a more definitely limited

* Loc. cit., p. 9.

oval or rounded figure (fig. 1 c). Some of the heaps attain $\frac{1}{1000}$th of an inch or more in diameter, while others have not more than a third or a fourth of that size. The smallest granules are rounded; of the larger, many are biconcave oval discs, others are rod-like,* the largest are irregular.

Solution of iodine stains the granules yellow, while it does not affect the matrix. Dilute acetic acid rapidly dissolves all but the finest and some of the coarsest granules, but apparently has no effect on the matrix. Moderately strong solution of caustic soda causes the matrix to swell up. The granules are little affected by weak alkalies, but are dissolved by strong solutions of caustic soda or potash.

I have been unable to discover any nucleus in the midst of the heaps of granules, and they exhibit no trace of a membranous envelope. It occasionally happens that a granule-heap contains nothing but granules (fig. 1 a), but, in the majority of cases, more or fewer coccoliths lie upon, or in the midst of, the granules. In the latter case the coccoliths are almost always small and incompletely developed (fig. 1 b, c).

The *coccoliths* are exceedingly singular bodies. My own account of them, quoted above, is extremely imperfect, and in some respects erroneous. And though Mr. Sorby's description is a great improvement on mine, it leaves much to be said.

I find that two distinct kinds of bodies have been described by myself and others under the name of coccoliths. I shall term one kind *Discolithus*, and the other *Cyatholithus*.

The *Discolithi* (fig. 2) are oval discoidal bodies, with a thick, strongly refracting rim, and a thinner central portion, the greater part of which is occupied by a slightly opaque, as it were, cloud-like patch. The contour of this patch corresponds with that of the inner edge of the rim, from which it is separated by a transparent zone. In general, the discoliths are slightly convex on one side, slightly concave on the other, and the rim is raised into a prominent ridge on the more convex side, so that an edge view exhibits the appearance shown in fig. 2 d.

The commonest size of these bodies is between $\frac{1}{7000}$th and $\frac{1}{5000}$th of an inch in long diameter; but they may be found, on the one hand, rising to $\frac{1}{2700}$th of an inch in length, (fig. 2 f), and, on the other, sinking to $\frac{1}{11000}$th (fig. 2 a). The last mentioned are hardly distinguishable from some of

---

* These apparent rods are not merely edge views of disks.

the granules of the granule-heaps. The largest discoliths are commonly free, but the smaller and smallest are very generally found imbedded among the granules.

The second kind of coccolith (fig. 4 *a—m*), when full grown, has an oval contour, convex upon one face, and flat or concave upon the other. Left to themselves, they lie upon one or other of these faces, and in that aspect appear to be composed of two concentric zones (fig. 4 *d*, 2, 3) surrounding a central corpuscle (fig. 4 *d*, 1). The central corpuscle is oval, and has thick walls; in its centre is a clear and transparent space. Immediately surrounding this corpuscle is a broad zone (2), which often appears more or less distinctly granulated, and sometimes has an almost moniliform margin. Beyond this appears a narrower zone (3), which is generally clear, transparent, and structureless, but sometimes exhibits well-marked striæ, which follow the direction of radii from the centre. Strong pressure occasionally causes this zone to break up into fragments bounded by radial lines.

Sometimes, as Dr. Wallich has already observed, the clear space is divided into two (fig. 1 *e*). This appears to occur only in the largest of these bodies, but I have never observed any further subdivision of the clear centre, nor any tendency to divide on the part of the body itself.

A lateral view of any of these bodies (fig. 4 *f—i*) shows that it is by no means the concentrically laminated concretion it at first appears to be, but that it has a very singular and, so far as I know, unique structure. Supposing it to rest upon its convex surface, it consists of a lower plate, shaped like a deep saucer or watch-glass; of an upper plate, which is sometimes flat, sometimes more or less watch-glass-shaped; of the oval, thick-walled, flattened corpuscle, which connects the centres of these two plates; and of an intermediate substance, which is closely connected with the under surface of the upper plate, or more or less fills up the interval between the two plates, and often has a coarsely granular margin. The upper plate always has a less diameter than the lower, and is not wider than the intermediate substance. It is this last which gives rise to the broad granular zone in the face view.

Suppose a couple of watch-glasses, one rather smaller and much flatter than the other; turn the convex side of the former to the concave side of the latter, interpose between the centre of the two a hollow spheroid of wax, and press them together —these will represent the upper and lower plates and the central corpuscle. Then pour some plaster of Paris into the interval left between the watch-glasses, and that will take the place of the intermediate substance. I do not wish to imply, however, that the intermediate substance is something totally distinct from the upper and lower plates. One would naturally expect to find protoplasm between the two plates; and the granular aspect which the intermediate substance frequently possesses is such as a layer of protoplasm might assume. But I have not been able to satisfy myself completely of the presence of a layer of this kind, or to make sure that the intermediate substance has other than an optical existence.

From their double-cup shape I propose to call the coccoliths of this form *Cyatholithi*. They are stained, but not very strongly, by iodine, which chiefly affects the intermediate substance. Strong acids dissolve them at once, and leave no trace behind; but by very weak acetic acid the calcareous matter which they contain is gradually dissolved, the central corpuscle rapidly loses its strongly refracting character, and nothing remains but an extremely delicate, finely granulated, membranous framework of the same size as the cyatholith.

Alkalies, even tolerably strong solution of caustic soda, affect these bodies but slowly. If very strong solutions of caustic soda or potash are employed, especially if aided by heat, the cyatholiths, like the discoliths, are completely destroyed, their carbonate of lime being dissolved out, and afterwards deposited usually in hexagonal plates, but sometimes in globules and dumb-bells.

The *Cyatholithi* are traceable from the full size just described, the largest of which are about $\frac{1}{10000}$th of an inch long, down to a diameter of $\frac{1}{8000}$th of an inch. Their structure remains substantially the same, but those of $\frac{1}{3000}$th of an inch in diameter and below it are always circular instead of oval; the central corpuscle, instead of being oval, is circular, and the granular zone becomes very delicate. In the smallest the upper plate is a flat disc, and the lower is but very slightly convex (fig. 1 *f*). I am not sure that in these very small cyatholiths any intermediate substance exists, apart from the under or inner surface of the upper disc. When their flat sides are turned to the eye, these young cyatholiths are extraordinarily like nucleated cells; and it is only by carefully studying side views, when the small cyatholiths remind one of minute shirt-studs, that one acquires an insight into their real nature. The central corpuscles in these smallest cyatholiths are often less than $\frac{1}{40000}$th of an inch in diameter, and are not distinguishable optically from some of the granules of the granule-heaps.

The *coccospheres* occur very sparingly in proportion to the coccoliths. At a rough guess, I should say that there is not one of the former to several thousand of the latter. And owing to their rarity, and to the impossibility of separating them from the other components of the Atlantic mud, it is very difficult to subject them to a thorough examination.

The coccospheres are of two types—the one compact, and the other loose in texture. The largest of the former type which I have met with measured about $\frac{1}{1300}$th of an inch in diameter (fig. 6 e). They are hollow, irregularly flattened spheroids, with a thick transparent wall, which sometimes appears laminated. In this wall a number of oval bodies (1), very much like the "corpuscles" of the cyatholiths, are set, and each of these answers to one of the flattened facets of the spheroidal wall. The corpuscles, which are about $\frac{1}{4500}$th of an inch long, are placed at tolerably equal distances, and each is surrounded by a contour line of corresponding form. The contour lines surrounding adjacent corpuscles meet and overlap more or less, sometimes appearing more or less polygonal. Between the contour line and the margin of the corpuscle the wall of the spheroid is clear and transparent. There is no trace of anything answering to the granular zone of the cyatholiths.

Coccospheres of the compact type of $\frac{1}{1700}$th to $\frac{1}{7000}$th of an inch in diameter occur under two forms, being sometimes mere reductions of that just described, while, in other cases (fig. 6, c), the corpuscles are round, and not more than half to a third as big ($\frac{1}{11000}$th of an inch), though their number does not seem to be greater. In still smaller coccospheres (fig. 6 a, b) the corpuscles and the contour lines become less and less distinct and more minute until, in the smallest which I have observed, and which is only $\frac{1}{4500}$th of an inch in diameter (fig. 6 a) they are hardly visible.

The coccospheres of the loose type of structure run from the same minuteness (fig. 7 a) up to nearly double the size of the largest of the compact type, viz. $\frac{1}{700}$th of an inch in diameter. The largest, of which I have only seen one specimen (fig. 7, d), is obviously made up of bodies resembling cyatholiths of the largest size in all particulars, except the absence of the granular zone, of which there is no trace. I could not clearly ascertain how they were held together, but a slight pressure sufficed to separate them.

The smaller ones (fig. 7 b, c, and a) are very similar to those of the compact type represented in figs. 6, c and d; but they are obviously, in the case of b and c, made up of bodies resembling cyatholiths (in all but the absence of the granular zone), aggregated by their flat faces round a common

centre, and more or less closely coherent. In *a*, only the corpuscles can be distinctly made out.

Such, so far as I have been able to determine them, then, are
the facts of structure to be observed in the gelatinous matter
of the Atlantic mud, and in the coccoliths and coccospheres.
I have hitherto said nothing about their meaning, as in an
inquiry so difficult and fraught with interest as this, it seems
to me to be in the highest degree important to keep the questions of fact and the questions of interpretation well apart.

I conceive that the granule-heaps and the transparent
gelatinous matter in which they are imbedded represent
masses of protoplasm. Take away the cysts which charac-
terise the *Radiolaria*, and a dead *Sphærozoum* would very
nearly resemble one of the masses of this deep-sea "Urschleim," which must, I think, be regarded as a new form of
those simple animated beings which have recently been so well
described by Haeckel in his 'Monographie der Moneren.'*
I proposed to confer upon this new "Moner" the generic
name of *Bathybius*, and to call it after the eminent Professor of Zoology in the University of Jena, *B. Haeckelii*.

From the manner in which the youngest *Discolithi* and
Cyatholithi are found imbedded among the granules; from
the resemblance of the youngest forms of the *Discolithi* and
the smallest "corpuscles" of *Cyatholithus* to the granules;
and from the absence of any evident means of maintaining
an independent existence in either, I am led to believe that
they are not independent organisms, but that they stand in
the same relation to the protoplasm of *Bathybius* as the
spicula of Sponges or of *Radiolaria* do to the soft part of
those animals.

That the coccospheres are in some way or other closely
connected with the cyatholiths seems very probable. Mr.
Sorby's view is that the cyatholiths result from the breaking
up of the coccospheres. If this were the case, however, I
cannot but think that the coccospheres ought to be far more
numerous than they really are.

The converse view, that the coccospheres are formed by
the coalescence of the cyatholiths, seems to me to be quite as
probable. If this be the case, the more compact variety of
the coccospheres must be regarded as a more advanced stage
of development of the loose form.

On either view it must not be forgotten that the components of the coccospheres are not identical with the free
cyatholiths; but that, on the supposition of coalescence, the
disappearance of the granular layer has to be accounted for;

* 'Jenaische Zeitschrift,' Bd. iv, Heft 1.

while, on the supposition that the coccospheres dehisce, it must be supposed that the granular layer appears after dehiscence; and, on both hypotheses, the fact that both coccospheres and cyatholiths are found of very various sizes proves that the assumed coalescence or dehiscence must take place at all periods of development, and is not to be regarded as the final developmental act of either coccosphere or cyatholith.

And, finally, there is a third possibility—that the differences between the components of the coccospheres and the cyatholiths are permanent, and that the coccospheres are from the first independent structures, comparable to the wheel-like spicula associated in the wall of the "seeds" of *Spongilla*, and perhaps enclosing a mass of protoplasm destined for reproductive purposes.

In addition to *Bathybius* and its associated discoliths, cyatholiths, and coccospheres, the Atlantic mud contains—

a. Masses of protoplasm surrounded by a thick but incomplete cyst, apparently of a membranous or but little calcified consistence, and resembling minute *Gromiæ*. It is possible that these are unfinished single chambers of *Globigerinæ*.

b. *Globigerinæ* of all sizes and ages, from a single chamber $\frac{1}{7500}$th of an inch in diameter, upwards. I may mention incidentally that very careful examination of the walls of the youngest forms of *Globigerina* with the $\frac{1}{12}$th leads me to withdraw the doubt I formerly expressed as to their perforation.

In the absence of any apparent reproductive process in *Globigerinæ*, is it possible that these may simply be, as it were, offsets, provided with a shell, of some such simple form of life as *Bathybius*, which multiplies only in its naked form?

c. Masses of protoplasm enclosed in a thin membrane.

d. A very few *Foraminifera* of other genera than *Globigerina*.

e. *Radiolaria* in considerable numbers.

f. Numerous *Coscinodisci* and a few other Diatoms.

g. Numerous very minute fragments of inorganic matter.

The *Radiolaria* and Diatoms are unquestionably derived from the surface of the sea; and in speculating upon the conditions of existence of *Bathybius* and *Globigerina*, these sources of supply must not be overlooked.

With the more complete view of the structure of the cyatholiths and discoliths which I had obtained, I turned to the chalk, and I am glad to have been enabled to verify Mr. Sorby's statements in every particular.   The chalk contains cyatholiths and discoliths identical with those of the Atlantic soundings, except that they have a more dense look and coarser contours (figs. 3 and 5).   In fact, I suspect that they are fossilized, and are more completely impregnated with carbonate of lime than the recent coccoliths.

I have once met with a coccosphere in the chalk; and, on the other hand, in one specimen of the Atlantic soundings I met with a disc with a central cross, just like the body from the chalk figured by Mr. Sorby (fig. 8).

Micr. Journ. Vol. VIII. N.S. Pl. IV

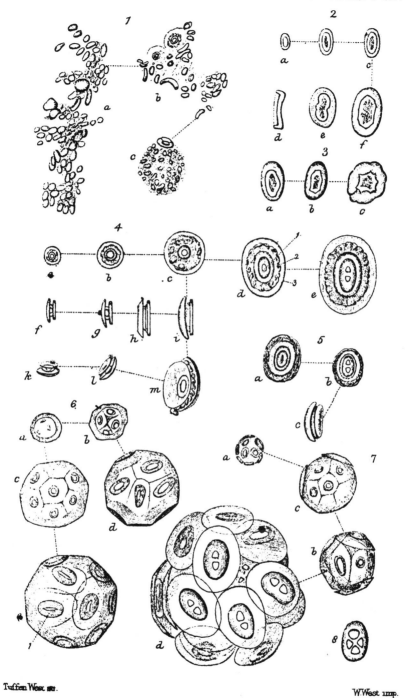

Tuffen West del.

W. West imp.

[*Editor's Note:* See next page for Plate description.]

JOURNAL OF MICROSCOPICAL SCIENCE.

DESCRIPTION OF PLATE IV,

Illustrating Prof. Huxley's paper on Organisms from Great Depths in the North Atlantic Ocean

Fig.

1.—Masses of the gelatinous substance.

2.—*Discolithi* from Atlantic mud.

3.—   ,,   from the chalk of Sussex.

4.—*Cyatholithi* from the Atlantic mud.

5.—   ,,   from the chalk of Sussex.

6.—Coccospheres of the compact type.

7.—   ,,   of the loose type.

8.—A crucigerous disk from Atlantic mud.

All the figures are drawn to the same scale, and are supposed to be magnified 1200 diameters.

4

Reprinted from pages 342–348 and 349–350 of *Ann. Mag. Nat. Hist.*, ser. 4, **19:**342–350 (1877)

OBSERVATIONS ON THE *COCCOSPHERE*

G. C. Wallich, M.D.

Surgeon-Major Retired List H.M. Indian Army

THE history of what may be termed the Coccosphere question is a remarkable one. Seventeen years ago I pointed out, as the result of actual observation, that the "coccoliths," which had been discovered three years previously by Professor Huxley in soundings from the Atlantic, are not independent structures, but merely cast-off appendages of the Coccosphere-cell. Yet, from that period to the present, the physiological relation existing between these two integral portions of one and the same organism has remained shrouded in mystery. Since 1868 a number of elaborate observations have been published, both here and abroad, on the characters and supposed affinities of the various forms of " coccolith." But, unfortunately, the value of these observations has been materially diminished, owing to their being based on one or other of the following essentially fallacious assumptions :—namely, that the " *coccolith* " itself is a " cell; " that it is an independently developed and independently living structure; and that, as a " *coccolith*," it is capable of taking part in any subsequent vital combination.

These assumptions have possibly had their origin in two statements made by Prof. Huxley :—the first, in 1858*, that "*coccoliths* somewhat resemble single *cells* of the plant *Protococcus*;" the second, ten years later, namely in 1868†, that the varieties of " *coccoliths* " named by him " *Discoliths* and *Cyatholiths* stand in the same relation to the protoplasm of *Bathybius* as the *spicula* of sponges or of *Radiolaria* do to the soft parts of these animals." It is true that in the same paper Prof. Huxley noticed three alternative "possibilities " in relation to the *cocco-*

---

* 'Deep-sea Soundings in the North Atlantic,' made in H.M.S. 'Cyclops,' Commander Dayman, in 1875. Appendix, Report on Soundings, by Prof. Huxley, p. 64.

† "On some Organisms living at Great Depths in the North Atlantic Ocean," by Prof. Huxley, F.R.S., 'Quart. Journ. Microsc. Science,' Oct. 1868, p. 210.

spheres. But any one who carefully studies his remarks must, I think, conclude that, on the whole, he was disposed to give "*Bathybius*" the benefit of the doubt, and to regard the *coccospheres* as subsidiary productions due to "the coalescence" of the "*coccoliths*"—a view, which then, as now, I venture most respectfully to contest. For although the supreme interest that centred in the "*coccoliths*" has waned since they ceased to constitute the bones of *Bathybius*, we must not forget the important part already played by them in the construction of certain rocks, and which they still continue to play in the construction of certain oceanic deposits. I may be pardoned therefore for seeking to redeem the coccosphere-question from the chaos into which it has drifted, and for suggesting that had the fact indicated by me in a paper "On the *Polycystina*" (read at the Royal Microscopical Society in 1865), namely that I "had met with *coccospheres* as free floating organisms in tropical seas" in 1857, been recognized as I think it ought, Sir Wyville Thomson would have abstained, in 1872 *, from casting unmerited doubts on my view regarding the true relation of the "*coccoliths*" to the *coccospheres*, and, in 1874, from adopting and publishing that view as a new and original observation made on board the 'Challenger' †.

From first to last in my published writings on the subject, I have never made the statement so persistently attributed to me (and which involves a contradiction of the opinion really entertained and expressed by me), namely that "sometimes the coccoliths are found aggregated into spheroids" (see 'Lay Sermons,' "On a Piece of Chalk," by Prof. Huxley, 5th edit. 1874, p. 186) ‡, but have invariably adhered to the opinions

* "Sometimes the '*Coccoliths*' are found aggregated on the surface of small transparent balls, and these, *which seemed at first to have something to do* with the production of the '*coccoliths*,' Dr. Wallich has called '*coccospheres*.'" (Sir Wyville Thomson, 'The Depths of the Sea,' 1872, p. 413.)

† "I need only say that I believe *our* observations have placed it beyond a doubt that the '*coccoliths*' are the separated elements of a peculiar calcareous armature which covers certain spherical bodies (the '*coccospheres*' of Dr. Wallich)." (Sir W. Thomson, 'Proceedings Roy. Soc.' vol. xxiv. No. 154, Nov. 1874, p. 38.)

‡ See also 'The Microscope,' 5th edit. 1875, p. 464, where Dr. Carpenter speaks of "the larger spherical *aggregations* first observed by Dr. Wallich, and designated by him as *coccospheres*;" and at p. 466, "The coccospheres are made up *by the aggregation of bodies* resembling cyatholiths." As (in the 'Introduction to the Study of the Foraminifera,' 1862, pp. 46–7) Dr. Carpenter quoted almost *in extenso* both the description and figures of "*coccoliths*" and *coccospheres* given by me in 'The Annals' of July 1861, it is difficult to see how he could so completely have misunderstood what I both described and figured.

of which a correct *résumé* is given in my paper " On Deep-sea
Protozoa " (' Monthly Micr. Journ.' Jan. 1869)—namely, that
after a careful and long-continued study of these organisms,
whether occurring as free floating inhabitants of the surface-
waters of the Indian Ocean and mid-Atlantic, as components
of the present deep-sea deposits, in a fossil condition in the
post-tertiary earths, or as living organisms in the British
Channel, I have never deviated from the opinion that the free
coccoliths are derived from their parent *coccospheres*.   In
some deep-sea deposits, as stated by Prof. Huxley, free *coccoliths* undoubtedly occur in overwhelming number as compared
with the *coccospheres*; but it is equally true that *coccospheres* are, at times, present in great abundance, whereas free
coccoliths are comparatively scarce.   Coupling these facts
with the very important one, that perfect *coccospheres* are to
be met with of every intermediate size between the $\frac{1}{5000}$ and
$\frac{1}{830}$ of an inch in diameter, I am induced to believe that the
free *coccoliths* are, *in every instance*, formed on, or *pari passu*
with, the spheroidal cells on which they rest, *their state of
attachment to these cells being their normal as well as pristine
condition.*   That they revert at any future stage of their
history, after once becoming free, to their original composite
state, there is no recorded evidence forthcoming to prove.
(In an appended footnote it was stated that " some of the *free-
floating coccospheres are oblong.*")   Lastly, I stated (*loc. cit.*)
(with reference to the " granular zone " which Prof. Huxley described as *possibly* forming a normal portion of the
coccolith), that " amongst the immense numbers of coccospheres which had been examined both in the recent state and
in the preserved though still recent material of the soundings,
I had never met with any proof that this zone exists as an integral portion of the structure; nor had any evidence presented
itself" that the " granular zone " is any thing more than an
accidental accretion, *or that its presence is due to any inherent
condition without which the organism would be incomplete.*
(" On Deep-sea Protozoa," ' Monthly Micr. Journ.' Jan. 1869,
pp. 35 and 36).

Having thus far shown that there is no reason to suppose
that the *Coccosphere* is a secondary formation, resulting in any
way from an " *aggregation* " of independently developed " *coccoliths*," but that the balance of evidence is altogether in
favour of the view that the " *coccoliths* " are normally developed
upon, and simultaneously with, their parent *coccosphere*, I have
now to state the grounds on which I base the opinion that
the " *coccolith* " presents none of the characters of a true
" *cell*."

Although Prof. Huxley, in his first brief notice of the coccoliths (already referred to as having appeared in 1858) described the "*coccolith*" as being somewhat like a single cell of the plant *Protococcus*," he has nowhere asserted that it is a cell. In his paper describing *Bathybius* ('Quart. Journ. Micr. Science,' Oct. 1868, p. 207) he alludes to "*a central corpuscule*," and says, "*there is in its centre a clear and transparent space*," adding that "sometimes, as Dr. Wallich has already observed, the clear space is divided into two. This appears to occur only in the largest of these bodies; but I have never observed any further subdivision of the clear centre, nor any tendency to divide on the part of the body itself." In the same paper Prof. Huxley pointed out, for the first time, the double or *shirt-stud*-like figure of the "*coccoliths*," a feature which I had altogether overlooked, owing, doubtless, to my attention having been chiefly directed towards the *Coccosphere* as a whole.

Now *every thing* depends on a correct interpretation of what Prof. Huxley describes as the *central corpuscule* and the *clear space* at its centre. He says, "Suppose a couple of watch-glasses, one rather smaller than the other; turn the convex side of the former to the concave side of the latter; interpose between the centre of the two a hollow spheroid of wax, and press them together: these will represent the upper and lower plates and the central corpuscule" (loc. cit. p. 207). This description is most closely borne out by Prof. Huxley's figures. To facilitate my explanation, I have reproduced three of his figures in the Plate which accompanies this paper—namely, figs. 13 H, 14 H, and 15 H. It will be seen from these, that if we apply his experimental illustration of the two watch-glasses and the hollow spheroid of wax, where there is *one* clear space in the centre of the central corpuscule, we should have to employ either two hollow spheroids of wax, or one spheroid with two cavities in it, to represent the *coccolith* in which two central clear spaces occur; and so on, whatever the number of central clear spaces may be. To my mind this does not by any means give a correct idea of the appearances; which, on the contrary, indicate that the central clear space or spaces are either single or double perforations in the external disk—its "markings," as it were—and nothing more. They have, therefore, no physiological significance, and certainly do not represent any thing that can be called a cell. See Plate XVII. fig. 10, which gives a diagrammatic sectional view of a coccolith. There is no evidence forthcoming, that I am aware of, to show whether the *stem* of the stud (*i. e.* the intermediate piece between the two disks), is or is not continuous with the disks. As the appearance of concentric rings is constant, being observable even in the fossil coccoliths, I presume the stem must be continuous with the disks.

Instead of the watch-glasses and hollow spheroid of wax, imagine a shirt-stud made of colourless glass, with a minute shallow hole drilled at the centre of the *larger* of its two disks, which (as in the case of the coccolith) would constitute the outer disk. Imagine this glass stud to be enveloped in transparent varnish or any glairy fluid. On looking down upon it we should see (fig. 5, *b*) a minute central ring formed by the edge of the minute central hollow ; external to, and at a little distance from this, a second ring (*c*), formed by the outline of the *stem* of the stud ; again, a little external to this, a third ring (*d*), formed by the outline of the smaller of the two disks of the stud (*e*) ; and lastly, the marginal outline. Of course the multiple " central clear-spaces " might be imitated by drilling a corresponding number of holes in the outer disk (see Plate XVII. fig. 7). Now here we should have precisely the same appearance of concentric rings and central spaces as we find in the " *coccolith* ; " and what is more, they would have a similar origin. Of course the only difference observable in looking down on the *coccolith* or the glass stud from the direction of the inner or smaller disk, would be that the " central clear space " would be somewhat less distinct, whereas the outline of the smaller disk would be more distinct.

I have now to refer to Mr. Carter's views as embodied in his paper on " *Melobesia unicellularis* " (Annals and Mag. Nat. Hist., Mar. 1871). Let me, however, at once confess that whilst I dissent, *in toto* (for reasons already assigned), from the view that the " *coccolith* " is, in any sense, " *a cell*," I am quite prepared to adopt Mr. Carter's opinion, if he will permit me, as applicable to the parent and entire structure, namely the *coccosphere* with its " *coccoliths*." The only difficulty I see in the way of regarding the *Coccosphere* as a protophyte, resides in the remarkable evidence of its relationship to certain Foraminifera, furnished by the discovery (at first in one or two specimens only, but afterwards in many) of shells so regularly studded with *coccoliths*, as to suggest the idea that the chambers originated as *coccospheres* *. One thing would seem certain, that this regularity is incompatible with the supposition that the *coccoliths* got into their position accidentally. How then, did they attain it ? I once asked Mr. Carter if he could explain the matter ; and he obligingly sent

* See my observations on this subject, and accompanying figures in 'The Annals,' for July 1861, p. 55; and in the 'Monthly Microscopical Journal,' for Jan. 1869, pp. 37, 38.

the best explanation I have as yet come across, though even this has a weak point in it. It was, that the animals of the Foraminifera probably employed the *coccoliths*, which abound in the mud, instead of sand or other particles for the strengthening of their shells, as we know to be the habit of a large number of the Foraminifera that live at the bottom of the deep sea. But, although the sparse kind of tessellation with large mineral particles here and there on the shell is undoubtedly characteristic of some species (as for example *Proteonina* and *Buliminia*; and, as I have elsewhere shown to be the case in certain deep-sea Foraminifera as well as freshwater testaceous Rhizopods, "the selective and adaptive power" exhibited in the material and workmanship of the shells is simply marvellous), in the shells now under notice the arrangement of the *coccoliths* appears almost *too* like that observable on the *coccospheres* to render it easily intelligible how the animal of the Foraminifer could have so exactly "mimicked" it. On the other hand, there is a piece of evidence which would seem to support Mr. Carter's view of unicellular algal affinity (supposing it to be extended to the *coccosphere*), namely, an appearance of "dehiscence" which presents itself not unfrequently in the large oblong *coccospheres* met with in tropical seas, and so invariably occurs, at one end only, as to negative the idea of its being accidental (See Plate XVII. fig. 4).

Mr. Carter suggests that the "loose type" of coccosphere described and figured by Prof. Huxley may "be a still more developed form of the *sporangium* or coccosphere, perhaps undergoing dehiscence" (*loc. cit.* p. 189). He will, however, I know, pardon me saying that it is going too far ahead of the evidence to assume that the coccosphere *is* a sporangium at all; for if it be, out of the multitudes I have seen, none has ever departed from the sporangial phase, either in those met with at the top or at the bottom of the ocean. But a glance at the curious object I have depicted (Plate XVII. fig. 18), which I have repeatedly met with in some parts of Bengal, will at once show that Unicellular Algæ do undoubtedly assume a sporangial condition in accordance with that which Mr. Carter must have had in his mind's eye when he suggested that the coccosphere might be a sporangium. My specimen is, I believe, the sporangial condition of a branching stipitate form of *Ankistrodesmus*, each of the kidney-shaped bodies being a frond.

Figures 1 to 4 (see Plate XVII.) represent the only two species of *Coccosphere* I have hitherto met with :—the spherical one being the ubiquitous oceanic form, which I propose to call *Coccosphæra pelagica*; the oblong species, which is not so common by any means, being, so far as my experience goes, confined to tropical or subtropical seas.   I propose to name it after Mr. Carter, *Coccosphæra Carterii*.

The following are the characters of the two species :—

Genus COCCOSPHÆRA (Wall.).

1. *Coccosphæra pelagica* (Wall.).

Cell spherical, hyaline, with a distinct membranous wall. Cell-contents, a perfectly colourless glairy protoplasm.   *Coccoliths* generally more or less elliptical, numbering from 16 to 36, arranged side by side, and, in the normal state, not over-lapping.   Central aperture of *Coccolith* single, margin of external disk finely and radially striate.   Internal disk plain.

Diameter of *Coccosphere* ranging from $\frac{1}{5000}$ to $\frac{1}{850}$, of an inch.   Length of *Coccoliths* from $\frac{1}{9000}$ to $\frac{1}{1000}$ of an inch.

Habitat. Free-floating, Indian Ocean and North Atlantic; and (dead) in North Atlantic muds.   Always most abundant where the *Globigerinæ* are in greatest profusion, and the deposit of the purest kind.

2. *Coccosphæra Carterii* (Wall.).

Cell oblong.   Long diameter about twice that of short diameter.   Cell as in *C. pelagica*.   *Coccoliths* varying in number from 16 to 38, more or less oblong, with two central apertures arranged lengthwise, margin finely and radially striate.   Internal disk plain.   Length of *Coccosphere* from $\frac{1}{1000}$ to $\frac{1}{800}$ of an inch.   Length of coccolith from $\frac{1}{5000}$ to $\frac{1}{1000}$ of an inch.

Habitat. Free-floating, Indian Ocean, and Mid-Atlantic. (N.B. I have not observed any intermediate form between the spherical and oblong.)

It only remains for me to add, that I have not referred in the course of the preceding observations to the highly important researches of Sorby, Oscar Schmidt, Haeckel, Gümbel, and others, simply because my own inquiries have been directed principally towards an aspect of the subject upon which they have hardly touched at all—my object having been to sustain the accuracy of my own observations, not to question that of others.

EXPLANATION OF PLATE XVII.

Fig. 1. *Coccosphæra pelagica* (Wall.), with its complement of *coccoliths.*

Fig. 2. Cell-wall of same, showing distinct membranous outline ; most of the *coccoliths* having been thrown off.

Fig. 3. *Coccosphæra Carterii* (Wall.).

Fig. 4. The same in the *dehiscent* (?) condition.

Fig. 5. *Coccolith* of *C. pelagica* seen from external aspect ; showing the radiate striation on margin of outer disk, and the central depression which constitutes the " central clear space " of Huxley.

Fig. 6. *Coccolith* of *C. Carterii*; side view, showing the *two* central depressions and radiate marginal striæ, together with the inner disk and intermediate piece.

Fig. 7. The same, as seen from its external aspect, this being, in short, a front view of the outer disk. Here also the two button-hole-like depressions are shown.

Fig. 8. Circular *coccolith* of *C. pelagica* occasionally met with.

Fig. 8 *a.* A specimen of a form of *coccolith* occasionally but rarely occurring, in which there is no central depression, but apparently an aperture close to the margin of the outer disk.

Fig. 9 D. Diagrammatic, enlarged, side view of coccolith of *C. pelagica.*

Fig. 10 D. Diagrammatic vertical section of same, showing the central depression (*a*), in external disk : *s*, the stem : *e d*, the inner disk.

Fig. 11 D. Diagrammatic front view of the outer disk of same : *a*, the central depression, the "central clear space" of Huxley, and "nucleus" of other writers ; *b*, the innermost ring, indicating the margin of this depression ; *c*, the ring indicating the outline of the intermediate piece, or stem uniting the two disks ; *d*, the ring indicating the margin of the inner disk ; *e*, the outline of the outer disk itself. *Possibly* these are the rings referred to in Prof. Huxley's Report of 1868, when describing the *coccoliths* as " curious rounded bodies, to all appearance consisting of several concentric layers surrounding a minute clear centre."

Fig. 12 S. This figure is copied from fig. 20, plate 16, appended to Prof. Oscar Schmidt's paper " On Coccoliths and Rhabdoliths " Annals & Mag. Nat. Hist. Nov. 1872, translated by W. S. Dallas, F.L.S. It is described in the text (p. 367) as " a decided coccolith with a dorsal shield, as may be ascertained by placing it on its edge, the dark non-granular part, *b*, representing the granular zone, and the clear spaces in it ; *a*, *divided medullar space without central granules.*"

Figs. 13 H, 14 H, and 15 H. Three figures copied from the plate accompanying Professor Huxley's paper ; described as " Cyatholiths from the Atlantic Mud." The central corpuscle with its clear space, *a*, in the centre is shown in figs. 13 and 14. The "granular zone," *g z*, is shown in fig. 15.

Fig. 16 represents a two-celled or chambered *coccosphere*—being apparently the first stage in the formation of the *coccolith*-covered *Textulariæ* and *Rotaliæ* which have been described by me in former papers, and of which mounted specimens are extant.

Fig. 17. A *coccolith* of *C. Carterii* as seen in preserved specimens, an aggregation of granules being observable around the stem between the outer and inner disks, the so-called " granular zone " of authors.

Fig. 18. Sporangium of a protophyte from Bengal, probably allied to *Ankistrodesmus* : *a.* the globular colourless and transparent sporangial cell ; *b b b*, the kidney-shaped fronds of same. These never have a *flagellum* or cilia, and are not zoospores.

N.B. In figs. 5, 7, and 11 D the letters indicate the same portions of the structure.

·Dr. G. C. Wallich *on the* Coccosphere.

Ann. & Mag. Nat. Hist. S. 4. Vol. 19. Pl. XVII.

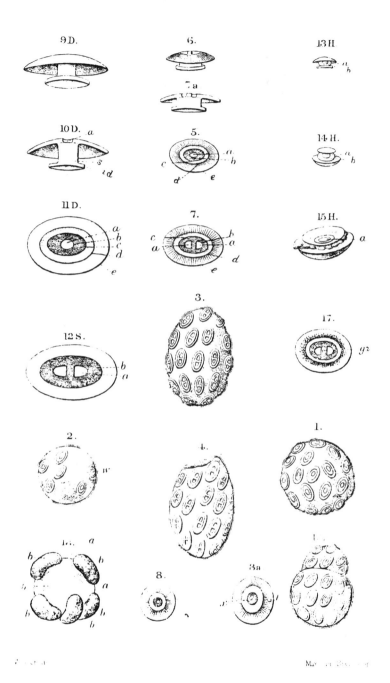

Part II

STRUCTURAL, BIOLOGICAL, BIOGEOGRAPHIC, AND PRESERVATIONAL ASPECTS OF NANNOPLANKTON

Editor's Comments
on Papers 5 Through 16

5 BLACK
The Fine Structure of the Mineral Parts of Coccolithophoridae

6 PAASCHE
Biology and Physiology of Coccolithophorids

7 WATABE and WILBUR
Effects of Temperature on Growth, Calcification, and Coccolith Form in Coccolithus Huxleyi *(Coccolithineae)*

8 MCINTYRE, BÉ, and ROCHE
Modern Pacific Coccolithophorida: A Paleontological Thermometer

9 OKADA and HONJO
The Distribution of Oceanic Coccolithophorids in the Pacific

10 BRAMLETTE
Significance of Coccolithophorids in Calcium-Carbonate Deposition

11 BERGER
Deep-Sea Carbonates: Evidence for a Coccolith Lysocline

12 ROTH, MULLIN, and BERGER
Coccolith Sedimentation by Fecal Pellets: Laboratory Experiments and Field Observations

13 HONJO
Coccoliths: Production, Transportation and Sedimentation

14 THIERSTEIN
Selective Dissolution of Late Cretaceous and Earliest Tertiary Calcareous Nannofossils: Experimental Evidence

15 WISE and KELTS
Abstract from *Inferred Diagenetic History of a Weakly Silicified Deep Sea Chalk*

16   NOËL and MANIVIT
Nannofaciès de "black shales" aptiennes et albiennes d'Atlantique sud (legs 36 et 40). Intérêt sédimentologique

This section includes a rather mixed selection of articles on the fine structure of coccoliths, a review of our knowledge of the biology of coccolithophores, their extant distribution, the postmortem dissolution of coccospheres in the water column, and their role in the deposition and preservation of carbonate sediments.

The prodigious contribution of E. Kamptner to the study of coccoliths included the first attempt at explaining the crystallographic details of coccolith plates. Based on the polarized light images of various types of coccoliths, Kamptner (1954) determined that their minute crystallites were arranged in a helicoid manner and the slight differences in this arrangement produced the characteristic extinction patterns of different coccolith species under cross-polarized light. This paper on the microscopic structure of coccoliths is a valuable contribution to the early research in nannopaleontology; however, due to its extensive length (90 pages) it could not be included in the present volume.

Black applied his petrological talents to the observation of nannofossil crystallography under a transmission electron microscope (see Paper 5) and demonstrated that three calcite crystal habits occur in nannoplankton. Unmodified hexagonal prisms are the most common crystal habit for holococcoliths (coccoliths constructed of crystallites of the same shape and size, associated with the motile phase of the life cycle). Rhombohedral prisms, however, also occur in some holococcoliths. Heterococcoliths (coccoliths made of crystallites of different shapes and sizes, occurring in the nonmotile phase of the life cycle) are always composed of modified rhombohedra. Discoasters, on the other hand, crystallize the tubular habit of calcite; each discoaster arm is made up of a single tubular crystal, joined with other similar crystals to form asteroliths. Black (1972) further demonstrated that discoasters are usually constructed of three radially stacked structural elements. After diagenetic changes, the tubular crystals can become rhombohedra arranged asymmetrically around the center.

Paasche's paper (Paper 6) contains a summary of our rather limited knowledge of the life histories of the coccolithophore species. In spite of their quantitative prominence among marine phytoplankton and their importance as primary producers, culture studies on coccolithophores have been rare, and have been hampered by the difficulties of keeping the species alive in the restrictive environment of the laboratory. Most of the species included in Paasche's synthesis are near-shore forms with a benthic stage in their life cycle, and are thus not representative of the oceanic species. *Emiliania huxleyi* and *Coccolithus pelagicus,* occurring both near-shore and in the open sea, are the only ubiquitous species that have been cultured successfully. [A detailed account of the biological aspects of coccolithophores can be found in Tappan (1980)].

The mineralization of coccoliths has been observed in the cultures of *Emiliania huxleyi* (Wilbur and Watabe, 1963) and *Coccolithus pelagicus* (Manton and Leedale, 1969). In *E. huxleyi,* Wilbur and Watabe observed that deposition of coccolith calcite within the protoplasm occurs near a reticular body next to the Golgi vesicle. Precipitation of calcite starts near the center and proceeds upwards and outwards to the edges of the shield. After calcification is complete, the shields are pushed to the surface of the cell, where they arrange themselves in an interlocking coccospheric envelope around the protoplasm. In *C. pelagicus,* the coccolith shields originate within the Golgi vesicle and remain attached to the underlying organic scales. Two concentric envelopes of morphologically different types of coccoliths around the same cell have also been observed in some species of *Syracosphaera* (Okada and McIntyre, 1977).

Watabe and Wilbur's study of the effects of temperature on growth and calcification in *Coccolithus huxleyi* (= *Emiliania huxleyi*) (see Paper 7) demonstrated that changes in temperature can significantly alter the rate of growth of coccoliths as well as the size of the resultant placoliths. In culture this species tolerates a temperature range of 7° to 27°C, whereas in nature varieties of *E. huxleyi* have been recorded in water temperatures as low as 2°C. Other studies have shown the influence of calcium, strontium, and magnesium concentrations on the calcification and growth of coccoliths. Blackwelder et al. (1976) amd Weiss et al. (1976) studied the near-shore species *Cricosphaera carterae* in culture. Their data indicated that in calcium-deficient media the addition of strontium stimulated both cell division and calcification, and that the presence of magnesium in the medium also enhanced coccolith formation, with up to 40% of $CaCO_3$ deposited as aragonite. In the absence of magnesium, only calcite was deposited. In the fossil assemblages, however, the concentrations of magnesium and strontium in coccoliths are very low—too low to be detected by X-ray diffraction and electron microprobe analysis (Shaffer, 1973; Siesser, 1977).

The biogeography of calcareous nannoplankton has been mapped both in the Atlantic (McIntyre and Bé, 1967) and in the Pacific Oceans

(Papers 8 and 9; Honjo and Okada, 1974; Geitzenauer et al., 1976). On the basis of mapped assemblages, McIntyre and Bé were able to distinguish tropical, subtropical, transitional, and subpolar nannofloral provinces in the Atlantic, with their boundaries corresponding to the surface isotherms. The highest diversities were recorded in the tropical and subtropical zones. In the Pacific, a similar relationship between nannofloral zones and surface water temperatures were also demonstrated (Papers 8 and 9).

In a north-south transect study in the Pacific Ocean between 50°N and 15°S, Okada and Honjo (Paper 9) found high diversities in the tropical and subtropical latitudes, but the highest numbers of cells per liter were found in the convergence zone of the subarctic and the narrow equatorial belt only. Honjo and Okada (1974) also found a vertical stratification of coccolithophores in the photic zone, with an abrupt change at about 130 m. depth, separating a middle photic zone dominated by delicate species from a lower photic zone dominated by heavily calcified and fortified forms. Okada and McIntyre (1979) have also studied the seasonal variation of coccolithophores in the western North Atlantic. As expected, they found that coccoliths are less populous during winters at high latitudes and during summers at low (subtropical) latitude stations. The average standing crop in the surface waters, however, was consistent throughout all climatic regions. They also found that populations decrease with depth and towards higher latitudes, and that many species showed seasonally biased occurrences.

Papers 10 through 16 deal with aspects of production and sedimentation of coccoliths, including dissolution and diagenesis. Bramlette's paper (Paper 10) was the first article that recognized the dominant role of coccolithophorids in carbonate deposition, their significance to the geochemical cycle and $CaCO_3$ budget in the oceans, and their promise in biostratigraphy. Bramlette also pointed out that pre-Quaternary oozes were much richer in coccoliths than the Quaternary ones, although the reason for this still remains unexplained.

Berger (Paper 11) has attempted to rank dissolution resistant vs. susceptible coccolith species and point out the problems associated with identifying a coccolith lysocline (a facies boundary between well-preserved and poorly-preserved coccolith assemblages) in deep sea carbonates. Such ranking provides an index of the degree of dissolution, and Roth and Berger (1975) conclude that coccoliths are better dissolution indicators than Foraminifera for depth above the carbonate lysocline. Dissolution ranking of numerous extant and fossil species have been given by Schneidermann (1973) and by Roth and Berger (1975).

Papers by Roth et al. (Paper 12) and Honjo (Paper 13) address the question of the settling of coccolithophore particles through the water column. It has long been recognized that individual coccospheres were too small to settle rapidly enough to avoid excessive dissolution in the water column. A mechanism of accelerated sinking had to be found to account for the extensive accumulation of coccolith oozes on the sea floor. Transportation via the fecal pellets provided the answer (Honjo, 1975). Coccospheres pass relatively undamaged through the guts of grazing zooplankton, such as copepods, and are delivered to the water column as fecal pellets, held together and protected by the mucose material. This packaged settling accelerates sinking and reduces the residence time in the water column from many years (as required for individual coccoliths) to a few days for the fecal pellet-size particles, thereby delivering relatively unaffected coccoliths to the sea floor.

Adelseck et al. (1973) subjected coccolith samples to high temperatures and pressure to simulate diagenesis in the laboratory. From the results of their experiment, they concluded that the size of coccolith surface area was an important factor in determining whether dissolution or reprecipitation of calcite takes place. Smaller, more delicate, forms are solution susceptible, whereas robust and more massive forms are resistant to dissolution and tend to show signs of calcite overgrowth. Thierstein (Paper 14) studied the effects of dissolution on nannofloral composition by deploying late Cretaceous and earliest Tertiary assemblages in deep moorings in the Sargasso Sea. He reaffirmed the importance of available surface area in the dissolution-precipitation process and established a dissolution ranking for the late Cretaceous and earliest Tertiary nannofossils. Other studies have produced similar results (e.g., Hill, 1975).

The abstract of the paper by Wise and Kelts (Paper 15) and the article by Noël and Manivit (Paper 16) also deal with the dissolution and preservational aspects of coccoliths, the modification of the sediment after diagenesis, or its preservation in a confined, anoxic environment. Wise and Kelts take up the question of calcite reprecipitation during early diagenesis of chalk. Noël and Manivit use the nannofacies to interpret the paleoenvironment of the mid-Cretaceous black shales recovered in Deep Sea Drilling Project sites in the South Atlantic.

REFERENCES

Adelseck, C. G., Jr., G. W. Geehan, and P. H. Roth, 1973, Experimental Evidence for the Selective Dissolution and Overgrowth of Calcareous Nannofossils during Diagenesis, *Geol. Soc. Am. Bull.* **84:**2755–2762.

Black, M., 1972, Crystal Development in Discoasteraceae and Braarudosphaeraceae (Planktonic Algae), *Paleontology* **15:**476–489.

Blackwelder, P. L., R. E. Weiss, and K. M. Wilbur, 1976, Effects of Calcium, Strontium and Magnesium on the Coccolithophorid, *Cricosphaera (Hymenomonas) carterae.* I. Calcification, *Marine Biology* **34:**11-16.

Geitzenauer, K. R., M. B. Roche, and A. McIntyre, 1976, Modern Pacific Coccolith Assemblages: Derivation and Application to Late Pleistocene Paleotemperature Analyses, *Geol. Soc. America Mem. 145* p. 423-428.

Hill, M. E., III, 1975, Selective Dissolution of Mid-Cretaceous (Cenomaniam) Calcareous Nannofossils, *Micropaleontology* **21**(2):227-235.

Honjo, S., 1975, Dissolution of Suspended Coccoliths in the Deep Sea Water Column and Sedimentation of Coccolith Ooze, in *Dissolution of Deep Sea Carbonates,* W. V. Sliter, A. W. H. Bé, and W. H. Berger, eds., Cushman Foundation Foraminiferal Research Special Pub. No. 13, pp. 114-128.

Honjo, S., and H. Okada, 1974, Community Structure of Coccolithophores in the Photic Layer of the Mid-Pacific, *Micropaleontology* **20:**209-294.

Kamptner, E., 1954, Untersuchungen über den Feinbau der Coccolithen, *Arch. Protistenkd.***100:**1-90.

Manton, J., and G. F. Leedale, 1969, Observations on the Microanatomy of *Coccolithus pelagicus* and *Cricosphaera carteri,* with Special Reference to the Origin and Nature of Coccoliths and Scales, *Marine Biol. Assoc. United Kingdom Jour* **49:**1-16.

McIntyre, A., and A. W. H. Bé, 1967, Modern Coccolithophoridae of the Atlantic Ocean—I. Placoliths and Cyrtoliths, *Deep-Sea Research.* **14:**561-597.

Okada, H., and A. McIntyre, 1977, Modern Coccolithophores of the Pacific and North Atlantic Oceans, *Micropaleontology* **23:**1-55.

Okada, H., and A. McIntyre, 1979, Seasonal Distribution of Modern Coccolithophores in the Western North Atlantic Ocean, *Marine Biology* **54:**319-328.

Roth, P. H., and W. H. Berger, 1975, Distributions and Dissolution of Coccoliths in the South and Central Pacific, in *Dissolution of Deep-Sea Carbonates,*W. V. Sliter, A. W. H. Bé, and W. H. Berger, eds., Cushman Foundation Foraminiferal Research, Special Pub. No. 13, pp. 87-113.

Schneidermann, N., 1973, Deposition of Coccoliths in the Compensation Zone of the Atlantic Ocean, in *Proceedings of Symposium on Calcareous Nannofossils,* L. A. Smith and J. Hardenbol, eds., Houston, Gulf Coast Section, Soc. Econ. Paleontologists and Mineralogists, pp. 140-151.

Shaffer, B. L., 1973, Electron-probe Microanalysis of Calcareous Nannofossils, *Geoscience and Man* **7:**89-83.

Siesser, W. G., 1977, Chemical Composition of Calcareous Nannofossils, *South African Jour. Science* **73:**283-285.

Tappan, H., 1980, Haptophyta, Coccolithophores and other Calcareous Nannoplankton, in *Paleobiology of Plant Protists,* Freeman and Co., San Francisco, pp. 678-803.

Weiss, R. E., P. L. Blackwelder, and K. M. Wilbur, 1976, Effects of Calcium, Strontium and Magnesium on the Coccolithophorid *Cricospahera (Hymenomonas) carterae.* II. Cell-Division. *Marine Biology* **34:**17-22.

Wilbur, K. M., and N. Watabe, 1963, Experimental Studies in Calcification in Molluscs and the Alga *Coccolithus huxleyi, New York Acad. Sci. Ann.* **109:**82-112.

5

Reprinted from *Linnean Soc. London Proc.* **174**:41–46 (1963), by permission of the Council of The Linnean Society of London

THE FINE STRUCTURE OF THE MINERAL PARTS OF COCCOLITHOPHORIDAE

By Maurice Black

Before the last decade, our knowledge of the fine structure of coccoliths depended almost entirely upon inferences drawn from the appearance of these bodies in polarized light. Because calcium carbonate, crystallized either as aragonite or calcite, has an exceptionally strong power of double refraction, even the very minute crystals present in coccoliths have an appreciable effect upon polarized light, and the interference patterns observable between crossed nicols are so characteristic that they can be used to differentiate between species which look alike in ordinary light.

By such methods much can be discovered about the general pattern of crystal arrangement, which in many species is quite complicated. Except in the largest coccoliths, the crystals themselves are too small to be resolved by a light-microscope, and the patterns observed are actually arrangements of optical directions (axes or bisectrices), which are not necessarily parallel with the length of the crystals, as has sometimes been assumed. Another source of uncertainty is the difficulty of distinguishing optically between sub-microscopic crystals of calcite and aragonite, since the only property it is practicable to use with such small particles is birefringence, and this has similar values for both minerals. With the higher magnifications obtainable under an electron microscope, it is usually easy to distinguish the two minerals by their different crystal forms, and if necessary a diffraction picture can be obtained from individual crystals. Also, by making use of carbon replicas, the crystal patterns can be directly observed and photographed.

In recent work with the electron microscope, the mineral coating of coccolithophorid scales has always proved to be in the form of calcite, and earlier statements that aragonite is present in some genera have not been confirmed. Calcite crystals can adopt any one of four distinct habits : rhombohedral, prismatic, tabular and scalenohedral. A large majority of coccolithophorids secrete rhombohedral crystals (fig. 1c, d) and a smaller number use hexagonal prisms (fig. 1a, b). The tabular habit, which is characteristic of the discoasteridae, has not been seen in the coccolithophoridae, and scalenohedral crystals have not yet been recognised in either group. When the individual crystals can be clearly seen, and their angles measured, they are invariably found to adopt the simplest form possible : the primitive rhombohedron on the one hand, or a hexagonal prism on the other.

If we glance over the whole range of structures to be seen in coccoliths, at one extreme we find types in which these very simple crystal forms are retained unmodified, and at the other, a host of more elaborately designed coccoliths in which the calcite is moulded by the organism into shapes quite foreign to inorganic crystallography. The fundamental principles of construction can be worked out in the simpler forms where the crystal faces are easily seen and identified, and the visible arrangement of crystals can be correlated with the interference patterns seen in polarized light. There is then no great difficulty in extending these principles to more complicated forms, since in the absence of well-defined faces, crystal orientation may still be determined optically.

The simplest, and at the same time the most orderly, crystal arrangement consists of a regular packing of hexagonal prisms, discovered by Halldal & Markali (1954, Pl. II, figs. c and d) in *Calyptrosphaera papillifera* Halldal, and represented diagrammatically in fig. 1b. A similar structure has also been reported from half a dozen other species with calyptroform coccoliths (Halldal & Markali, 1955, Pls. I–VI).

In the great majority of species which have so far been examined the coccoliths consist of rhombohedral crystals, and any fibrous structures that may be present are built of elongate rhombohedra and not of hexagonal prisms. The two structures are easily distinguished from each other in polarized light, since the optic axis is parallel with the length of a hexagonal prism, but is oblique to a rhombohedral fibre (Fig. 1*d*).

The peculiar geometry of a rhombohedral crystal lends itself to the development of a great variety of elaborate patterns when groups of such crystals are forced to fit into a circular or elliptical framework, and the oblique direction of the optic axis leads to very distinctive interference effects in polarized light. Dense crowding of crystals is apt to result in modification of the simple rhombohedral shape, or alternatively to cause a spiral imbrication. In order to understand the more complicated structures, it is advisable to start by considering coccoliths in which the crystals are well shaped and not too crowded.

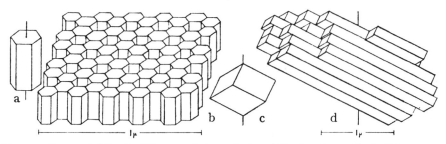

Fig. 1.—The two habits of calcite crystals found in Coccolithophoridae. Vertical lines represent the direction of the optic axis.
(*a*) Prismatic habit: the prism edges are parallel with the optic axis, and prismatic fibres give straight extinction between crossed nicols.
(*b*) Hexagonal packing of prisms in the crown of *Calyptrosphaera papillifera* Halldal. The central prism from each hexagonal group is missing, leaving a regular sieve structure.
(*c*) Rhombohedral habit: the edges are all oblique to the optic axis, so that rhombohedral fibres give oblique extinction.
(*d*) Rhombohedral fibres in the stem of *Rhabdosphaera stylifer* Lohmann, drawn with the optic axis vertical, as in the previous figures.

In the Crystallolithus phase of *Coccolithus pelagicus* (Wallich) the calcite is in the form of minute unmodified rhombohedra, scattered over the surface of the oval scale without completely covering it (Gaarder & Markali, 1956 : Pl. I, figs. 5–8 ; Parke & Adams, 1960 : Pl. I, fig. 3). At the margin there is a rim, built up into a wall of rhombohedra, one crystal wide and two crystals high. The whole coccolith is very simple and delicate, but nevertheless displays in rudimentary form the essential structures found in more robust and heavily calcified discoliths.

A more crowed arrangement, leaving no spaces between the crystals, is seen in the coccoliths of *Discolithus phaseolus* Black and Barnes (Pl. 1, fig. 6). In their general plan of construction, these resemble Crystallolithus plates. They have a raised rim made of small rhombs, and within this the disk is completely covered by rhombohedral crystals, but instead of being similar in size and shape to the rim-units, these are expanded to form plates. Towards the centre, the plates become more crowded, and tend to pile up into a conical mound or short spike, acquiring a spiral twist as they do so. In other species of *Discolithus*, similarly developed spikes rise to a considerable height (Pl. I, fig. 2), and the spirally arranged crystals become elongated until they resemble fibres. The same structure is seen in the stem of *Rhabdosphaera*, where it arises from the basal plate in a similar manner.

Most coccoliths whose essential feature is a single disk appear to be constructed upon a plan of this same kind : a wall of rhombohedral crystals arises at the circumference, and from the foot of this, crystals grow inwards towards the centre of the disk. There is, however, a considerable amount of variation in the extent to which the rather haphazard grouping of crystals seen in *Discolithus phaseolus* is replaced by more orderly patterns. In many species of *Syracosphaera* the wall grows up to a substantial height, giving barrel-shaped forms, and the inward-growing crystals on the surface of the disk take the form

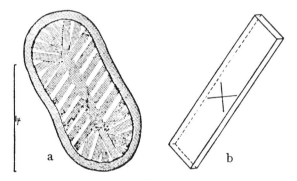

Fɪɢ. 2.—Arrangement of calcite crystals in the ordinary coccoliths of *Calciopappus caudatus* Gaarder & Ramsfjell.
(*a*) Ordinary coccolith, after Gaarder, Markali & Ramsfjell,1954 : Pl. IIc.
(*b*) Rhombohedral crystal of calcite drawn parallel with the lateral ribs in (*a*) and showing principal optical directions.

of elongated rhombohedral bars, giving an appearance like the spokes of a wheel. In the species illustrated on Pl. I fig. 5, instead of leaving the rim in a strictly radial direction, they are slightly oblique, and are all inclined in the same rotary direction as they are followed round the margin of the shield. A similar rotary arrangement is noticeable in many discoliths and related types ; in circular forms this results in a simple rotary symmetry about the centre of the circle, but in oval coccoliths the situation is complicated by the change in curvature. When the individual rays each make a constant angle with the circumference, a highly characteristic pattern is produced, which turns up repeatedly in a large number of species. The ordinary coccoliths of *Calciopappus caudatus* Gaarder and Ramsfjell illustrate this pattern with beautiful simplicity (Gaarder, Markali and Ramsfjell, 1954 : Pl. IIc). Such crystal arrangements can be recognized under a petrological microscope by the peculiar and very distinctive extinction cross which they give between crossed nicols. It was from these remarkable optical effects that Kamptner (1952 : 379) worked out the ultra-fine structure of *Pontosphaera scutellum* Kamptner and obtained a result agreeing with crystal arrangements that have actually been seen under the electron microscope in other species.

In all the forms discussed above, the crystal arrangement is closely related to the circumference of the coccolith, and gives the impression that crystal growth started there, and spread later to the centre. When we turn our attention to the placoliths, which consist of two shields united by a cylindrical column, we find that this simple plan no longer holds good (Pl. 1, figs. 3, 4, 7). There is here no peripheral wall, and the key to the structure lies near the top of the central column rather than at the circumference. The placoliths of *Umbilicosphaera mirabilis* Lohmann illustrate the essential features of this structural plan (Black & Barnes, 1961: Pl. 25, figs. 4 and 5). In the outer shield, a wide central pore is surrounded by a ring of rhombohedral crystals, tapered slightly towards the pore, as is necessary to make them fit together in a ring, and expanded towards the outer margin of the shield. A similar pattern of rhombohedral crystals builds up the outer shields of *Coccolithus pelagicus* (Wallich) and *C. sarsiae* Black, and can also be traced in a modified form in *C. leptoporus* (Murray and Blackman) and in several other species with similar circular shields (Black & Barnes, 1961: Pl. 24, fig. 3; Black, 1962: Pls. VIII and IX). In all of these, there is a well-defined circle or ellipse near the centre of the shield along which the crystals preserve their characteristic geometrical form, and it is a reasonable hypothesis that crystal growth starts here and extends towards the circumference, where each unit expands and to some extent modifies its shape.

The structure of the tubular pillar which joins together the two shields of placoliths is not always easy to make out, but in those species where it can be clearly seen, the wall of the tube consists of rhombohedral crystals either packed together like the staves of a barrel, or more commonly running obliquely down the wall of the tube to give a spiral structure (Black & Barnes, 1961: Pl. 20).

We have so far been concerned with those coccoliths in which the calcite crystals in large measure retain their characteristic edges and faces. When

---

EXPLANATION OF PLATE

PLATE 1

FIG. 1.—*Syracosphaera* sp. A coccolith in which crystal faces are completely suppressed. No. 3,427, ×18,000.

FIG. 2.—*Discolithus* sp. A broken discolith, showing details of construction. The rim is of simple rhombohedra, as in fig. 6. The radial elements of the coccolith-floor have rounded edges and are without crystal outlines; in the centre they assume a spiral twist and turn upwards to make a spine. This preparation has been shaded from two directions, giving a second shadow from the single spine. No. 3,349, ×10,000.

FIG. 3.—*Umbilicosphaera mirabilis* Lohmann. An abnormally developed placolith, showing the large central pore surrounded by interpenetrating calcite crystals. The rays in the upper right-hand part of the figure are in the normal condition for this species, and do not show crystal faces at the outer margin; in the lower part, the crystals are well developed as far as the margin, and reveal the interpenetrant relationship between adjacent crystals. No. 3,372, ×16,000.

FIG. 4.—*Gephyrocapsa oceanica* Kamptner. A replica of a placolith, showing the inner shield through the outer. In this specimen the oblique bar across the central pore is incomplete, being represented by two groups of crystals grown from opposite sides. In normal specimens, the two halves of the arch are united at the centre to form a continuous bridge. No. 9,516, ×16,000.

FIG. 5.—*Syracosphaera* sp. A lopadolith, seen from the internal surface. The floor consists of elongate rhombohedra. The marginal rim is all that can be seen of a vertical wall, constructed of lath-shaped crystals, whose ends are visible in the rim. The height of the wall can be judged from the breadth of the shadow, the angle of shading being 60° to the horizontal. No. 3,353, ×16,000.

FIG. 6.—*Discolithus phaseolus* Black and Barnes. Discolith with a rim of simple rhombohedra, and a floor of rhombic plates. No. 2,580, ×16,000.

FIG. 7.—*Coccolithus pelagicus* (Wallich). External view of a placolith. The radial elements are overlapping plank-shaped crystals. The central area is covered by thin rhombic plates, which conceal the two large pores characteristic of this species. No. 10,021, ×5,000.

The numbers quoted before the magnification are those of the original negatives in the reference collection of electron micrographs at the Sedgwick Museum, Cambridge.

Specimens illustrated are all from modern Globigerina Ooze, those in figs. 4 and 7 from Discovery Station 4269 (Biscay Abyssal Plain), the remainder from Challenger Station 338 (South Atlantic).

All figures are electron micrographs of chromium-shaded carbon replicas.

PLATE 1, FIGS. 1–7.

we pass on to consider other types, such as *Coccolithus huxleyi* (Lohmann), in which these crude outlines are suppressed and give way to more delicately moulded shapes, we still find many species in which the crystals, despite their much-altered shape, are fitted together according to one of the designs we have already considered (Pl. 1, fig. 1 : Black & Barnes, 1961 : Pls. 20–23). This is often evident from the appearance of the coccolith under the electron microscope ; in more obscure cases it is made apparent by the interference effects seen in polarized light, which sometimes reveal the presence of internal crystal patterns different from those seen at the surface.

The variety of structural patterns, great though it may be, appears to be limited by certain restrictions. If, as seems probable from recent work on the scale-bearing Chrysophyceae, all coccoliths of the discolith type are formed by accretion of mineral matter upon a primarily organic scale, some of these restrictions become more understandable. The crystals commonly lie with one rhombohedron face resting on the scale, and crystal growth appears to be virtually confined to directions parallel with the surface of the scale, except at the margin and centre. In all the discoliths so far examined, crystal growth never extends outwards beyond the margin of the scale, and once an outer rim or wall has been built in this position, further extension of crystals can only take place inwards. After the floor of the disk has been completely covered by crystalline plates, continued growth leads to imbricate overlap, like the leaves of an iris diaphragm, since the faces of adjoining rhombohedra are bevelled at supplementary angles (75° and 105°). It is probably because of this circumstance that very few, if indeed any, coccoliths have a strictly bilateral or radial symmetry : there is almost invariably some rotary or spiral element involved. The possible exceptions are those calyptroliths which consist of prismatic crystallites to the exclusion of rhombohedra.

The coccolith may be regarded as a part of a compound skeleton, in which the shape of the ultimate component crystals can be clearly seen, and the way in which they are fitted together can be easily studied. So few crystals are employed in building the smaller coccoliths that they can be counted without trouble. In these favourable circumstances, it is possible to study the varying extent to which inorganic crystal growth has been brought under control by different species. In the simplest coccoliths, the calcite crystals are no different in shape from those that can be precipitated in a test-tube, but their position on the primary organic scale is clearly under control, and no growth takes place beyond the margin of the scale. In a multitude of slightly more elaborate forms, the crystals still retain their geometrical faces and angles, but are flattened to form plates, elongated into bars, or extended in other ways, and then fitted together into intricate designs without masking their fundamental rhombohedral habit. Finally, there are the very elaborately constructed coccoliths in which crystal faces and angles are suppressed, and the moulding of the calcite units is completely controlled by the organism. It may be that these developmental types could be arranged in some sort of evolutionary series, but the geological record offers us no encouragement to attempt this, for the simplest of all coccoliths are borne by plants living today in the North Atlantic, and some of those found fossil in the Chalk must be ranked amongst the most elaborate.

REFERENCES

BLACK, M. 1962. Fossil coccospheres from a Tertiary outcrop on the continential slope. *Geol. Mag.*, **99** : 123.

BLACK, M. & BARNES, B. 1961. Coccoliths and discoasters from the floor of the South Atlantic Ocean. *J.R. micr. Soc.*, **80** : 137.

DEFLANDRE, G. & FERT, C. 1954. Observations sur les Coccolithophoridés actuels et fossiles en microscospie ordinaire et électronique. *Ann. Paleont.*, **40** : 117.

GAARDER, K. R., MARKALI, J. & RAMSFJELL, E. 1954. Further observations on the Coccolithophorid *Calciopappus caudatus*. *Avh. norske VidenskAkad.*, 1954 (1), 1.

GAARDER, K. R. & MARKALI, J. 1956. On the Coccolithophorid *Crystallolithus hyalinus* n. gen., n. sp. *Nytt Mag. Bot.*, **5** : 1.

HALLDAL, P. & MARKALI, J. 1954. Observations on coccoliths of *Syracosphaera mediterranea* Lohm., *S. pulchra* Lohm. and *S. molischi* Schill. in the electron microscope. *J. Cons. int. Explor. Mer.*, **19** : 329.

—— —— 1955. Electron microscope studies on coccolithophorides from the Norwegian Sea, the Gulf Stream and the Mediterranean. *Avh. norske VidenskAkad.*, 1955 (1), 1.

KAMPTNER, E. 1952. Das mikroskopische Studium des Skelettes des Coccolitheen (Kalkflagellaten). *Mikroskopie*, **7** : 232, 375.

—— 1954. Untersuchungen über den Feinbau der Coccolithen. *Arch. Protistenk.*, 100 : 1.

LOHMANN, H. 1902. Die Coccolithophoridae, eine Monographie der Coccolithen bildenden Flagellaten, zugleich ein Beitrag zur Kenntnis der Mittlemeerauftriebs. *Arch. Protistenk.*, **1** : 89.

PARKE, MARY. 1961. Some remarks concerning the class Chrysophyceae. *Brit. Phyc. Bull.*, **2** : 47.

PARKE, MARY & ADAMS, IRENE. 1960. The motile (*Crystallolithus hyalinus* Gaarder and Markali) and non-motile phases in the life-history of *Coccolithus pelagicus* (Wallich) Schiller. *J. mar. biol. Ass. U.K.*, **39** : 263.

6

BIOLOGY AND PHYSIOLOGY OF
COCCOLITHOPHORIDS[1]

BY E. PAASCHE

Institute of Marine Biology, University of Oslo, Norway

CONTENTS

SELECTION OF SPECIES FOR CULTURE STUDIES............................ 71
LIFE CYCLES.. 73
SOME ASPECTS OF GROWTH AND NUTRITION.............................. 75
COCCOLITH FORMATION... 77

The coccolithophorids are a group of mostly unicellular algae of the flagellate type. In recent systems of classification (20, 79), they have been included in the class Haptophyceae since some of them have been shown to possess the flagellar apparatus and the organic surface scales characteristic of that class (59, 76, 94). What distinguishes the coccolithophorids from other Haptophyceae and from all other algae, is their ability to form surface scales or bodies of a second type, known as coccoliths. The coccoliths consist of calcium carbonate crystals which are elaborated in a variety of ways, and each coccolithophorid species exhibits its own characteristic coccolith morphology.

The coccolithophorids form an important component of the marine phytoplankton, and the coccoliths may at times make up a large fraction of recent and fossil sediments. There is a considerable literature on the taxonomy, morphology, and distribution of living and fossil coccolithophorids. Almost all of this is outside the scope of the present review; the reader is referred to the brief introductory account by Black (8), and to the index by Loeblich & Tappan (55), which covers most of this literature. By comparison, laboratory studies on coccolithophorids are few in number although it is apparent from recent reviews on algal nutrition (38) and on the physiology of coccolith formation (100) that physiologists are growing increasingly aware of these organisms. The emphasis of the present review is on selected aspects of reproduction, growth physiology, and coccolith formation. Much of the pertinent information has been obtained from culture studies that have been undertaken during the past ten years.

SELECTION OF SPECIES FOR CULTURE STUDIES

Laboratory studies on life histories have been restricted almost entirely to the following species: *Cricosphaera carterae* and allied forms, *Ochros-*

[1] The survey of literature pertaining to this review was concluded in December 1967.

phaera neapolitana, Coccolithus pelagicus, and *Coccolithus huxleyi.* The only coccolithophorids that have been used to any extent in physiological work are *Cricosphaera* species and *Coccolithus huxleyi,* which are easier to grow in culture than other species.

The *Cricosphaera* form group includes a variety of isolates which have been referred to a number of different taxa by the original investigators, but all of which appear to possess ring-shaped coccoliths of the "cricolith" type, considered by Braarud (13) to be characteristic of the genus *Cricosphaera.* Pending a taxonomic revision of this group, many investigators have continued to use the names that were originally applied. Matters have been simplified in this review by treating all these forms as *Cricosphaera* species, with the understanding that no formal taxonomic transfer of species is intended. The "species" referred to in this article, some of which may be identical, are listed below under their valid names (when such exist) or the names most frequently used in the recent literature. The references following each name contain electron micrographs or other evidence that cricoliths are present. P.C.C. No. denotes the number given to an isolate in the Plymouth Culture Collection.

1. *Cricosphaera* (= *Syracosphaera* = *Hymenomonas*) *carterae.* (13).
2. *Cricosphaera* (= *Syracosphaera*) *elongata.* (13).
3. *Cricosphaera sp.,* P.C.C. No. 336. (11).
4. *Hymenomonas* sp., P.C.C. No. 156. (11, 22).
5. cf. *Hymenomonas coccolithophora,* isolated by Valkanov. (96).
6. *Pleurochrysis scherffelii.* (76, 77).
7. *Syracosphaera* sp., isolated by von Stosch. (13).
8. *Syracosphaera* sp., P.C.C. No. 181. (77).

The information published on the origin of these strains (11, 17, 25, 81, 93, 96), and of others of a similar morphology (e.g., 19, 24, 58), indicates that the natural distribution of *Cricosphaera*-like forms is, on the whole, restricted to extreme inshore or estuarine areas (marine or brackish pools, shore-line localities, etc.).

Ochrosphaera neapolitana has been isolated from sea-water aquaria (57, 87) and probably has a littoral distribution.

Coccolithus pelagicus is a typical plankton alga but, unlike the majority of coccolithophorids, it is found mainly in cold seas (14, 63).

Coccolithus huxleyi is the most common of all plankton coccolithophorids and is widely distributed in offshore (14, 63) as well as inshore (1, 6) marine areas. There is some indication that physiologically different strains occur in nature (84).

This list is deplorably short considering that there may be as many as 250 coccolithophorid species in existence (55). Moreover, the selection is hardly representative: a large number of coccolithophorids occur only in the tropical or subtropical offshore marine plankton, and there are also some fresh-water species. Both of these categories may pose interesting physiological problems which have not been touched so far.

Life Cycles

All the *Cricosphaera* species appear to be capable of multiplying by binary fission in the motile coccolith-bearing stage. In addition to this, however, a number of strains have been shown to produce benthic (bottom-dwelling), frequently multicellular and filamentous stages in which coccoliths are absent. According to von Stosch (93), these nonmotile stages may result from the germination of flagellate zoospores which arise in a number of four from the cell contents of the ordinary coccolithophorid cell. Alternatively, as described by Parke (76), the first event in the transition process may be a loss of motility in the coccolithophorid cell, followed by a double cell division, resulting in a tetrad. Further cell divisions then lead to a succession of forms which greatly resemble various benthic chrysophycean algae of the genera *Chrysosphaera, Apistonema, Chrysonema, Thallochrysis,* and *Chrysotila* (76). Valkanov (96) observed that both of these modes of transition from the motile to the nonmotile stage could occur in cultures of one and the same *Cricosphaera* strain. In some strains at least, the first motile cells to reappear in cultures of the filamentous stages are naked (10, 76, 93, 96). However, in two strains studied by Boney & Burrows (11), normal coccolithophorids were observed within 24 hours of immersion of the filamentous stages in fresh media. Rayns (82) has demonstrated that the coccolithophorid stage of *C. carterae* is diploid and the benthic stages haploid. Meiosis probably occurs during the formation of tetrads (or zoospores), while the exact point of sexual fusion remains uncertain. The environmental factors governing the release of motile cells from benthic stages have been studied in some detail (10, 11).

In *Ochrosphaera neapolitana,* cell fission appears to be restricted to the coccolithophorid stage, which is nonmotile (57, 76, 87, 88). At times, flagellate-like swarmers are liberated but they immediately give rise to a new generation of nonmotile coccolithophorids (76, 88). A sexual reproductive cycle in which the swarmers functioned as gametes was described by Schwarz (88) but could not be confirmed by Parke (76). According to Parke (76), this species produces cysts containing wall elements reminiscent of those found in certain coccolithophorid-like forms (*Thoracosphaeraceae, Discoasteraceae*) of uncertain systematic position.

A very remarkable life cycle has been described for *Coccolithus pelagicus* by Parke & Adams (78). Here, the motile stage (previously known from nature as a separate species, *Crystallolithus hyalinus*) produces coccoliths which are entirely different from those of the much larger, nonmotile though likewise unicellular stage. The motile stage normally propagates by binary fission, but in old cultures the nonmotile stage will appear, possibly as a result of sexual fusion. The nonmotile cells may go through a double cell division and then give rise to a new generation of the motile stage (78). However, the nonmotile stage can also be made to multiply as such for indefinite periods (77). If a sexual cycle is involved, the motile stage would represent the haploid phase rather than the diploid phase as in *Cricosphaera carterae.* The two stages are frequently encountered together in the plankton (14), indicating that there is a ready transition from one to the other in the natural environment.

Information on the reproduction of *Coccolithus huxleyi* is very incomplete and partly contradictory. The coccolith-forming stage, which is the only one known with certainty from nature, is typically nonmotile although flagella have been observed on occasion (67, 85). In a strain investigated in the laboratory by Braarud (15), cell division took place only after the cell contents had escaped from the coccolith envelope in the form of a naked, fusiform, flagella-less swarmer, a complete new set of coccoliths being acquired only gradually by the two daughter cells. These cultures were grown with a natural illumination cycle. In more recent work on a different strain, the cells were found to divide by simple fission whereby the original coccoliths were retained by the daughter cells; this was true in continuous (70) as well as in intermittent (74) light. Naked nonmotile cells appeared in old cultures of this strain; when isolated in clonal culture, these naked cells could be subcultured as such for indefinite periods (70). Fusiform swarmers as well as naked, biflagellate cells, which were also described by Braarud (15), were observed in the permanently naked clones on some occasions (70). A complicated life cycle may well be present in *C. huxleyi* though the exact interrelations between the various cell types remain to be elucidated.

Observations on natural populations indicate that other coccolithophorids as well may reproduce in more complicated ways than by binary fission. According to Bernard (3), a number of different cell types (including flagellates, cysts and palmelloid stages) have been observed in *Cyclococcolithus fragilis* in sea-water samples. The life cycle proposed by Bernard must be considered hypothetical until confirmed by culture studies. In a number of other species, the cell contents reportedly divide into 4 to 16 swarmers whose further fate is not known (2, 52, 86). Resting spores have been observed in some instances (31, 47, 48). There are several reports of "hybrid" cells in which part of the coccolith envelope contains coccoliths of one species while the rest is made up by coccoliths that would be classified as belonging to an entirely different species (48, 53, 56). It is possible that such cells are transitional stages in complex life cycles.

Further culture studies may confirm life history relationships that can already be surmised, and reveal others that are totally unexpected. Thus, many algae described as benthic Chrysophyceae may turn out to be stages in life cycles of *Cricosphaera*-like coccolithophorids, as has already been demonstrated for some of them (11, 76). Life cycles analogous to the one discovered in *Coccolithus pelagicus* may prove to be quite common among the planktonic coccolithophorids. However, it is not certain that the motile stage will always prove to be identical to a second coccolithophorid "species": a "naked" haptophycean flagellate of the genus *Chrysochromulina* has been suspected of filling a similar role in a coccolithophorid life cycle

(59). What has already been achieved in this field suffices to show that the present classification of coccolithophorid species by coccolith morphology alone is unsatisfactory (76).

SOME ASPECTS OF GROWTH AND NUTRITION

Investigations of the growth of *Cricosphaera* species and *Coccolithus huxleyi* in the laboratory have been largely motivated by a desire to know more about the factors governing the natural distribution of coccolithophorids. The following brief discussion is limited to three aspects which at present seem particularly relevant to ecology: the effect of temperature, the effect of salinity, and the relative importance of different modes of nutrition (autotrophy, heterotrophy).

The fastest growth rates recorded for coccolithophorids in culture appear to be 2.25 divisions per day for *Cricosphaera elongata* (29) and 1.85 divisions per day for *Coccolithus huxleyi* (29, 74), which would place these species among the faster-growing plankton algae.

There is little information on the temperature tolerance of motile stages of *Cricosphaera* species. Boney & Burrows (11) have shown experimentally that the nonmotile benthic stages survive temperatures of 35° to 40° C for an hour or more, as well as deep-freezing for periods up to four days. The criterion for viability in these experiments was the ability of the nonmotile stages to release motile cells upon return to sea water at 16° C. In nature, benthic stages of *Cricosphaera* frequently grow near the high-water mark where they are presumably exposed to similarly great variations in temperature. Growth of *Coccolithus huxleyi* in culture exhibits a strikingly narrow optimum in the temperature range of 20° to 23° C (66, 74, 98). At about 10° C, the growth rate is only about one fourth of the optimum rate, and it is somewhat surprising that by far the greatest concentrations of *C. huxleyi* ever observed in nature were found growing at this temperature (1). Mjaaland (66) found that 7° C was the lower temperature limit for growth in culture, but sizable populations have been observed in nature at 2° to 3°C (35). The upper temperature limit established for one strain was about 27° C (74), which agrees with field data from the tropics, showing that *C. huxleyi* tends to disappear from the plankton once the temperature rises much above this value (37, 63, 91).

Oceanic sea water has a salinity of about 3.5 per cent, but in littoral and estuarine areas much higher or lower values may be encountered. In keeping with their occurrence in such environments, various species or strains of *Cricosphaera*, grown in the motile phase in the laboratory, have been shown to tolerate salinities as low as 0.4 to 0.8 per cent (12, 25, 64, 80) and as high as 4.5 to 5 per cent (12, 80) or even 9 per cent (11, 25). The nonmotile stages are able to withstand a salinity of 23.6 per cent or even evaporation of the sea water to the point of crystallization (11). *Coccolithus huxleyi*, on the other hand, cannot be grown at salinities below 1.6 per cent or above 4.5 per cent (33, 66, 80). Very large natural populations have been observed near the lower of these salinity limits (6). The upper limit appears to be considerably in excess of any salinity that this species is likely to encounter in the open sea.

Cricosphaera species and *Coccolithus huxleyi* contain the same photosynthetic pigments that are found within the Haptophyceae in general (45, 83), and they carry out an ordinary photosynthetic carbon assimilation (45, 46). As far as is known, all strains of these species will grow photoautotrophically in the absence of organic carbon sources. Available data indicate that their growth response to variations in light intensity or day length is much the same as that of other autotrophic marine plankton algae (45, 66, 74).

Organic compounds have sometimes been found to stimulate the growth of coccolithophorids in the light. This is true of *Cricosphaera* species in particular, although there are exceptions (27). The growth of two *Cricosphaera* strains was considerably promoted by lactate, pyruvate, acetate, and other potential sources of organic carbon (80). More recently, one of these strains, P.C.C. No. 156, has been stated to utilize a large number of different carboxylic acids, amino acids, etc., as carbon sources in the light (44). Neither of these reports contains evidence of a more direct kind that would indicate that the organic compounds were actually metabolized by the cells. Crenshaw (22) demonstrated that lactate produced an increased cell yield in the last-mentioned strain only when the culture medium was allowed to become depleted in carbon dioxide and bicarbonate. Under such conditions, part of the lactate disappeared from the medium. This strain grows poorly or not at all on lactate in the dark (44, 80), and it seems possible that a light-dependent assimilation of lactate took place in Crenshaw's experiments. The growth of *Coccolithus huxleyi* in the light is only slightly affected by the presence of organic substances (80), and attempts to demonstrate directly a significant light-dependent uptake of glucose, acetate, and glutamate in this alga yielded negative results (90). The capacity for photoassimilation of dissolved organic substances would seem to offer no ecological advantage to a true plankton alga such as *C. huxleyi;* on the other hand, it might perhaps be of use to *Cricosphaera* species inhabiting rockpools or similar localities, where organic carbon may at times be relatively plentiful and carbon dioxide depleted by algal photosynthesis.

The ability of some strains of *Cricosphaera* to utilize organic sources of nitrogen instead of nitrate (25, 44, 80) has been considered by Pintner & Provasoli (80) to be additional proof that heterotrophic tendencies are present in these littoral coccolithophorids. Utilization of organic monophosphates instead of orthophosphate, which is equally pronounced in the planktonic *Coccolithus huxleyi* and in *Cricosphaera* strains (49, 50, 80), cannot be regarded in the same way, since it has been shown that these compounds are hydrolyzed by phosphatases located at the cell surface so that only the phosphate moiety is assimilated (50).

Efforts to grow coccolithophorids on organic substrates in the dark have usually met with failure (27, 80, 90). One *Cricosphaera* strain, P.C.C. No. 156, is reportedly capable of very slow growth in the dark in the presence of lactate or various amino acids (44). *Cricosphaera* species lack the ability to ingest solid food particles (76), but there are indications that in some of the planktonic coccolithophorids, phagotrophy may serve at least as a supplement to photosynthesis. Parke & Adams (78) have shown that the motile stage of *Coccolithus pelagicus* will ingest bacteria and small algae in the same way as do other haptophycean flagellates. It is not known whether this mode of nutrition can substitute fully for photosynthesis and so permit growth in the dark.

In the sea, coccolithophorids are frequently observed at depths of several hundred metres, where no light is available for photosynthesis (4, 5, 30, 35, 36, 61). A heterotrophic or phagotrophic mode of nutrition has sometimes been invoked to account for such observations, in particular those relating to *Cyclococcolithus fragilis* (4, 5). However, the cells found in deep water may no doubt have settled from overlying strata in many cases, and there appears to be no really satisfactory evidence that coccolithophorids grow actively in the nonilluminated parts of the sea.

The capacity for autotrophy is not absolute in coccolithophorids, since all strains that have been examined require at least one vitamin. *Coccolithus huxleyi* needs thiamine, whereas various strains of *Cricosphaera* require either thiamine or vitamin B_{12}, or both (26, 33, 80). Thus, differences in vitamin requirements do not match other physiological and ecological differences between *Cricosphaera* and *Coccolithus huxleyi*. Nevertheless, the lack of a vitamin B_{12} requirement in *C. huxleyi* is thought to be ecologically important in that it enables this alga to grow regularly in the offshore tropical waters of the Sargasso Sea (65).

COCCOLITH FORMATION

Most of the experimental work reviewed in this section was performed on either *Coccolithus huxleyi* (22, 68–75, 97–99) or *Cricosphaera* sp., P.C.C. No. 156 (22, 23, 39–44, 51).

The crystallography and fine structure of coccoliths have been admirably reviewed by Black (7). The morphology of the crystals that make up the coccoliths, as well as electron diffraction data, indicates that the calcium carbonate is quite generally present in the form of calcite. This has been confirmed by X-ray diffraction studies of cultured material of *Cricosphaera* species (22, 42, 54) and of *Coccolithus huxleyi* (99). It is of interest, therefore, that in the latter species, the calcite is largely replaced by aragonite and vaterite when the cells are subjected to nitrogen starvation (99). Very little strontium is present in coccoliths (54).

Some coccoliths, known as "holococcoliths" (16), of which those of the motile stage of *Coccolithus pelagicus* are an example, are made up of a large number of regularly packed calcite crystals of simple hexagonal-prismatic or rhombohedral shape (7). The majority of coccoliths are of the "heterococcolith" type (16), in which the crystals are basically rhombohedral though their shape may be modified to such an extent that the characteristic calcite crystal faces become partly or wholly obliterated (7). Examples of heterococcoliths are the ring-shaped coccoliths of *Cricosphaera* (18, 22, 24), and the "placoliths" of *Coccolithus huxleyi* and of nonmotile cells of *C. pelagicus* (18, 63, 97). The placoliths consist of flattened elements arranged radially in two levels and joined by a short central tube. According to Watabe (97), the crystals which make up the *C. huxleyi* coccolith are very elaborate in that each of them may consist of one upper and one lower radial element in addition to a sector of the central tube. Electron microscope investigations of holo- and heterococcoliths of different species [e.g., (8, 9, 34, 63)] have, on the whole, revealed an astonishing variety of shape and morphological detail. In many species, each cell carries two, in some even three, distinct types of coccoliths. Apart from this, coccolith morphology is remarkably constant within any given species, although laboratory experiments on *C. huxleyi* by Watabe & Wilbur (98) and observations on natural populations of the same species by McIntyre & Bé (63), have shown that the shape of the constituent crystals may be influenced to some extent by the temperature at which the cells are growing.

Recent laboratory studies on *Cricosphaera* (22, 42, 51), *Coccolithus huxleyi* (68, 99), and the nonmotile stage of *C. pelagicus* (78) have demonstrated that the heterococcoliths produced by these species are formed inside the cell and only subsequently deposited on the cell surface. Prior to these investigations, coccoliths had been observed inside the cells of a number of other species (19, 52, 56, 58) although the significance of this was not always recognized. However, there is still some doubt as to whether all coccoliths are formed inside the cell. In a very thorough investigation of the motile stage of *C. pelagicus,* Manton & Leedale (59) failed to demonstrate the presence of intracellular coccoliths, and they concluded that the coccoliths may be formed outside the cytoplasmic membrane though inside an outer "skin" which forms a loose casing around the cell in this particular coccolithophorid. In this case, the coccoliths are holococcoliths composed of simple calcitic rhombohedra, the formation of which may require a less precise environmental control than is needed to produce the elaborately modified crystals found in the heterococcoliths (7). Yet it is difficult to see how the orderly crystal pattern which is manifest in any coccolith, no matter how simple, could result unless some sort of cytoplasmic influence were involved.

According to Wilbur & Watabe (99), the coccoliths of *C. huxleyi* are formed in an unstructured portion of the cytoplasm, termed the "matrix region," which has an outline corresponding to the shape of a completed coccolith. Crystal growth starts near the base of the central tubular part of the coccolith. There is indirect evidence that in other types of coccoliths, crystal growth is initiated near the coccolith periphery (7). Very recently it has been shown that the coccolith-forming matrix in the nonmotile stage of *C. pelagicus* is derived from the Golgi body (60). A second cytoplasmic organelle, the "reticular body," is thought to be also involved in the formation of coccoliths, as it is present in coccolith-forming but not in naked cells of *C. huxleyi* (99). A similar structure has been observed in *Cricosphaera* (39). Isenberg and his collaborators (23, 51) have presented evidence obtained by histochemical and autoradiographic techniques that the coccolith-forming matrix in this species contains mucopolysaccharide-like material.

In several species, the formation of coccoliths appears to be intimately connected with the production of organic surface scales. The reader is reminded that such scales, which are known to be formed inside the Golgi apparatus and which are now thought to consist mainly of polysaccharide (32), are present in other Haptophyceae as well as the coccolithophorids. In the nonmotile stage of *Coccolithus pelagicus*, the basal crystals of each coccolith are laid down on a preformed scale in a Golgi cisterna (60). The scales remain attached to the undersides of the completed coccoliths on the cell surface, as is also the case in other species carrying large placoliths [Plate 11A in (63); (62)]. A similar association of internal and external coccoliths with scales was seen in *Cricosphaera* (22). There may be no universal rule, however; for, in the motile stage of *Coccolithus pelagicus*, the scales are formed intracellularly even though the coccoliths apparently are not (59). It should also be noted that an excess of scales, not associated with coccoliths, is produced by this species (59) as well as by *Cricosphaera* (22), and probably by others.

A very elaborate model for the morphological events leading to coccolith formation in *Cricosphaera* has been presented by Isenberg et al. (39, 40). It is possible that this model rests on a misinterpretation of the electron micrographs presented in the papers in question. The structures identified as striate "fibres" [e.g., Fig. 9 in (39)] appear to be indistinguishable from organic scales as seen in corresponding electron micrographs of sections of the same *Cricosphaera* strain (22) or of the motile stage of *Coccolithus pelagicus* (59). An important role in coccolith formation was assigned to these "fibres" but the question of their exact nature merits further investigation.

The coccoliths remain coated with an organic membrane even after they have been extruded by the cell (18, 22). Isenberg and collaborators (23, 42, 51) have shown that the organic material present in coccoliths may be derived from the intracellular organic matrix region in which they are formed. The same investigators have subjected the organic components of the extracellular coccoliths isolated from *Cricosphaera* to a very detailed chemical analysis (39, 41). One fraction extractable from the coccoliths with distilled water was shown to consist of protein and polysaccharide. The amino acid composition of the protein component was remarkable because of the presence of hydroxyproline, known primarily from structural proteins. Since "fibre" material was apparently included in the analyses

(39), it seems possible that some of these chemical data relate to organic scales rather than to the bounding membranes of coccoliths.

In long-term growth experiments on *Cricosphaera*, Isenberg et al. (44) investigated the effects of a number of non-nitrogenous organic compounds and amino acids on the quantitative development of coccoliths. The ratio of coccolith calcium to cellular nitrogen varied within wide limits, depending on the particular compound present in the growth medium. The organic substances were thought to influence the synthesis of cellular structures associated with coccolith formation. An interesting experiment by Wilbur & Watabe (99) on *Coccolithus huxleyi* likewise demonstrates that the capacity for coccolith formation is influenced by the nutritional state of the cells. A strain which does not form coccoliths under normal conditions was shown to do so when grown in a medium strongly deficient in nitrate.

The external coccoliths can be removed by lowering the pH of the culture medium. Cells of *Coccolithus huxleyi* decalcified in this manner may acquire a complete coccolith envelope (about 15 coccoliths) within 15 hours of their being transferred back to a normal medium in the light (72). Complete recalcification in *Cricosphaera* may require 40 hours (22). In both instances, cell division was shown not to be a prerequisite for the formation of new coccoliths.

The effect of temperature on the quantity of coccoliths produced by *Coccolithus huxleyi* has been investigated by Watabe & Wilbur (98). Increased proportions of naked cells were found near the extremes of the temperature growth range. Some doubt remained whether this was due to temperature effects on the calcification process as such or to preferential growth of (permanently) naked cells which were present initially in the inocula. In a recent study on a clone where all cells formed coccoliths, it was found that the amount of coccolith calcium produced per unit cell volume was fairly independent of temperature (75).

In the upper strata of the oceans where most coccolithophorids grow, the sea water is thought to be generally supersaturated with respect to calcite. In artificial sea water, coccoliths dissolve if the product of the concentrations of calcium and carbonate ions is appreciably smaller than the solubility product of calcite (70, 95). Experiments on *C. huxleyi* suggest that coccoliths continue to be formed inside the cells even when the calcium content of the medium is reduced to levels where external coccoliths dissolve (70). However, coccolith production is considerably retarded at calcium concentrations less than half of that of sea water (22, 70). The inorganic carbon deposited in coccoliths apparently enters the cell in the form of bicarbonate ions (70), and it seems that very little, if any, carbonate is taken up as such from the medium (22, 70). More precise information on the mechanisms of calcium and carbon uptake is lacking, but it is of some interest to note that pools of exchangeable calcium (22), and probably of inorganic carbon as well (69, 70), may be present in the cells. *Cricosphaera* does not form coccoliths when grown in the presence of an inhibitor of carbonic anhydrase (43). This enzyme may somehow be involved in the intracellular regulation of the carbon dioxide-bicarbonate-carbonate system at the sites of coccolith formation. It should be evident from all this that precipitation of calcium carbonate from sea water by coccolithophorids is only very indirectly dependent on the factors [supersaturation, influence of magnesium ions, etc. (21, 89)] governing nonbiological carbonate precipitation in the sea.

Short-term experiments involving the use of isotopic and other methods have demonstrated that light plays an essential role in coccolith formation in *Coccolithus huxleyi* (22, 68–70) as well as in *Cricosphaera* (22). Crenshaw (22) found no indication that coccoliths could be produced in the dark by either species. More recently, it has been demonstrated that at least one strain of *C. huxleyi* will form coccoliths in the dark, although the dark rate is only about one tenth of the rate obtained at light saturation (72). The direct association of coccolith formation with some photochemical process is evident from experiments showing that there is an immediate sharp drop in the rate of coccolith formation when the light is turned off, and that the basal dark rate is re-established within an hour of this (72). Coccolith formation in *C. huxleyi* becomes light-saturated at a light intensity somewhat lower than the one required for saturation of photosynthesis (70). The slope of the light saturation curve of coccolith formation appears to be steeper below the photosynthetic compensation point than above it (70). This may possibly be due to some sort of interaction between photosynthesis, respiration, and coccolith formation.

The exact nature of the light-dependent reaction(s) involved in coccolith formation has yet to be established. If the inorganic carbon required for photosynthesis were derived from the same source of bicarbonate from which the coccolith carbonate is drawn, photosynthetic carbon dioxide assimilation might provide the driving force of coccolith formation by shifting the equilibrium

$$2 \; HCO_3^- \rightleftharpoons CO_2 + CO_3^{--} + H_2O$$

to the right. Crenshaw (22) adopted this explanation mainly because he found that photosynthesis and coccolith formation in *Cricosphaera* were inhibited to the same extent by the photosynthetic inhibitor 3-(3,4-dichlorophenyl)-1,1-dimethylurea (DCMU). However, results obtained with *C. huxleyi* have suggested that light-dependent coccolith formation can at best be only partially accounted for in this manner. In some isolates of this species, the ratio of carbon precipitated in coccoliths to carbon assimilated in photosynthesis normally exceeds unity (70), which is the highest ratio permitted by the above formulation. More important, light-dependent coccolith formation was shown to continue for hours, albeit at a reduced rate, in the presence of concentrations of 3-(*p*-chlorophenyl)-1,1-dimethylurea (CMU) sufficient to block photosynthetic carbon assimilation and oxygen evolution completely (70, 71). Some forms of photosynthetic phosphorylation are known to be relatively insensitive to the substituted ureas, and it was speculated that coccolith formation might depend on chemical energy supplied through photo-phosphorylation. This energy would be consumed in the uptake of calcium or bicarbonate ions, for example. Alternatively, some nonphotosynthetic photochemical reaction might be involved. The action spectrum of coccolith formation in *C. huxleyi* (73) does not give an unequivocal answer to this. It exhibits peaks at those wavelengths in the blue and red at which maximum cell absorption and photosynthetic carbon assimilation occur, thereby implicating a chloroplast reaction (photophosphorylation or carbon assimilation, or both); but there is also an additional stimulatory effect of blue light, not seen in photosynthesis. Thus, the effect of light may be complex. It seems very likely that the role of light is similar in all coccolithophorids exhibiting a light-dependent coccolith formation. Further work will undoubtedly show that the apparent differences between *Cricosphaera* and *Coccolithus huxleyi* are less fundamental than the experiments with substituted ureas would seem to suggest.

The relationship between coccolith formation and photosynthesis has also been examined from a somewhat different angle, namely, with a view to the possibility that the uptake of carbon in connection with coccolith formation might be a prerequisite for efficient photosynthesis and thereby for growth. It is of interest in this connection that cultures of *Cricosphaera* can be grown to quite high densities even when the production of coccoliths is abolished by the presence of a carbonic anhydrase inhibitor (43) or by the substitution of strontium for calcium in the growth medium (42). Growth of *Coccolithus huxleyi* is inhibited at low calcium concentrations (22), but from short-term experiments on this alga, it appears that photosynthetic carbon assimilation is not reduced at calcium concentrations sufficiently low to preclude coccolith formation (70). Other data from the same investigation suggested that naked clones of this species utilize a mixture of bicarbonate and molecular carbon dioxide in photosynthesis in the same way as do coccolith-forming clones, and this was considered an additional, though less direct, indication that photosynthesis is not modified in the absence of coccolith formation (70). Steemann Nielsen (92) has reinterpreted these data to the effect that the naked cells take up only carbon dioxide from the medium, whereas coccolith-forming cells obtain additional carbon dioxide from bicarbonate transported into the cell by some light-dependent mechanism (driven by photosynthetic phosphorylation?). Coccolith formation would be essential to this uptake mechanism since it would serve to neutralize the hydroxyl ions left over from the bicarbonate during photosynthesis. Steemann Nielsen's argument rests on the observation (70) that photosynthesis per cell in naked clones is only about half of that observed in coccolith-forming clones under comparable conditions. However, this is not necessarily an indication that photosynthesis is less efficient in terms of carbon assimilated relative to carbon present in cells, since the cell volume is normally smaller in the naked clones [compare data given in (29) and (74);

see also footnote p. 1088 in (28)]. Also, Steemann Nielsen's hypothesis predicts much slower growth of naked than of coccolith-forming cultures at light saturation (92), whereas growth rates measured in experiments at light saturation and under otherwise comparable conditions are the same in the two cell types (29, 74). Thus, there is still some uncertainty whether the carbon uptake during photosynthesis is at all connected with the uptake of inorganic carbon during coccolith formation.

LITERATURE CITED

1. Berge, G., *Sarsia*, **6**, 27–40 (1962)
2. Bernard, F., *Arch. Zool. Exptl. Gen.*, **81** (Notes et Revue), 33–44 (1939)
3. Bernard, F., *J. Conseil, Conseil Perm. Intern. Exploration Mer*, **15**, 177–88 (1948)
4. Bernard, F., in *Symposium on Marine Microbiology*, 215–28. (Oppenheimer, C. H., Ed., C. C Thomas, Springfield, Ill., 769 pp., 1963)
5. Bernard, F., *Rappt. Comm. Intern. Exploration Sci. Mer Mediterranee*, **18**, 341–44 (1965)
6. Birkenes, E., Braarud, T., *Avhandl. Norske Videnskaps-Akad. Oslo, I. Mat.-Naturv. Kl.*, **2**, 1–23 (1952)
7. Black, M., *Proc. Linnean Soc. London*, **174**, 41–46 (1963)
8. Black, M., *Endeavour*, **24**, 131–37 (1965)
9. Black, M., Barnes, B., *J. Roy. Microscop. Soc.*, **80**, 137–47 (1961)
10. Boney, A. D., *J. Exptl. Marine Biol. Ecol.*, **1**, 7–33 (1967)
11. Boney, A. D., Burrows, A., *J. Marine Biol. Assoc. U.K.*, **46**, 295–319 (1966)
12. Braarud, T., *Physiol. Plantarum*, **4**, 28–34 (1951)
13. Braarud, T., *Nytt Mag. Botan.*, **8**, 211–12 (1960)
14. Braarud, T., *J. Oceanog. Soc. Japan*, 20th Anniversary Vol., 628–49 (1962)
15. Braarud, T., *Pubbl. Staz. Zool. Napoli*, **33**, 110–16 (1963)
16. Braarud, T., Deflandre, G., Halldal, P., Kamptner, E., *Micropaleontology*, **1**, 157–59 (1955)
17. Braarud, T., Fagerland, E., *Avhandl. Norske Videnskaps-Akad. Oslo, I. Mat.-Naturv. Kl.*, **2**, 1–10 (1946)
18. Braarud, T., Gaarder, K. R., Markali, J., Nordli, E., *Nytt Mag. Botan.*, **1**, 129–34 (1952)
19. Chadefaud, M., Feldmann, J., *Bull. Museum Natl. Hist. Nat. (Paris)*, Ser. 2, **21**, 617–21 (1949)
20. Christensen, T., in *Botanik*, **2**, No. 2, 1–178. (Böcher, T. W., Lange, M., Sörensen, T., Eds., Munksgaard, Copenhagen, 1962)
21. Cloud, P. E., Jr., in *Chemical Oceanography*, **2**, 127–58. (Riley, J. P., Skirrow, G., Eds., Academic Press, London and New York, 508 pp., 1965)
22. Crenshaw, M. A., *Coccolith formation by two marine coccolithophorids*, Coccolithus huxleyi *and* Hymenomonas sp. (Doctoral thesis, Duke Univ., Durham, N.C., 1964)
23. Douglas, S. D., Isenberg, H. D., Lavine, L. S., Spicer, S. S., *J. Histochem. Cytochem.*, **15**, 285–91 (1967)
24. Dragesco, J., *Cahiers Biol. Marine*, **6**, 83–115 (1965)
25. Droop, M. R., *J. Marine Biol. Assoc. U.K.*, **34**, 233–45 (1955)
26. Droop, M. R., *J. Marine Biol. Assoc. U.K.*, **37**, 323–29 (1958)
27. Droop, M. R., McGill, S., *J. Marine Biol. Assoc. U.K.*, **46**, 679–84 (1966)
28. Eppley, R. W., Sloan, P. R., *J. Fisheries Res. Board Can.*, **22**, 1083–97 (1965)
29. Eppley, R. W., Sloan, P. R., *Physiol. Plantarum*, **19**, 47–59 (1966)
30. Fournier, R. O., *Science*, **153**, 1250–52 (1966)
31. Gaarder, K. R., *Rept. Sci. Results "Michael Sars" N. Atlantic Deep-Sea Expedition 1910*, **2**, No. 4, 1–20 (1954)
32. Green, J. C., Jennings, D. H., *J. Exptl. Botany*, **18**, 359–70 (1967)
33. Guillard, R. R. L., in *Symposium on Marine Microbioloy*, 93–104. (Oppenheimer, C. H., Ed., C. C Thomas, Springfield, Ill., 769 pp., 1963)
34. Halldal, P., Markali, J., *Avhandl. Norske Videnskaps-Akad. Oslo, I. Mat.-Naturv. Kl.*, **1**, 1–30 (1955)
35. Hasle, G. R., *Nytt Mag. Botan.*, **8**, 77–88 (1960)
36. Hentschel, E., *Wiss. Ergeb. Deut. Atlantischen Expedition "Meteor" 1925–1927*, **11**, 1–344 (1936)
37. Hulburt, E. M., Ryther, J. H., Guillard, R. R. L., *J. Conseil, Conseil Perm. Intern. Exploration Mer*, **25**, 115–28 (1960)
38. Hutner, S. H., Provasoli, L., *Ann. Rev. Plant Physiol.*, **15**, 37–56 (1964)
39. Isenberg, H. D., Douglas, S. D., Lavine, L. S., Spicer, S. S., Weissfellner, H., *Ann. N.Y. Acad. Sci.*, **136**, 155–88 (1966)
40. Isenberg, H. D., Douglas, S. D., Lavine, L. S., Weissfellner, H., in *Proc. Intern. Congr. Tropical*

Oceanography, 155–77. (Bayer, F. M., Ed., Univ. Miami Inst. Marine Sci., Miami, Fla., 1967)

41. Isenberg, H. D., Lavine, L. S., Mandell, C., Weissfellner, H., *Nature*, **206**, 1153–54 (1965)

42. Isenberg, H. D., Lavine, L. S., Moss. M. L., Kupferstein, D., Lear, P. E., *Ann. N.Y. Acad. Sci.*, **109**, 49–64 (1963)

43. Isenberg, H. D., Lavine, L. S., Weissfellner, H., *J. Protozool.*, **10**, 477–79 (1963)

44. Isenberg, H. D., Lavine, L. S., Weissfellner, H., Spotnitz, A., *Trans. N.Y. Acad. Sci., Ser. 2*, **27**, 530–45 (1965)

45. Jeffrey, S. W., Allen, M. B., *J. Gen. Microbiol.*, **36**, 277–88 (1964)

46. Jeffrey, S. W., Ulrich, J., Allen, M. B., *Biochim. Biophys. Acta*, **112**, 35–44 (1966)

47. Kamptner, E., *Sitz.-Ber. Akad. Wiss. Wien, Math.-Naturw. Kl., Abt. I*, **146**, 67–76 (1937)

48. Kamptner, E., *Ann. Naturhist. Museum Wien*, **51**, 54–149 (1941)

49. Kuenzler, E. J., *J. Phycology*, **1**, 156–64 (1965)

50. Kuenzler, E. J., Perras, J. P., *Biol. Bull.*, **128**, 271–84 (1965)

51. Lavine, L. S., Isenberg, H. D., Moss, M. L., *Nature*, **196**, 78 (1962)

52. Lecal-Schlauder, J., *Ann. Inst. Oceanog. (Paris)*, **26**, 255–362 (1951)

53. Lecal-Schlauder, J., *Bull. Soc. Hist. Nat. Afrique Nord*, **52**, 63–66 (1961)

54. Lewin, R. A., Chow, T. J., *Plant Cell. Physiol. (Tokyo)*, **2**, 203–8 (1961)

55. Loeblich, A. R., Jr., Tappan, H., *Phycologia*, **6**, 81–216 (1966)

56. Lohmann, H., *Arch. Protistenk.*, **1**, 89–165 (1902)

57. Magne, F., *Rev. Gen. Botan.*, **59**, 231–40 (1952)

58. Magne, F., *Rev. Gen. Botan.*, **61**, 389–415 (1954)

59. Manton, I., Leedale, G. F., *Arch. Mikrobiol.*, **47**, 115–36 (1963)

60. Manton, I., Leedale, G. F. (Personal communication, 1967)

61. Marshall, H. G., *Limnol. Oceanog.*, **11**, 432–35 (1966)

62. McIntyre, A. (Personal communication, 1967)

63. McIntyre, A., Bé, A. W. H., *Deep-Sea Res.*, **14**, 561–97 (1967)

64. McLachlan, J., *Can. J. Microbiol.*, **7**, 399–406 (1961)

65. Menzel, D. W., Spaeth, J. P., *Limnol. Oceanog.*, **7**, 151–54 (1962)

66. Mjaaland, G., *Oikos*, **7**, 251–55 (1956)

67. Norris, R. E., *New Zealand J. Botan.*, **2**, 258–78 (1964)

68. Paasche, E., *Nature*, **193**, 1094–95 (1962)

69. Paasche, E., *Physiol. Plantarum*, **16**, 186–200 (1963)

70. Paasche, E., *Physiol. Plantarum*, Suppl. 3, 1–82 (1964)

71. Paasche, E., *Physiol. Plantarum*, **18**, 138–145 (1965)

72. Paasche, E., *Physiol. Plantarum*, **19**, 271–78 (1966)

73. Paasche, E., *Physiol. Plantarum*, **19**, 770–79 (1966)

74. Paasche, E., *Physiol. Plantarum*, **20**, 946–56 (1967)

75. Paasche, E., *Limnol. Oceanog.*, **13** (In press, 1968)

76. Parke, M., *Brit. Phycol. Bull.*, **2**, 47–55 (1961)

77. Parke, M. (Personal communication, 1967)

78. Parke, M., Adams, I., *J. Marine Biol. Assoc. U.K.*, **39**, 263–74 (1960)

79. Parke, M., Dixon, P. S., *J. Marine Biol. Assoc. U.K.*, **44**, 499–542 (1964)

80. Pintner, I. J., Provasoli, L., in *Symposium on Marine Microbiology*, 114–21. (Oppenheimer, C. H., Ed., C. C Thomas, Springfield, Ill., 769 pp., 1963)

81. Pringsheim, E. G., *Arch. Mikrobiol.*, **21**, 401–10 (1955)

82. Rayns, D. G., *J. Marine Biol. Assoc. U.K.*, **42**, 481–84 (1962)

83. Riley, J. P., Wilson, T. R. S., *J. Marine Biol. Assoc. U.K.*, **47**, 351–62 (1967)

84. Ryther, J. H., Kramer, D. D., *Ecology*, **42**, 444–46 (1961)

85. Schiller, J., *Arch. Protistenk.*, **51**, 1–130 (1925)

86. Schiller, J., *Arch. Protistenk.*, **53**, 326–42 (1926)

87. Schussnig, B., *Oesterr. Botan. Z.*, **79**, 164–70 (1930)

88. Schwarz, E., *Arch. Protistenk.*, **77**, 434–62 (1932)

89. Simkiss, K., *Nature*, **201**, 492–93 (1964)

90. Sloan, P. R., Strickland, J. D. H., *J. Phycol.*, **2**, 29–32 (1966)

91. Smayda, T. J., *Inter-Am. Trop. Tuna Comm. Bull.*, **11**, 353–612 (1966)

92. Steemann Nielsen, E., *Physiol. Plantarum,* **19,** 232–40 (1966)
93. Stosch, H. A. von, *Naturwissenschaften,* **42,** 423 (1955)
94. Stosch, H. A. von, *Naturwissenschaften,* **45,** 140–41 (1958)
95. Swift, E., 5th, Taylor, W. R., *J. Phycol.,* **2,** 121–25 (1966)
96. Valkanov, A., *Rev. Algologique,* **6,** 220–26 (1962)
97. Watabe, N., *Calcified Tissue Res.,* **1,** 114–21 (1967)
98. Watabe, N., Wilbur, K. M., *Limnol. Oceanog.,* **11,** 567–75 (1966)
99. Wilbur, K. M., Watabe, N., *Ann. N.Y. Acad. Sci.,* **109,** 82–112 (1963)
100. Wilbur, K. M., Watabe, N., in *Proc. Inter. Congr. Tropical Oceanography,* 133–54. (Bayer, F. M., Ed., Univ. Miami Inst. Marine Sci., Miami, Fla., 1967)

ERRATA

Page 79, line 5 should read: "*C. huxleyi* (99). A similar structure has been observed in *Cricosphaera* . . . ."

Page 82, line 8 should read: ". . . chloroplast reaction (photophosphorylation) or . . . ."

7

Reprinted from Limnology and Oceanography **11**:567–575 (1966)

EFFECTS OF TEMPERATURE ON GROWTH, CALCIFICATION, AND COCCOLITH FORM IN *COCCOLITHUS HUXLEYI* (COCCOLITHINEAE)

Norimitsu Watabe and Karl M. Wilbur

Department of Zoology, Duke University, Durham, North Carolina   27706

INTRODUCTION

The Coccolithineae deposit structures of calcium carbonate called coccoliths that have a form characteristic of the species. In *Coccolithus huxleyi*, the coccolith is a complex body made up of many elements (Braarud et al. 1952; Kamptner 1956), each of which is a single crystal of calcium carbonate (Watabe, unpublished data). The study of Wada (1961) on mollusc shell led us to consider that the influences of temperature on calcification might be explored using the coccoliths of *C. huxleyi*. Wada had observed that crystal size in the shell of the pearl oyster *Pinctada martensii* was greater during growth in colder months. If crystal size in *C. huxleyi* was similarly altered by changes of temperature, the form and dimensions of the coccoliths might be changed. If such were the case, coccoliths of a single species might be found to have differing forms and dimensions in natural waters of differing temperatures. The recent finding of Bé and McIntyre (1964) that the morphology of coccoliths of *C. huxleyi* is different in warm and cold waters is of special pertinence in this respect.

Coccoliths are found in fossil deposits beginning with the Jurassic and also occur in more recent ocean sediments (Bramlette 1958; Black and Barnes 1961; Noël 1965*a*, *b*). It would be of considerable interest to know whether coccolith morphology is influenced by temperature to the extent that coccolith form could give an index of environmental temperatures, past and recent.

Using pure cultures of *C. huxleyi* maintained at constant temperatures, the studies reported here have examined 1) the ability of cells to grow and deposit coccoliths over a wide temperature range, and 2) the influence of temperature on coccolith dimensions and abnormalities of form. It has been found that this species will grow and calcify over the range 7–27C and that coccolith dimensions are distinctly different according to the temperature at which the coccoliths are formed.

We wish to express our sincere thanks to Miss Suzanne B. Collings, Mr. Carlie W. Powell, and Mrs. Edna Green for their excellent technical assistance. We are also grateful to Miss Collings for taking the electron micrograph in Fig. 5-C.

This study was supported by Public Health Service Research Grant DE 01382-06 from the National Institute of Dental Research, National Institutes of Health.

MATERIALS AND METHODS

Culture conditions

C. huxleyi, strain BT-6, was originally obtained from Dr. John H. Ryther and Dr.

TABLE 1. *Culture conditions for cells of* Coccolithus huxleyi *in the experimental media*

	Medium A	Medium B
Temperatures	7, 18, 24, and 27C	7, 12, 18, 24, and 27C
Temperature range	± 0.5C	± 0.2C
Temperature control	3 BOD incubators and a plant growth chamber	5 water baths with heaters and thermoregulators in 4C constant temperature room
Illumination	Culture flasks illuminated from below by 6 "cool white" 40-w fluorescent lamps giving 5,380 lux at the bottom of the flasks	Culture flasks were illuminated from above by 4 "cool white" 40-w fluorescent lamps giving 3,540 lux at the water level
Salinity	35‰	35‰
pH	pH 8.15	pH 8.00

Robert R. L. Guillard and has been maintained in liquid culture for five years. The culture medium was medium f-1 of Guillard and Ryther (1962), including artificial seawater described by Lyman and Fleming (1940). This will be designated Medium B used in studies of growth, calcification, and coccolith form. The growth and calcification studies were repeated in Medium B to which 2mM Na_2CO_3 were added (designated Medium A).

Five or 10 ml of stock cultures were inoculated into 100 ml of Medium A or Medium B in 1,000-ml erlenmeyer flasks. Four flasks were maintained at each of the experimental temperatures. The stock cultures in Medium A and Medium B were in logarithmic growth at the time of use and contained 90 and 35% calcified cells, respectively.

The culture conditions for cells in the two media are given in Table 1.

Determination of growth of calcified and uncalcified cells

At intervals of one or two days after inoculation, numbers of calcified and uncalcified cells were counted using a hemocytometer with a polarizing microscope. Calcified cells were easily discernible by the strong birefringence of the coccoliths.

Coccolith morphology

At the end of the growth period at the various temperatures, the cultures were centrifuged at 140 × g for 10 min; the pellets were resuspended in distilled water and centrifuged. The washing procedure was repeated three times, and the final pellets were diluted 20-fold with distilled water. Single drops of the suspension were placed on electron microscope specimen grids coated with collodion and carbon. Ten grids were prepared from cultures at each temperature. Electron micrographs were printed at a final magnification of 16,600 diam.

Carbon replicas of coccoliths were prepared in some cases. The grids were shadowed with platinum at an angle of about 30°, followed by coating with carbon. The grids, with mounted side down, were floated on 0.1 N HCl for 20 min to decalcify the coccoliths and were then washed.

The number of coccoliths of normal and abnormal form were counted. Observations were also made on the number of well-formed coccoliths with fused upper elements and with closed central region of lower discs.

Only well-formed coccoliths were used for the study of dimensions. The over-all length and width of whole coccoliths were measured directly from prints. The width at half length of the upper elements was also measured from prints, using a dissecting microscope and ocular micrometer. Approximately 20 elements on each coccolith were measured. The total number of coccoliths observed was 3,036. Of these, measurements were made on 528. The number of upper elements measured was 10,332.

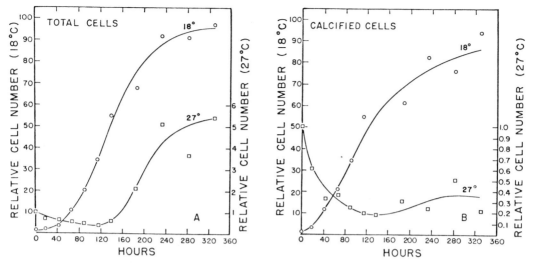

FIG. 1. A. Growth curves of cells of *Coccolithus huxleyi* at 18 and 27C. The curves were drawn by inspection. B. Growth curves of calcified cells in the same culture shown in Fig. 1A. The curves were drawn by inspection.

As the upper elements are curved downward, accurate measurements of their lengths were not possible. On each coccolith the apparent lengths of two elements on the long axis were measured and averaged.

RESULTS

Growth

The range of temperatures permitting substantial growth was first investigated and found to be 7 to 27C. No attempt was made to determine the extremes of temperature. Cultures in Medium A (high Na_2CO_3) and Medium B were then incubated at 7, 12 (Medium B only), 18, 24, and 27C, and the increase in total cells and calcified cells was determined during periods of 14 to 34 days.

At 18 and 24C, growth followed a logarithmic course and then ceased (Fig. 1A, 18C; *see also* Paasche 1964). Some cultures later showed a decrease in cell number. Cultures at 7, 12, and 27C exhibited a lag phase not seen at 18 and 24C, and during this time there was commonly a decrease in cell number (Fig. 1A, 27C). The length of the lag phase was longest in Medium B at 7C and lasted 11–18 days.

The effect of temperature on growth rate is shown in Fig. 2. The patterns of temperature effects on growth rate in both media were similar, but not identical. However, the rates in Medium A were more than double those in B at 18 and 24C. Growth rates at 18C were essentially the same as at 24C. The doubling time at these temperatures in Medium A was about 19 hr. The highest rate judged from the shape of the curves, may occur at a temperature between 18 and 24C as shown by Mjaaland (1956). Above 24 and below 18C, growth rates decreased. The Q_{10} value of growth rates for the interval 7–18C was 2.95 for Medium A and 1.66 for Medium B. Differences in growth rates were statistically significant for cultures grown at 7 and 18C and at 24 and 27C in Medium A, and at 18 and 27C in Medium B.

Variability in growth rate between individual cultures, illustrated by the wide limits standard deviations of samples in Fig. 2, is a curious and common finding. It occurred even though all cultures were inoculated from the same stock into a common batch of medium maintained in the same temperature bath or chamber and with the positions of the culture flasks rotated to

83

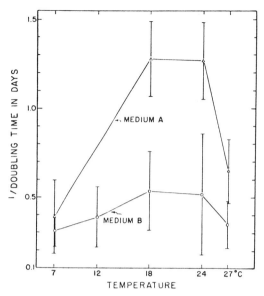

FIG. 2. Rates of division of cells as a function of temperature. Each point gives the reciprocal of the mean doubling time, in days, of four cultures. The ranges indicate standard deviations of the samples.

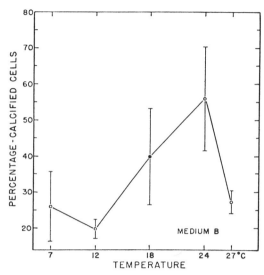

FIG. 3. Maximum percentages of calcified cells at different temperatures. Each point represents the mean value of the maximum percentages observed in four cultures. The ranges indicate standard deviations of the samples.

equalize any small differences in light intensities. The observed variability serves to emphasize the importance of unknown factors on growth.

Calcification

Not all cells of *C. huxleyi* do calcify. The growth curves of calcified cells were compared with those for all cells at each temperature. At 18 and 24C (Medium A), most of the curves were parallel. However, at 7 and 27C, where fewer cells were calcified, the curves are similar in some cultures and not in others. The extreme difference is seen in Fig. 1A and B at 27C.

The quantitative relationship of temperature to the percentage of cells bearing coccoliths was next considered. No single method of representation is completely satisfactory. This is due to two factors: The percentage of calcified cells changes with age of the culture, and cultures probably do not "age" at the same rate at different temperatures. Data on percentage cell calcification have been represented as 1) means of maximum percentages of calci-

fied cells attained in individual cultures and 2) as mean percentages of new cells calcified during the first period of approximate doubling of cell number following inoculation of the cultures. The percentage of new cells calcified, r, was calculated from the relationship:

$$r = \frac{zy - x}{z - 1},$$

where x is percentage of cells calcified at time t_o (just before the marked increase in cell number), y is percentage of cells calcified at time t_e (when approximate doubling of cell number has taken place), and z is the ratio of total cells at t_e to total cells at t_o.

Application of both criteria 1) and 2) demonstrated that more cells produced coccoliths at 18 and 24C than at lower or higher temperatures. The percentage of calcified cells in this intermediate range was twofold to threefold that at 7 and 27C, both for the maximal calcification attained during extended periods (Fig. 3) and for new cells calcified during the first doubling of cell number (Fig. 4).

Temperature and coccolith form

A well-formed coccolith of *C. huxleyi* has a placolith-type structure consisting of an upper and a lower disc connected by a hollow cylinder (Braarud et al. 1952; Kamptner 1956; Wilbur and Watabe 1963). The upper disc is made up of a radial arrangement of elongated elements (designated upper elements) usually enlarged at their outer ends where they may be fused with neighboring elements or unfused (Fig. 5C). The lower disc also has radial elements, wider than the upper elements and extending both outward and centrally from the base of the cylinder.

The sections which follow present an analysis of temperature effects on various aspects of coccolith growth and form.

Abnormalities   Most coccoliths of cells grown at 18C had a normal form (Fig. 6, curve A). At higher and lower temperatures, the percentage of abnormal coccoliths increased as the temperature difference from 18C increased. The abnormal forms had a lower disc plus a few upper elements (Fig. 5D), or irregularities of upper and lower elements (Fig. 5E).

Incomplete growth   In well-formed coccoliths, determinations were made at the several temperatures of the percentage which had completed growth and possessed fused upper elements and a closed central region of the lower disc. The maximum percentage occurred at 18C (Fig. 6, curve D). The values were lower at 24 and 27C and lowest at 7 and 12C. The curves for separate determinations of percentage of coccoliths having fused upper elements (Fig. 6, curve C) and a closed central region (Fig. 6, curve B) were similar except that the latter had a high value at 27C. The similarity of curves C and D indicates that coccoliths with fused upper elements usually had a closed central region also.

The number of upper elements varied little with temperature (30 ± 2 at 7C, and 32 ± 1 at 18, 24, and 27C). At 12C, the number was slightly, but significantly, lower (29 ± 1) than at higher temperatures.

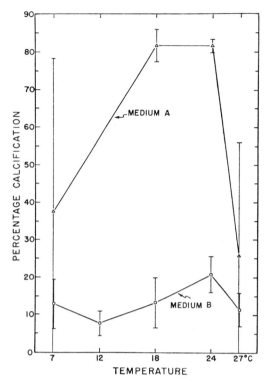

FIG. 4. Percentage of new cells calcified during the first period of approximate doubling of cell number following inoculation of the cultures. Each point is the mean value of the percentages in four cultures. The ranges indicate standard deviations of the samples.

Width and length of coccoliths   There were no significant differences in coccolith length or width at 7, 12, and 18C, but within the intervals 18–24C and 24–27C, both length and width decreased significantly (Fig. 7). The length : width ratio remained nearly the same (range, 1.19–1.22) at all temperatures.

Width and length of upper elements   Temperature affected the length and width of coccolith elements quite differently. The width of the upper elements was similar at 7 and 12C (Fig. 7). It increased progressively at 18 and 24C and then decreased slightly at 27C. All differences were significant except those among coccoliths developing between 7 and 12C.

The apparent lengths were similar at 7, 12, and 18C and decreased at higher tem-

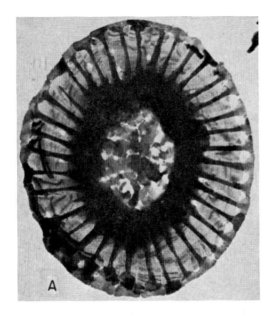

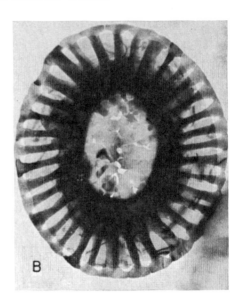

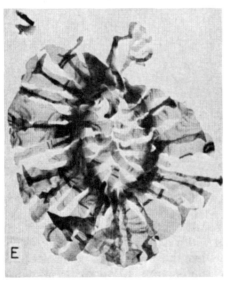

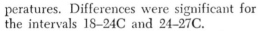

FIG. 5. A. A normal coccolith from a cell cultured at 18C. B. A normal coccolith from a cell cultured at 27C. Note that the overall length and width are less and that the upper elements are wider than in coccolith shown in Fig. 5A. C. A carbon replica of a normal coccolith shadowed with platinum. The upper elements with hammerlike heads and the lower elements connected with a hollow central cylinder. D. An abnormal coccolith from a cell cultured at 7C. The coccolith has only lower disc and one upper element indistinctly seen at lower center. E. An abnormal coccolith from a cell cultured at 12C. Note irregularities of the lower and upper elements. All electron micrographs are × 17,800.

peratures. Differences were significant for the intervals 18–24C and 24–27C.

The measurements of coccolith dimensions at various temperatures can be briefly summarized. Changes in coccolith length, width, and the apparent length of the upper elements followed parallel curves as a function of temperature. Over the range 7–18C, there was no significant change; at higher temperatures the dimensions decreased pro-

gressively. In the absence of a change in diameter of the central cylinder of the coccoliths, we should expect the observed parallelism in these dimensions because changes in apparent length of the upper elements would govern the overall changes in dimensions of the coccolith. In contrast to changes in length, the width of the upper elements increased from 12C to a maximum at 24C. Differences in coccolith form as affected by temperature are illustrated in Fig. 5A and B.

DISCUSSION

The results have demonstrated that cultures of C. huxleyi will grow over the

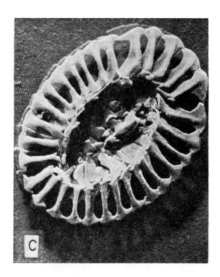

range 7–27C. This finding correlates with the distribution of this species in both cold and warm ocean waters (Bé and McIntyre 1964; *see also* Kamptner 1956). The rate of division was twofold to threefold higher at 18 and 24C than at 7 and 27C (Fig. 2). The effects of temperature on growth were similar in a general way to those reported by Mjaaland (1956). However, he found that the rate of growth at 20C was about 10 times higher than at 7 and 25C. Also, the maximum temperature permitting growth was found to be 25C, in contrast to our finding that the cells grow at 27C. Guillard (personal communication) found most rapid growth between 16 and 18C. Differences in temperature effects observed in the three laboratories are substantial and may involve differences in strains and conditions of culture.

The low *percentage* of calcified cells at upper and lower temperature extremes (Figs. 3 and 4) indicates a marked suppression of total calcification. The suppression could result from two phenomena: 1) Absence of calcification in cells which at intermediate temperatures would calcify, and 2) a preferential increase in constitutionally naked cells lacking the capacity to calcify (Paasche 1964). The data do not permit us to distinguish between the two mechanisms. If the temperature relationship found in cul-

tures also holds in the sea, we can expect that a disproportionately low percentage of cells will be identified at temperatures near 7 and 27C because they lack identifying coccoliths. The true abundance of *C. huxleyi* would thus be greater than the apparent abundance estimated from coccolith-bearing cells.

Three general processes are involved in coccolith production: 1) cellular intake of calcium and carbon, 2) synthesis of the organic matrix in which the coccolith is formed (Wilbur and Watabe 1963), and 3) nucleation and crystal growth. Data are not available to indicate which may be limiting coccolith formation at high and low temperatures in the event that temperature affects calcification directly. We do know, however, that the availability of inorganic carbon can be limiting. Crenshaw (1964) observed that supplementation of the Guillard and Ryther (1962) medium with sodium bicarbonate increased the amount of calcium carbonate deposited in *Hymenomonas* in which all cells calcify. In the present study of *C. huxleyi*, sodium carbonate added to the same medium increased the percentage of calcified cells, but here again the possibility of a preferential increase in constitutionally naked cells must be recognized in this species.

For the purpose of our study, calcified

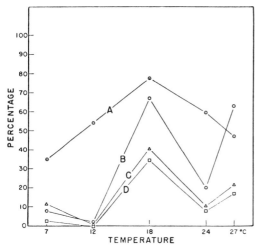

FIG. 6. Changes in coccolith form as a function of temperature. Curve A. Normal form. Curve B. Lower disc closed at the central region of the coccoliths. Curve C. Coccoliths with fused upper elements. Curve D. Coccoliths with lower disc closed at the central region and with fused upper elements.

cells were simply distinguished from uncalcified cells under polarized light which makes the coccoliths clearly visible. This method obviously does not provide information on total calcium carbonate deposition. Further, the complete loss of coccoliths adhering to the cell would result in an apparent reduction of percentage of cells calcified. Clearly, measurements of the rate of calcium carbonate formation in cells at various temperatures would be valuable in this regard.

Coccolith form permits one to distinguish populations grown at temperatures differing by 3 to 6C. From the width of the upper elements, populations at 12, 18, 24, and 27C were readily differentiated. At the three higher temperatures, populations can be distinguished by comparing any of three other dimensions: coccolith length, coccolith width, or upper element length. These findings lead to the question of the possible utilization of coccoliths of *C. huxleyi* as temperature indicators. The possibility has a special interest in that this species is present in deposits at least as early as the Tertiary (Black and Barnes 1961). Certainly, the findings of Bé and McIntyre

(1964) that the forms of coccoliths from Arctic and tropical waters differ in a manner that parallels that of the present findings on laboratory cultures is strongly suggestive. But before coccoliths can be used as temperature indicators with any assurance, we shall need information on the influence of light, nutrition, and ionic environment on their morphology.

The apparent length of the upper elements (*see* below) decreased with increased temperature. On the other hand, the width of the elements increased with increased temperature. Therefore, the rates of linear crystal growth along different axes were differently influenced by temperature. The shape (habit) of crystals growing *in vitro* is modified by the degree of supersaturation, impurities, pH, temperature, and so on. In the case of coccolith formation, these factors, other than temperature, are obviously under cellular control, and the complexity of the intracellular environment prohibits the analyses of possible factors altered by temperature.

Because the rates of linear growth in different axes of the upper elements were affected differently by temperature, the temperature effects should also be determined in terms of crystal volume. As the cross-sectional form of the elements is not uniform, and has not been studied, volume estimations cannot be made.

The measurements of crystal length require comment. Because the upper elements curve downward, measurements of their apparent length do not represent actual length. Variations in apparent length could result from differences in actual length, or the angle between the elements and the central cylinder, or both. The number of pictures showing side views of coccoliths was not sufficient to calculate the magnitudes of these two variables. However, if the angle is relatively constant (about 30°), as several pictures indicate, then the apparent length measured is directly proportional to the actual length.

The width : length ratio of the coccolith was found to remain constant even though the length of the upper elements was al-

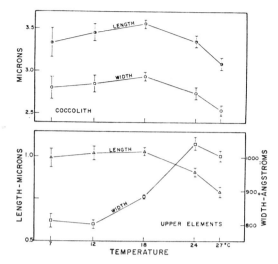

FIG. 7. The length and width of coccoliths and upper elements as a function of temperature. The ranges indicate 95% confidence limits of the mean values.

tered by temperature. This means that the central cylinder, made up of the same number of elements as the upper disc, must also change its size to a degree that maintains the shape constant.

Mechanisms governing coccolith form operated most favorably at 18C; at that temperature almost 80% of the coccoliths were normal. At higher and lower temperatures, the number of abnormalities increased. The normal form of the coccolith depends upon a nicely balanced growth of crystals that begins from a circle of growth centers in the basal region of the central cylinder (Wilbur and Watabe 1963). It is clear that abnormal coccoliths will result if crystal growth does not begin simultaneously in all centers in the basal region or if the crystals of the upper or the lower disc grow at different rates. Probably, most of the abnormalities are to be interpreted as an acceleration or an inhibition of growth of individual crystals resulting in coccolith asymmetry. The conditions concerned

would include local differences in calcium carbonate concentration and inhibitory substances.

REFERENCES

BÉ, A. W. H., AND A. McINTYRE. 1964. Recent coccoliths of the Atlantic Ocean. Geol. Soc. Am. Spec. Papers, **82**: 8.

BLACK, M., AND B. BARNES. 1961. Coccoliths and discoasters from the floor of the South Atlantic Ocean. J. Roy. Microscop. Soc., **80**: 137–147.

BRAARUD, T., K. GAARDER, J. MARKALI, AND E. NORDLI. 1952. Coccolithophorids studied in the electron microscope. Observations on *Coccolithus huxleyi* and *Syracosphaera carterae*. Nytt Mag. Botan., **1**: 129–134.

BRAMLETTE, M. N. 1958. Significance of coccolithophorids in calcium-carbonate deposition. Bull. Geol. Soc. Am. **69**: 121–126.

CRENSHAW, M. A. 1964. Coccolith formation by two marine coccolithophorids, *Coccolithus huxleyi* and *Hymenomonas sp.* Ph.D. Thesis, Duke University, Durham, N.C.

GUILLARD, R. R. L., AND J. H. RYTHER. 1962. Studies of marine planktonic diatoms. I. *Cyclotella nana* Hustedt, and *Detonula confervacea* (Cleve) Gran. Can. J. Microbiol., 8: 229–239.

KAMPTNER, E. 1956. Das Kalkskelett von *Coccolithus huxleyi* (LOHM). KPT. und *Gephyrocapsa oceanica* KPT. (Coccolithineae). Arch. Protistenk., **101**: 171–202.

LYMAN, J., AND R. H. FLEMING. 1940. Composition of sea water. J. Marine Res., **3**: 134–146.

MJAALAND, G. 1956. Some laboratory experiments on the Coccolithophorid, *Coccolithus huxleyi*. Oikos, **7**: 251–255.

NOËL, D. 1965a. Note preliminaire sur des coccoliths jurassiques. Arch. Doc., Centre Natl. Rech. Sci., **408**: 1–12.

———. 1965b. Sur les coccolithes du jurassique européen et d'Afrique du nord. Centre Natl. Rech. Sci., Paris. 209 p.

PAASCHE, E. 1964. A tracer study of the inorganic carbon uptake during coccolith formation and photosynthesis in the coccolithophorid *Coccolithus huxleyi*. Physiol. Plantarum, Suppl. 3, p. 1–82.

WADA, K. 1961. Crystal growth of molluscan shells. Bull. Natl. Pearl Res. Lab., **7**: 703–828.

WILBUR, K. M., AND N. WATABE. 1963. Experimental studies on calcification in molluscs and the alga *Coccolithus. huxleyi*. Ann. N.Y. Acad. Sci., **109**: 82–112.

8

Reprinted from *New York Acad. Sci. Trans.*, ser. 2, **32**:720-731 (1970)

MODERN PACIFIC COCCOLITHOPHORIDA:
A PALEONTOLOGICAL THERMOMETER[*][†]

Andrew McIntyre,[‡] Ph.D., Allan W. H. Bé,[§] Ph.D.,
and Michael B. Roche,[‡] B.S.

*Lamont-Doherty Geological Observatory
Palisades, N.Y.*

INTRODUCTION

The skeletal remains of the Coccolithophorida, a group of marine planktonic algae, play an increasingly important role in Cenozoic and Mesozoic biostratigraphy. Since their appearance in the Jurassic period, they have been characterized by rapid evolutionary change. This, combined with their planktonic habit, makes them excellent time-stratigraphic indicators. The rapidity of age determination of core samples aboard the drilling vessel *Glomar Challenger* of the Joint Oceanographic Institutions for Deep-Earth Sampling Program is due to the use of coccoliths.[1]

These golden-brown algae are restricted to the euphotic zone, approximately the upper 150 meters of the water column. Thus, they are under more direct climatic control than the other planktonic forms important in the fossil record. This is probably the cause of their complex evolutionary record and should make them excellent paleoclimatic indicators. The minute size of the living organisms (2–25 μ) and their skeletal plates or coccoliths (0.25–15 μ) prevented any detailed study, particularly of living floras, until the general availability of the electron and scanning-electron microscopes.[2,3] Size has also been an obstacle to oceanic sampling. Only recently, with the use of membrane surface filters with pore sizes below 1 μ, has a rapid and efficient shipboard technique been available.

An understanding of modern biogeographic distribution is required for paleoclimatic and stratigraphic use for two reasons. First, if the species composition in various water masses is known, the species can be grouped into floras representative of particular climates, so that a worldwide correlation will be meaningful and distinguishable from climatic zonation. Second, the optimum and maximum temperature ranges of the species must be ascertained to provide the paleontological thermometer needed to determine Pleistocene climatic fluctuations and help explain their cause.

The authors began such a study during 1963 in the Atlantic Ocean[3] which showed that definite coccolithophorid floral zones could be recognized; as a result, tentative temperature ranges were established for all paleontologically important species. The Pacific Ocean has been under study since 1967 to

[*]This paper was presented at a meeting of the Section on March 2, 1970; Lamont-Doherty Geological Observatory contribution no. 1574.
[†]This research has been supported by grants nos. GA-1205 and GA-14177 from the National Science Foundation, Washington, D.C.
[‡]Department of Geology, Queens College, Flushing, N.Y.
[§]Department of Biology, City College, New York, N.Y.

refine our knowledge of their ecological parameters. Our initial results are presented in this paper.

The most ubiquitous species in today's seas, *Coccolithus huxleyi,*‖ has the widest biogeographic and temperature range. Its biogeographic boundaries are also the limits for all coccolithophorids, reaching from the Arctic to the Antartic Convergences; coccolithophorids are not found in polar water masses. Presumably, such a species would be of little value for paleoclimatic work; however, *C. huxleyi* shows ecophenotypic variation with cold- and warm-water morphotypes that are clearly distinguishable from each other.[3,4] In the Pacific, the cold-

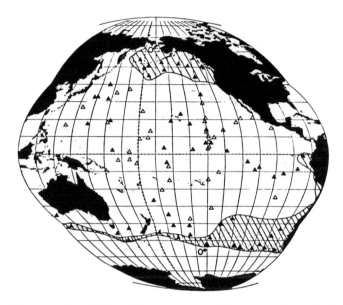

FIGURE 1. Biogeography of *Coccolithus huxleyi,* cold-water variety. Hachured area represents maximum concentration; triangles show sampling points, and solid symbols indicate the presence of the organism. Poleward boundary is the summer position of the 0° C surface-water isotherm.

water variety of *C. huxleyi* (FIGURE 1) dominates the floras in the subpolar waters, comprising 50–100% of the forms present, depending upon temperature and season. In surface waters below 6° C, it constitutes 90–100%. The distribution of this particular organism clearly delimits the subpolar-polar boundary, and as such can be used to locate it in late Pleistocene sediments. The cold-water form is found throughout the biogeographic range of *C. huxleyi,* but at temperatures above 14° C it never exceeds a small percentage of all *C. huxleyi* present.

‖*Note added in proof*—The following changes in nomenclature were recently made: *Coccolithus huxleyi* to *Emiliania huxleyi,*[8] and *Rhabdosphaera stylifera* to *Rhabdosphaera clavigera* (probably synonymous).

At the opposite end of the ecological spectrum are *Coccolithus pelagicus* and *Umbellosphaera irregularis* (FIGURE 2), representing subpolar and tropical waters, respectively, the most restricted distributions of the paleontologically common species. *Coccolithus pelagicus* at present has a narrow disjunct distribution; it is found only in the North Atlantic and North Pacific. Its temperature range is 6°–14° C, and its highest concentrations recorded to date occur between 9° and 12° C. Although its life cycle is quite complex,[5] the placolith stage is restricted to cold-temperate waters in the Northern Hemisphere, although southward-flowing boundary currents may carry this species to lower latitudes. The difficult problem with *C. pelagicus* is to explain its present distributional pattern. Although only found living in the Northern Hemisphere, it is found in modern surface sediments underlying the subpolar water masses in both the Northern and Southern Hemisphere. Since the age of these sediments is post-glacial, less than 12,000 years before the present, the disappearance of this species in the oceans of the Southern Hemisphere cannot be due to late Pleistocene glaciation.

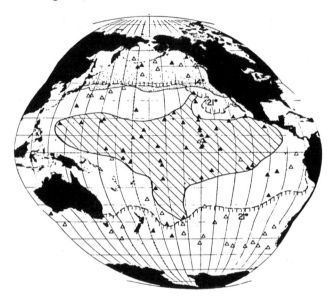

FIGURE 2. Biogeography of (a) *Umbellosphaera irregularis*, which is present between the summer 21° C isotherms (hachured area represents maximum concentration); and (b) *Coccolithus pelagicus*, found north of the winter position of the 14° C isotherm.

The region of temperate waters with surface temperatures between 6° and 14° C in the southern oceans is quite narrow, so that in many areas waters of subtropical character abut directly upon cold-subpolar waters. Thus, the ecological niche in which *C. pelagicus* could live is restricted. The easiest way to remove a species is to obliterate its environment. A post-glacial warming, if sufficiently strong, could result in a migration southward of subtropical water into the present area of subpolar water (with this would come an increase in circulation of the polar water, and a steepening of the surface-water temperature gradient), squeezing out the ecological habitat of *C. pelagicus*. A short time period would be sufficient to interrupt the life cycle of the species, with consequent regional extinction. It would not be necessary for all the temperate waters to be depressed below the euphotic zone, since in the southern oceans this zone tends to rotate with the West Wind Drift; thus, if at one point in the global gyre the narrow warm-subpolar water mass were to be forced below the subtropical waters and out of the euphotic zone, only the time for one complete circulation cycle would be required to kill all the *C. pelagicus*.

The climatic record derived from oxygen isotopes in the Greenland Icecap[6] and in the oceanic sediments[7] indicates that there was a very warm interval in the post-glacial period, approximately 8,000 years before the present, which could account for the current distributional pattern of *C. pelagicus*. The distribution was continuous between the Northern and Southern Hemispheres during the glacial epochs of the Pleistocene, particularly in the eastern Atlantic, where at times surface waters lower than 14° C crossed the equator.

The tropical character of *U. irregularis* is shown in the concentration of this species in a zone that surrounds the thermal equator in the Pacific, parallel to 8° N latitude. This tropical species is carried poleward from the tropical water mass by western boundary currents, the Kuroshio, East Australian and East New Zealand. Its northern boundary coincides with the Subtropical Convergence, but its southern boundary is more diffuse, owing to the seasonally irregular character of the Subtropical Convergence in the South Pacific.

Umbellosphaera tenuis (FIGURE 3), the only other species in the genus *Umbellosphaera*, prefers cooler waters (subtropical). Again, the well-defined water mass of the North Pacific is mirrored in the narrow latitudinal distributional pattern of this species, with borders roughly parallel to lines of latitude. The triangular pattern of maximum concentration of subtropical species in the South Pacific, with the apex pointing east toward South America, is a result

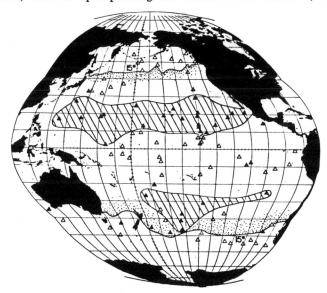

FIGURE 3. Biogeography of *Umbellosphaera tenuis*. Present between the summer 15° C isotherms; hachured area represents maximum concentration.

of the cold Peru current moving northward along the coast. The cold waters form a gyre that displaces the subtropical waters to the northwest, and is clearly seen in the distribution of *U. tenuis*. Both in the North and South Pacific, *U. tenuis* is found further poleward in the boundary currents than is *U. irregularis*. These two species are complementary in that a comparison of their distributional patterns in older sediments indicates the position and direction of the thermal equator and the intensity of the Subtropical Convergence.

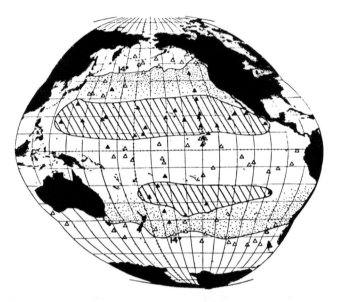

FIGURE 4. Biogeography of *Discosphaera tubifera*. Present between the summer 14° C isotherms; hachured area represents maximum concentration.

Perhaps the best indicators of the location of the subtropical water mass are *Discosphaera tubifera* and *Rhabdosphaera stylifera*¶ (FIGURE 4). These two closely related species prefer cool subtropical waters. Their poleward boundary coincides exactly with the subpolar boundary. *Discosphaera tubifera* shows a preference for slightly warmer surface waters, having an optimum temperature of about 1° C higher than that of *R. stylifera*. The distributional patterns for both of these species in today's seas appear disjunct, in that neither is found in equatorial waters.

For the paleontologist, the species of the *Gephyrocapsa* genus and *Cyclococcolithus leptoporus* and its variations may be the key to deciphering early Pleistocene climates (FIGURE 5). Unfortunately, these are not the most common forms in today's seas, and consequently our knowledge of them, despite the high density of ocean sampling, is incomplete. In addition, some problems in the taxonomy of these species remain, mainly whether the differences found represent ecophenotypic variations of the same species or separate species, particularly with respect to *C. leptoporus*. In this paper, we treat the mappable members of the genus *Gephyrocapsa* as separate species and those of *C. leptoporus* as varieties of the same species. Despite these difficulties, we can make

¶See footnote on p. 721.

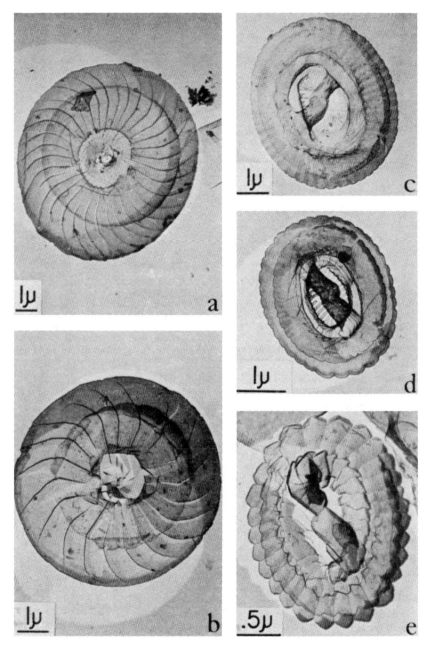

FIGURE 5. Distal surface of a single coccolith in: (a) *Cyclococcolithus leptoporus* var. B (Murray and Blackman) Kamptner; (b) *Cyclococcolithus leptoporus* var. C (Murray and Blackman) Kamptner; (c) *Gephyrocapsa oceanica* Kamptner (the diagnostic feature is the position of a bridge subparallel to the minimum diameter); (d) *Gephyrocapsa caribbeanica* Boudreux and Hay (the diagnostic feature is the bridge making an angle of less than 45° with the maximum diameter of the coccolith); (e) *Gephyrocapsa ericsonii* McIntyre and Bé (the diagnostic feature is its small size and tall bridge).

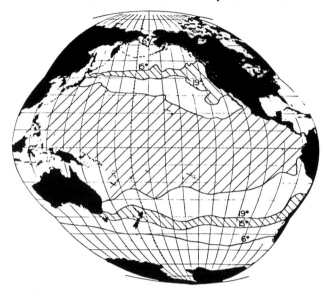

FIGURE 6. Generalized biogeography of *Gephyrocapsa oceanica*. Present between the summer 19° C isotherms, with maximum concentration in the hachured area centered on the equator. *Gephyrocapsa ericsonii* has its maximum concentration in the hachured area between the 19° C and the 15° C isotherms, the latter being its poleward boundary. *Gephyrocapsa caribbeanica* has the summer 6° C isotherms as its maximum poleward boundary.

some useful generalizations about biogeography and temperature ranges, which, when compared with known climatic fluctuations in the late Pleistocene, correlate perfectly.

Three species of *Gephyrocapsa, G. oceanica, G. caribbeanica,* and *G. ericsonii* (FIGURE 6), are presently living in Pacific waters. *Gephyrocapsa caribbeanica,* described initially as a fossil form from Caribbean cores of Pleistocene age,[8] is now known to be living in today's seas. It is the cold-preferring form (FIGURE 7) that ranges into the subpolar regions; we actually know more of this form and its relatives from studies in the Atlantic, since cooler water masses have been more densely sampled over a longer period of time in this ocean. Augmenting our distributional data are at-a-station productivity studies based on bimonthly sampling at different water depths. These samples also yield data of seasonal variation and temperature ranges for individual species. At present, this program is a joint study with the United States Coast Guard Oceanographic Office, involving weather-station ships occupying stations Bravo, Charlie, and Delta in the western North Atlantic. Of the three species, *G. caribbeanica* is the most cosmopolitan, ranging from tropical into subpolar seas, with a preference for cool waters. It has been found concentrated in the Pacific in samples from higher latitudes in water masses with surface temperatures between 5° and 15° C.

Gephyrocapsa ericsonii is in all probability a species that should be split, since we have found a bimodal distribution in both its biogeographic and temperature ranges. The cool-water form exists in subtropical waters ranging from 14° to 21° C, with an optimum temperature of 19° C. The rare warm-water form extends into tropical waters at temperatures from 23° to 30° C. *Gephyrocapsa oceanica* has perhaps the most restricted range, being limited to tropical and warm subtropical waters where it quite often constitutes a significant percentage of the total coccolithophorid flora. All three species occur in tropical seas, with *G. oceanica* dominant; in cool subtropical waters, *G. oceanica* is absent; and in subpolar waters, only *G. caribbeanica* is left.

Cyclococcolithus leptoporus (FIGURE 8) was previously reported to have three varieties dependent upon the number of skeletal units forming the coccoliths.[9] This view has been modified, and the largest form (variety A) has been given a new species name.[10] The remaining varieties now grouped under *C. leptoporus* appear to differ only in the number of elements forming the placolith plate and may be ecophenotypes. The distinguishing feature of *C. leptoporus* var. C is an average element count of 20. It is the eurythermal form, with its maximum poleward boundary in warm subpolar seas. While not found at latitudes as high as *Gephyrocapsa caribbeanica* or *Coccolithus huxleyi,* it is nevertheless paleontologically useful, particularly in the Southern Hemisphere where it occupies today the same water mass as *Coccolithus pelagicus* does in the Northern Hemisphere. This variety can, of course, be found in all water bodies between its maximum poleward boundaries. Thus, like *G. caribbeanica,* it is only useful when associated with certain other forms. *Cyclococcolithus leptoporus* var. B, with an element count averaging 30, prefers warm waters, tropical and warm subtropical. Since it is not as common as variety C, we cannot be as sure of its absolute biogeographic boundaries or temperature range until more sampling is done.

There are a number of other living coccolithophorids that have good paleontological records but, perhaps because of restricted depth habitat within the euphotic zone (e.g., *Cyclococcolithus fragilis*) or general rarity (e.g., *Helico-*

FIGURE 7. Biogeography of *Rhabdosphaera stylifera.* Present between the summer 14° C isotherms; hachured area represents maximum concentration.

FIGURE 8. Biogeography of *Cyclococcolithus leptoporus* var. B. Present between the summer 18° C isotherms; with maximum concentration in the hachured area centered on the equator. *Cyclococcolithus leptoporus* var. C has the summer 8° C isotherm as its poleward boundary; hachured area in the South Pacific is the maximum concentration area recorded to date.

sphaera carteri), we cannot as yet draw meaningful boundaries for them beyond the generalizations already published.

DISCUSSION

Although the Pacific Coccolithophorida biogeographic distributions presented are useful and can be correlated with Atlantic and Indian Ocean data, they are nevertheless biased. The samples collected over an 8-year period aboard the research vessels *Vema* and *Conrad* of the Lamont-Doherty Geological Observatory, while numbering in the thousands and covering all the water bodies in the Pacific Ocean, still are representative of only a finite time. Ideally, each water mass should be sampled bimonthly, as we are presently doing in the North Atlantic, to give a synoptic picture of the ecological parameters of each Coccolithophorida species.

There is a simple substitute for this involved and costly type of sampling program. The surface sediments of the oceans below each water mass should record its total flora over a long period of time, thus damping out seasonal variation and patchiness. In our study of the Atlantic Ocean,[3] the coccolith distribution in the sediments equated with living observations and demonstrated that contamination by deep currents or diffusion across surface boundaries was unimportant. Unfortunately, the bottom sediments in the Pacific Ocean are not the equal of those in the Atlantic. The floor of the North Pacific north of 12° N latitude is below the compensation zone for calcium carbonate and barren of coccoliths. South of 12° N latitude, the bottom sediments are carbonate-bearing, with the exception of an area in the southern Central Pacific. But even in this area the low sedimentation rate results in a long residence time in the surface layer where biological and chemical activity results in species-selective solution and disaggregation.

Species maps drawn on the basis of Pacific Ocean sediment data are more indicative of rates of destruction than of biogeographic distribution. As such, they are of more geochemical than paleoclimatic value. A single case illustrates this point: *Cyclococcolithus leptoporus* is a more resistant coccolith than is *Coccolithus huxleyi*.[11] In subtropical waters in the area of 20° S latitude and 110° E longitude, *C. huxleyi* constitutes 35% of the flora and *C. leptoporus* 5%. In the sediments below this area, the values are 25% for *C. huxleyi* and 55% for *C. leptoporus*. Across the whole South Pacific *C. leptoporus* was found in only about 30% of the surface-water samples, *C. huxleyi* in all. In the sediment samples, *C. leptoporus* was found in more cores than was *C. huxleyi*.

CONCLUSIONS

Although our Pacific Ocean study is still in progress, sufficient temperature data have been obtained to construct a temperature-range graph (FIGURE 9) of the species discussed in this paper. The additional data from the Pacific have allowed us to differentiate the temperature ranges for the *Gephyrocapsa* species and the ecophenotypes of *Cyclococcolithus leptoporus* and *Coccolithus huxleyi*, which we were unable to do in our first paper on the Atlantic Ocean.[3] Maximum

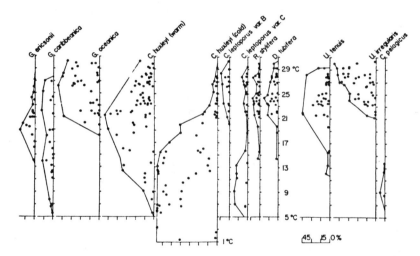

FIGURE 9. Graph of relative abundance of each species plotted against surface temperatures from over 200 samples throughout the Pacific Ocean.

and optimum ranges of the species are in general agreement with Atlantic and Indian Ocean data, both published and obtained from our more recent study of the United States Coast Guard Weather Ship samples. There have been some changes, e.g., the lowering of the low-temperature boundaries of *Discosphaera tubifera* from 18° to 14° C and of *Rhabdosphaera stylifera* from 15° to 14° C, and we can now assign a low-temperature boundary of 6° C to *Coccolithus pelagicus*.

The ten species we have discussed, with their included varieties, delineate four water masses based on both single and overlapping species temperature ranges. Surface waters of 0°–6° C are typified by a monospecific flora of *Coccolithus huxleyi*, cold-water variety. The flora increases in compexity with increasing temperature as *Gephyrocapsa caribbeanica*, *Cyclococcolithus leptoporus* var. C, and *Coccolithus pelagicus* join with *Coccolithus huxleyi*, cold-water variety, to characterize the waters at 6°–14° C. *Coccolithus pelagicus* is, as we have noted, only found in the Northern Hemisphere today, where its temperature range brackets this zone.

The addition of *Gephyrocapsa ericsonii*, *Rhabdosphaera stylifera*, *Discophaera tubifera*, and *Umbellosphaera tenuis* characterizes water temperatures of 14°–21° C. At temperatures above 21° C, the flora becomes less complex, with *Umbellosphaera irregularis*, *Gephyrocapsa oceanica*, and, less commonly, *Cyclococcolithus leptoporus* var. B as the predominant organisms, although the other forms, except *Coccolithus pelagicus*, are occasionally present in reduced numbers. Of all the species discussed in this paper, *Umbellosphaera irregularis*, *Gephyrocapsa oceanica*, and *Coccolithus huxleyi* are the only ones present above 29° C, constituting nearly 100% of the total flora.

These findings are of great value for Pleistocene paleoclimatic studies. Some of the species have already been used to show the migration of water bodies with late Pleistocene climatic fluctuations in the North Atlantic.[12] Now that similar distributions have been established for the Pacific and Indian Oceans, a new set of tools is available for general application in determining the mechanisms of glacial climatic changes.

ACKNOWLEDGMENTS

The authors wish to thank the many scientists aboard the research vessels *Vema* and *Conrad* of Lamont-Doherty Geological Observatory who collected the samples that made this study possible.

REFERENCES

1. BUKRY, D. & M. N. BRAMLETTE. 1969. Summary of coccolith biostratigraphy. *In* Initial Reports of the Deep Sea Drilling Project. I: 621–623. U.S. Govt. Printing Office. Washington, D.C.

2. HALLDAL, P. & J. MARKALI. 1955. Electron microscope studies on coccolithophorids from the Norwegian Sea, the Gulf Stream and the Mediterranean. Skrifter norske Videnskaps-Akad. 1955(1): 5–29.

3. MCINTYRE, A. & A. W. H. BÉ. 1967. Modern coccolithophoridae of the Atlantic Ocean. I. Placoliths and cyrtoliths. Deep Sea Res. 14: 561–597.

4. WATABE, N. & K. M. WILBUR. 1966. Effects of temperature on growth, calcification and coccolith form in *Coccolithus huxleyi* (Coccolithineae). Limnol. Oceanog. 11(4): 567–575.

5. PARKE, M. & I. ADAMS. 1960. The motile (*Crystallolithus hyalinus* Gaarder and Markali) and nonmotile phases in the life history of *Coccolithus pelagicus* (Wallich) Schiller. J. Mar. Biol. Ass. U.K. 39: 263.

6. DANSGAARD, W., S. J. JOHNSEN, J. MØLLER & C. C. LANGWAY, JR. 1969. One thousand centuries of climatic record from Camp Century on the Greenland Ice Sheet. Science 166: 377–381.

7. WOLLIN, G., D. ERICSON & M. EWING. 1970. Late Pleistocene climate Record in Atlantic and Pacific Deep Sea Sediments. *In* Silliman Lecture Volume Honoring Richard Foster Flint. Yale University Press. New Haven, Conn.

8. HAY, W. W., H. P. MOHLER, P. H. ROTH, R. R. SCHMIDT & J. E. BOUDREAUX. 1967. Calcareous nannoplankton zonation of the Cenozoic of the Gulf Coast and Carribbean: Antillean Area and transoceanic correlation. Trans. Gulf Coast Assoc. Geol. Soc. 17: 428–480.
9. MCINTYRE, A., A. W. H. BÉ & R. PREIKSTAS. 1967. Coccoliths and the Plio-Pleistocene boundary. Progr. Oceanog. 4: 3–24.
10. BUKRY, D. Personal communication.
11. MCINTYRE, A. & R. MCINTYRE. 1970. Coccolith concentrations and differential solution in oceanic sediments. *In* Micropaleontology of the Oceans. B. M. Funnel & W. R. Riedel, Eds. : 253–261. Cambridge University Press. Cambridge, England.
12. MCINTYRE, A. 1967. Coccoliths as paleoclimatic indicators of Pleistocene glaciation. Science 158: 1314–1317.

9

The distribution of oceanic coccolithophorids in the Pacific*

HISATAKE OKADA† and SUSUMU HONJO†

Abstract—Horizontal and vertical distributions of oceanic coccolithophorids were investigated along five traverses in the North and Central Pacific. The population was characterized by:
(1) a high standing crop in high latitude,
(2) an abrupt decrease in the temperate and subtropical areas except in the Kuroshio Water, and
(3) a significant increase in the equatorial and Kuroshio areas.
When most abundant, there were as many as 1.7×10^5 individuals per liter of water in the -5 m level at 50°N 155°W, but at some deeper localities in the photic layer there were no cells at all, particularly in the south Philippine Sea.
Approximately 90 species of the Coccolithophoridae were identified. Six coccolithophorid zones were established in the surface water along the 155°W meridian based on the distribution pattern of characteristic species: Subarctic, Transitional, Central North, Equatorial North, Equatorial South and Central South.
The Subarctic Zone corresponded to the Pacific Subarctic Current, where the specimens were almost exclusively a subarctic variety of *Emiliania huxleyi*. *Emiliania huxleyi*, a cold variety and *Rhabdosphaera clavigera* were dominant in the Transitional Zone, which coincided with the North Pacific Current. The Central zones were dominated by *Umbellosphaera irregularis*. The North Equatorial Current and Equatorial Countercurrent comprise the Central North Zone where the largest variety of species was observed. The Equatorial zones includes the South Equatorial Current and were characterized by an abundance of three placolith-type species: *Gephyrocapsa oceanica*, *Cyclococcolithina leptoporus* and *Cyclococcolithina fragilis*. The Central South Zone coincides with the northern portion of the South Pacific circulation.
Three vertical coccolithophorid zones were designated in the 200 m water column of the Transitional and Central zones, while the Subarctic and Equatorial zones had one and two vertical zones respectively. *Umbellosphaera irregularis* and *Rhabdosphaera clavigera* preferred shallow water while *Umbellosphaera tenuis* and *Cyclococcolithina fragilis* were mainly observed in the middle euphotic waters. *Florisphaera profunda* and *Thorosphaera flabellata* occurred only in lower euphotic waters.
Water temperature and light intensity appeared to be most influential factors in the distribution of coccolithophorid species, but currents also delimited the distribution of some species.

INTRODUCTION

SINCE WALLICH (1865) first observed living coccolithophorids, oceanic species have been described from the Mediterranean, Atlantic and Indian oceans. There have been several investigations in the Pacific and its marginal seas: Great Barrier Reef region (MARSHALL, 1933), Okhotsk Sea (SMIRNOVA, 1959), equatorial and subantarctic Pacific (HASLE, 1959; 1960), South Pacific (NORRIS, 1961), and Gulf of Panama and coast of Ecuador (SMAYDA, 1963; 1965; MARSHALL, 1970), as well as the global distribution in the Pacific of paleontologically significant coccolithophorid species (MCINTYRE, BÉ and ROCHE, 1970). The winter population in the western Pacific (OKADA and HONJO, 1970) and the oceanic coccolithophorids in the surface water of the North and Central Pacific (OKADA, 1970) have also been studied. Since LOHMANN's study (1902) of the Mediterranean, the vertical distribution of coccolithophorids has

---

*Contribution No. 2748 of the Woods Hole Oceanographic Institution, Woods Hole, Massachusetts 02543.
†Woods Hole Oceanographic Institution, Woods Hole, Massachusetts 02543.

been reported from various areas including relatively restricted parts of the Pacific (MARSHALL, 1933; HASLE, 1959, 1960; OKADA and HONJO, 1970; MARSHALL, 1970).

However, the wide spread, horizontal distribution of coccolithophorid species were hitherto but poorly known in the Pacific ocean. The present investigation was initiáted to clarify their geographic distribution and specific diversity in surface waters and throughout the water column. This study was conducted as a part of CSK (Cooperative Study of the Kuroshio and Adjacent Region) and IBP (International Biological Programme) cruises. Oceanographic data which were obtained at the same time as our sampling are also now available (TSUJITA, 1970; MARUMO, 1970).

SAMPLES AND TECHNIQUES

Approximately 600 water samples were obtained from 232 localities along five traverses in the North and Central Pacific (Fig. 1).* Surface water samples were

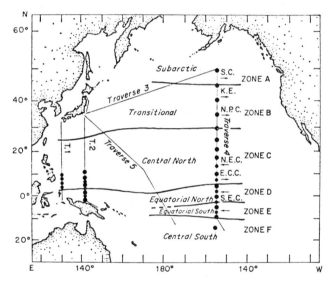

Fig. 1. Track chart of water sampling traverses and coccolithophorid zones. The large solid dots indicate major hydrographic stations where samples were collected from the entire water column, and the small solid dots indicate minor stations where samples were collected only from the upper 200 m of the column. Traverses 1 and 2 were conducted on board R.V. *Oshoro Maru*, Hokkaido University, November, 1968, to January, 1969, and traverses 3, 4 and 5 were conducted on board R.V. *Hakuho Maru*, Tokyo University, August to November, 1969.

normally collected at each degree of latitude, but in the tropical area along traverses 4 and 5, they were taken at half-degree intervals and at variable intervals along traverses 1, 2 and 3. Subsurface samples were also obtained along traverses 1, 2, and 4 with van Dorn-type samplers. At 19 stations along traverse 4, samples were collected at 10 intervals from the surface to 200 m level, and at approximately 10 intervals from 400 m to 5000 m.

*The sampling locality and hydrographic data were published by TSUJITA (1970) and MARUMO (1970) for traverses 1 and 2 and 3 to 5, respectively.

The water samples were filtered immediately after arrival on board using Millipore®️ filters with a 47 mm diameter (McIntyre and Bé, 1967), and a nominal pore size of 0·8 μ for the surface and photic water (upper 200 m) and 0·45 μ for the subphotic (200 m to bottom) water, but on traverses 1 and 2 filters with a pore size of 1·2 μ were used. Five to ten liters of water usually yielded a satisfactory number of coccolitho- phorids. After filtration, the filters were rinsed with approximately 20 ml of prefiltered distilled water, dried in an oven at 50°C for a few hours, and stored in a plastic con- tainer, for delivery to the laboratory ashore. For light microscopy, a 1 × 1 cm piece of the sampled filter was rendered transparent with type B Cargaille®️ immersion oil, and observed under a high magnification polarizing microscope. The coccolithophorid population per liter of water was obtained by computing the total number on a unit area and the water volume passed through the portion of the filter.

For electron microscopy the surface of another small piece of filter was coated uniformly with a thin carbon film in a vacuum evaporator after lightly shadowing with Pt–Pd (platinum–palladium) from a 30° intake angle. The filter was then dissolved by immersion in a series of acetone baths. Coccolithophorids covered with the carbon film were dissolved by bathing the film in a chromic acid mixture. After rinsing in distilled water, a piece of replicated film was mounted on a 150 mesh copper grid for transmission electron microscopy. Hitachi HU-11B and HU-12 electron micro- scopes were used for this investigation.

A Hitachi SSM-2 scanning electron microscope (SEM) was utilized where three dimensions were necessary to study the detailed morphology of the coccolithophorids. A gimbal stage was used to deposit a thin (approximately 300 Å) Pt–Pd film on the specimens (Honjo and Berggren, 1967).

The electron microscope was used almost exclusively to count the 200 individuals in each sample which were then identified to species. A specimen which appeared to be a perfect or a near perfect coccosphere under an optical or electron microscope was counted as one individual. The coccolith-free stages were disregarded.

RELATION BETWEEN COCCOLITHOPHORID DISTRIBUTION AND THE CURRENT SYSTEMS

Approximately 90 coccolithophorid species were identified, with the greatest diversity along traverse 4. The species assemblage and distribution of the standing crop allowed a preliminary description of six coccolithophorid zones, tentatively called Zones A to F from north to south (Figs. 1 and 2). The water masses appeared to have little relationship to the coccolithophorid zones, but the current pattern seemed to control the surface distribution of species (Table 1). The boundary between Zones A and B coincided with that of the Pacific Subarctic Current (eastward flowing) and the strong eastward flow of the Kuroshio Extension. The significant floral change at about 35°N is well defined by the southern limits of the Kuroshio Extension. The boundary between Zones B and C seems to differ somewhat from that of the North Pacific Current and North Equatorial Current. The species distribution is transitional between 30°N and 25°N (Fig. 2), where the currents are weak and unstable. The floral and current boundaries do not differ greatly since the 25°N station was apparently the boundary between the currents, and that between Zones B and C was established between 30°N and 29°N based on the abundance of *Umbellosphaera*

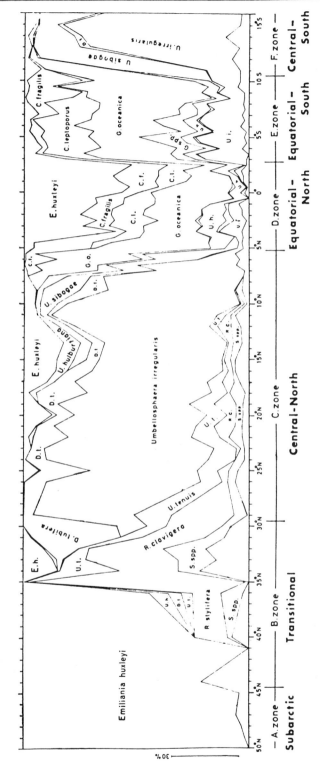

Fig. 2. Species assemblage of coccolithophorid in the surface water along traverse 4 (155°W). Abbreviations used for species names in Figs. 2, 4, 9, 10 and 11; A. q. *Anthosphaera quadricornu*. C. c. *Calciopappus caudatus*. C. f. *Cyclococcolithina fragilis*. C. l. *Cyclococcolithina leptoporus*. C. sp. A. *Coccolithus* sp. A. D. t. *Discosphaera tubifera*. E. h. *Emiliania huxleyi*. F. p. *Florisphaera profunda*. G. e. *Gephyrocapsa ericsonii*. G. o. *Gephyrocapsa oceanica*. G. sp. A. *Gephyrocapsa* sp. A. G. spp. *Gephyrocapsa* spp. S. spp. *Syracosphaera* spp. U. h. *Umbilicosphaera hulburtiana*. U. i. *Umbellosphaera irregularis*. U. s. *Umbilicosphaera sibogae*. U. t. *Umbellosphaera tenuis*.

Table 1. Coccolithophorid zones along the surface water of traverse 4 (155°W) compared with current, water mass and euphausiid zones.

	Coccolithophorid zones (Aug.–Oct., 1969)	Dominant species in surface waters	Current system (MARUMO, NAKAI and HASUMOTO, 1970) (Aug.–Oct., 1969)	Water mass (MARUMO, NAKAI and HASUMOTO, 1970) (Aug.–Oct., 1969)	Euphausiid zones (JOHNSON and BRINTON, 1963)
50°N	Subarctic Zone A (–45°N)	Emiliania huxleyi (subarctic variety)	Subarctic Current (–45°N)	Pacific Subarctic Water	Subarctic (ca. 57°N–41°N)
40°N	Transitional Zone B (45°N–30°N)	Emiliania huxleyi (cold variety) Rhabdosphaera clavigera	Kuroshio Extension (45°N–35°N)		Transition (41°N–33°N)
30°N			North Pacific Current (45°N–25°N)	Western North Pacific Central Water	
20°N	Central–North Zone C (30°N–5°N)	Umbellosphaera irregularis	North Equatorial Current (25°N–10°N)		Central (33°N–12°N)
10°N			Equatorial Countercurrent (10°N–5°N)		
0°	Equatorial–North Zone D (5°N–3°S)	Gephyrocapsa oceanica Emiliania huxleyi (warm variety)	South Equatorial Current (5°N–)	Equatorial Water	Equatorial (12°N–15°S)
	Equatorial–South Zone E (3°S–10°S)	Gephyrocapsa oceanica			
10°S	Central–South Zone F (10°S–)	Umbellosphaera irregularis	A part of the South Pacific Circulation (5°S–)		

106

irregularis. Zone C is composed of two current systems: the North Equatorial Current and the Equatorial Countercurrent. The North Equatorial Current is broad and flows westward. In contrast, the swift narrow Equatorial Countercurrent streams toward the east between the North and South Equatorial currents. Thus, there is a thorough mixing of the coccolithophorid flora between Zones C and D within the Equatorial Countercurrent; the closer to Zone D, the more scarce is *Umbellosphaera irregularis* and the more abundant *Gephyrocapsa oceanica*. Zones D and E coincide with the westerly South Equatorial Current. The division between these zones was less significant in the species assemblage than other boundaries. These two equatorial zones may almost be termed sub-zones.

South of 10°S Zone F was separated from Zone E in terms of the species assemblage, there was no clear boundary for the current system at this latitude. The species assemblage of Zone F is clearly identical with that in the southern part of Zone C,

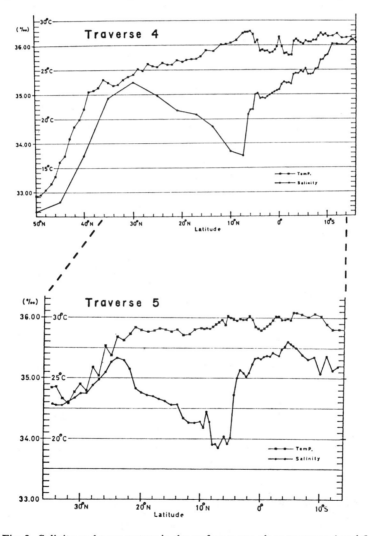

Fig. 3. Salinity and temperature in the surface water along traverses 4 and 5.

where the salinity record indicates a different water mass south of 10°S (Fig. 3). The South Equatorial Countercurrent* was not described by MARUMO, NAKAI and HASUMOTO (1970) but they did identify the South Pacific circulation south of 5°S. The species assemblage in the weak South Equatorial Countercurrent was very similar to that in the northern part of the South Equatorial Current, but it was transitional along traverse 5 between 5°S and 10°S (Fig. 4). Zones D and E then seem to correspond to the South Equatorial Current and the weak South Equatorial Countercurrent.

Many euphausiid species are limited to relatively narrow latitudinal zones, and four euphausiid zones have been established in the North Pacific with counterparts in the South Pacific (BRINTON, 1962). These zones roughly coincide with the coccolithophorid zonation (Table 1). It might, therefore, be useful to standardize the terminology to conform with Brinton's, i.e. Zone A, Subarctic; Zone B, Transitional; Zone C, Central (North); Zone D, Equatorial (North); Zone E, Equatorial (South) and Zone F, Central (South). The Transitional Zone extends to the Philippine Sea. The Central Zone occupies a broad area between 130°E and 155°W. The North and South Equatorial zones taper toward the west and become a single zone at around 180° (Fig. 1).

STANDING CROP OF OCEANIC COCCOLITHOPHORIDS

The distribution of the coccolithophorid standing crop was similar in the surface water along traverses 1, 2, 4, and 5 (Fig. 5)†, where they were numerous, particularly north of 45°N, with more than 100,000 individuals per liter of water, but they were usually scarcer in temperate and subtropical waters. In the Kuroshio and in the tropical waters of the Central Pacific, however, it was high. At the stations in the south Philippine Sea, none were detected in a liter of water.

The standing crop from the surface to 200 m was observed at 19 stations along traverse 4 (Fig. 6). It was large in the shallow layers in the north, but rapidly decreased in the deeper layers. At 50°N (Fig. 6, Sta. 1), the largest was 1.7×10^5 individuals/l. at 5 m, with an average in the upper 30 m of more than 120,000 individuals/litre. The depth of the densely populated layer increased southwards to 50 m at 45°N. The average population there decreased to 65,000 individuals/litre and disappeared at a depth of approximately 100 m level at 45°N. The distribution differed from that to the north between 40°N and 30°N (Fig. 6. Stas. 3, 4, and 5) where the standing crop increased rapidly with depth, with the greatest numbers existing at the 50–100 m level, and then rapidly decreasing with depth.

Between 25°N and 17°20′N (Fig. 6, Stas. 6, 7, and 8), the distribution was more or less even throughout the 200 m water column, particularly at 17°30′N (Sta. 8) where the count per litre showed very little fluctuation (3.8×10^3, 5.4×10^3 and 2.3×10^3/l. at the surface, 100 m and 200 m level, respectively). Deeper than 75 m coccolithophorids were scarce between 13°30′N and 7°30′N (Fig. 6, Stas. 9, 10, and 11, particularly Sta. 10, 10°N) perhaps due to the influence of the cold water channel there (Fig. 7). Between 5°N (Sta. 12) and 10°S (Sta. 18), the standing crop gradually increased down to the 50 or 75 m level, then decreased with depth. This was pronounced

*The South Equatorial Countercurrent has not been clearly identified, but the North Equatorial Countercurrent is distinct. It occurs in the South Equatorial Pacific between 5°S and 10°S (*National Geographic Atlas of the World*, Nat. Geog. Soc., Washington, 1970.)

†For coccolithophorid standing crop along traverses 1 and 2, see OKADA and HONJO (1970, Fig. 2).

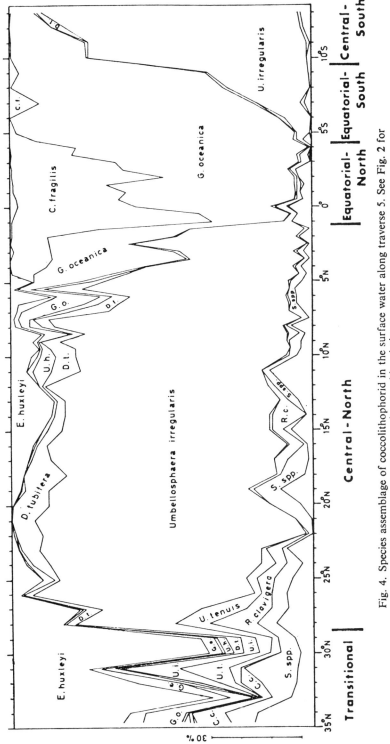

Fig. 4. Species assemblage of coccolithophorid in the surface water along traverse 5. See Fig. 2 for key to abbreviations.

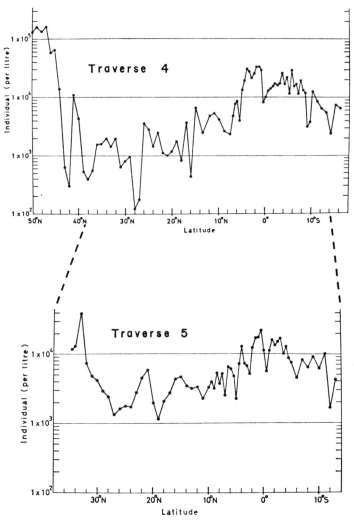

Fig. 5. Standing crop of coccolithophorid in the surface water along traverses 4 and 5.

at 100 to 125 m at 2°30'N (Sta. 13), possibly due to the cold Cromwell Current (Fig. 7). A south sub-tropical station (Sta. 19) had a relatively smaller standing crop for the several depth levels than did the equatorial stations. It increased at 200 m level, and an increase with greater depth also occurred at some equatorial stations owing to the dense population of species concentrated only in the lower euphotic layer.

The coccolithophorid population had a significant relation to the water temperature distribution in the 200 m water column, but salinity apparently had no influence on the population (Figs. 7 and 8). Two areas had a particularly high concentration of cells. One in the shallow waters off Alaska between 50°N and 45°N had the highest standing crop observed in this investigation, but it gradually diminished southward to 30°N where the greatest numbers occurred at 100 m. Deeper than 100 m, however, there were few cells, particularly at 45°N. This greatly populated area corresponded to a temperature range of approximately 10–19°C, and was limited to

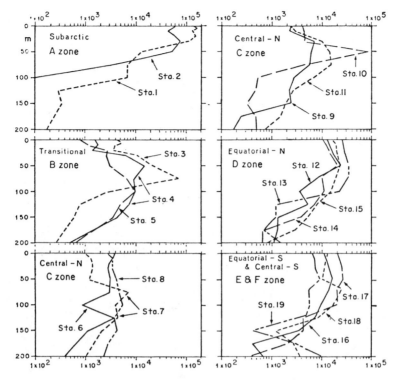

Fig. 6. Standing crop of coccolithophorids in the water column to 200 m at 19 stations along traverse 4 (155°W).

a depth of 100 m. The other area was between 10°N and 10°S, to a depth of approximately 100 m, with the maximum population at each station at 50 m. As in the high latitude area, the coccolithophorids were sparse in the lower levels of the 200 m water column; deeper than 100 and 150 m at 10°N and between 2°N and 7°30'S, respectively, where two cold underwater currents exist.

The total coccolithophorid number throughout the 200 m water column was as great as 6.0×10^7 individuals/2 m^3 (100 cm$^2 \times$ 200 m) in 200 m water column at 50°N (Sta. 1), then gradually decreased southward. The total number of cells in the 200 m water column was minimal (4.3×10^6 individuals/2 m^3) at 25°N (Sta. 6), but increased again near the tropics with a gradual decrease in the south sub-tropical area.

In the subphotic water column below 200 m, (Fig. 8), there were some coccolithophorids in suspension at most localities. In the deep layer there were more than 100 individuals/l. down to 4000 m in the equatorial area. An unusually large number of coccospheres (1.1×10^4/l.) occurred at 600 m at 5°N. Between 40°N and 35°N there was also a large number of cells.

SPECIES DISTRIBUTION OF OCEANIC COCCOLITHOPHORIDS

The greatest diversity of species and widest coverage of latitude were observed in the surface waters of traverse 4 (Figs. 1 and 2, Table 2). Three different species

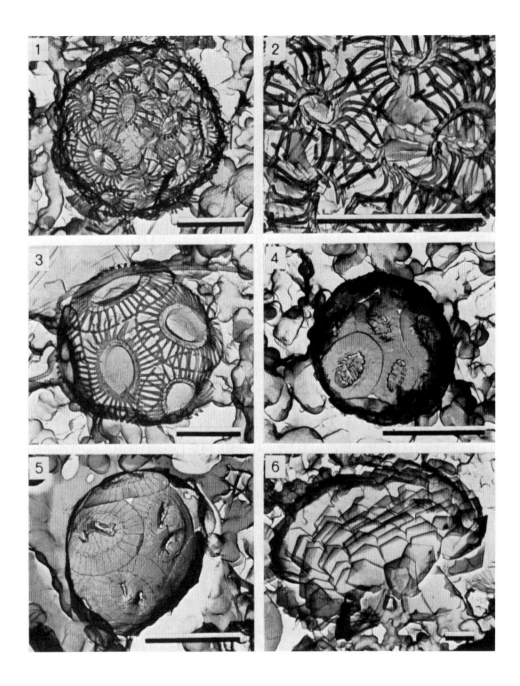

Plate 1.   Bars in pictures represent 3 μ.
No. 1. *Emiliania huxleyi*, subarctic water variety.
No. 2. *Emiliania huxleyi*, subarctic water variety.
No. 3. *Emiliania huxleyi*, cold water variety.
No. 4. *Coccolithus* sp. A.
No. 5. *Gephyrocapsa* sp. A.
No. 6. *Florisphaera profunda* var. B.

[*facing p.* 364]

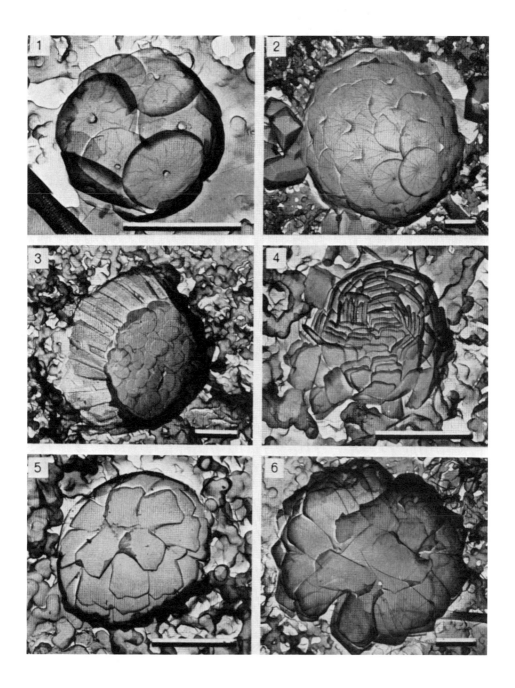

Plate 2. Bars in pictures represent 5 μ.
No. 1. *Cyclococcolithina fragilis.*
No. 2. *Cyclococcolithina fragilis* var. A.
No. 3. *Thorosphaera flabellata.*
No. 4. *Florisphaera profunda* var. A.
(Holotype for genus and species).
No. 5. *Florisphaera profunda* var. A.
No. 6. *Florisphaera profunda* var. B.

113

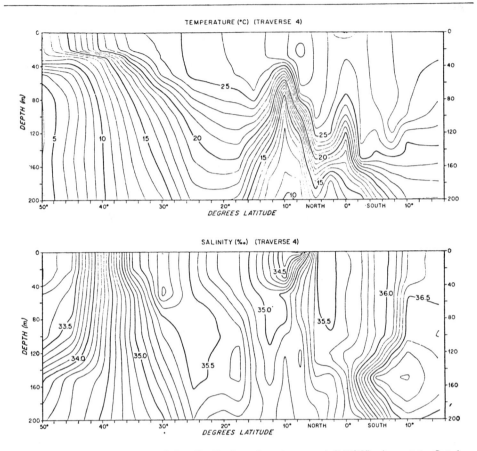

Fig. 7. Water temperature and salinity distribution along traverse 4 (155°W), August to October, 1969, based on the data obtained from the *Hakuho Maru* cruise (KH-69-4) conducted simultaneously with our sampling (MARUMO, 1970).

assemblages limited by the upper, middle and lower euphotic layers, respectively, were identified in the 200 m water column of the transitional and central zones (Figs. 9 and 10).

Zone A. Except for the 100 m level at Sta. 1 where *Calciopappus caudatus* was common, the subarctic ecophenotype of *Emiliania huxleyi* (Pl. 1, Nos. 1 and 2) monopolized the entire euphotic layer. This variety is characterized by a deformed dorsal shield, is poorly developed and formed of irregularly oriented 'T'-shaped elements. It seems to differ from the cold variety (MCINTYRE and BÉ, 1967) (Pl. 1, No. 3), which abruptly replaced it at 44°N. This then marks the boundary between zones A and B.

Zone B. The cold variety of *E. huxleyi* was still predominant at the surface but was replaced by *Rhabdosphaera clavigera* at 35°N. *Umbellosphaera tenuis* dominated the middle euphotic layer, and *Florisphaera profunda* n. gen., n. sp. (see p. 19), *Thorosphaera flabellata*, *Cyclococcolithina fragilis* and *Anthosphaera quadricornu* the lower euphotic layer, especially at the southern station (Sta. 5).

Zone C. *Umbellosphaera irregularis* was most abundant in the upper layer. The

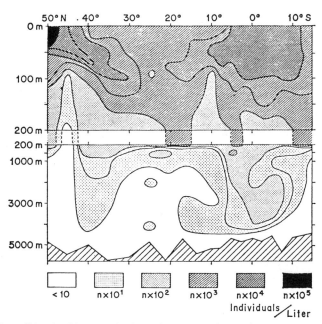

Fig. 8. Coccolithophorid counts in the entire water column along traverse 4 (155°W).

Table 2. Characteristic coccolithophorid species for zones and vertical layers in the Pacific.

	Zone A Subarctic	Zone B Transitional	Zone C Central-North	Zone D Equatorial-North	Zone E Equatorial-South	Zone F Central-South	Vert. dist. in photic layers
Emiliania huxleyi	○	○	●	●	○	○	
Gephyrocapsa oceanica			○	⊙	⊙	+	Throughout
Discosphaera tubifera			○			+	
Rhabdosphaera clavigera		⊙	○				Upper
Umbellosphaera irregularis			⊙	+	○	⊙	
Cyclococcolithina leptopolus				○	○		
Umbilicosphaera hulburtiana			+	○	+		Upper–middle
Umbilicosphaera sibogae			○			○	
Calciopappus caudatus	○						
Cyclococcolithina fragilis		●	○	●	●	○	
Gephyrocapsa sp. A			○		○		Middle
Umbellosphaera tenuis		●	○			○	
Anthosphaera quadricornu			○	○			
Florisphaera profunda		○	○	●	●	○	Lower
Thorosphaera flabellata		+	○	⊙	⊙	○	

+ 3–10%　　　○ 10–25%.　　● 25–50%.　　⊙ 50%–100%.

115

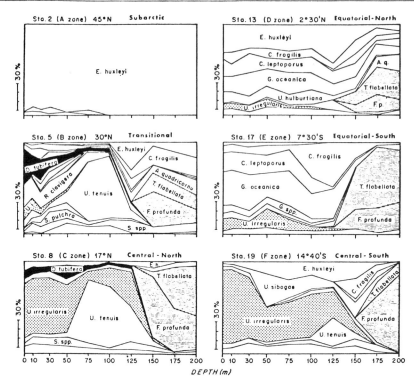

Fig. 9. Typical vertical assemblage of coccolithophorid species for Subarctic through Central–South zones. See Fig. 2 for key to abbreviations.

flora was similar in the middle and lower layers to Zone B except for Sta. 10 (10°N) where a cold undercurrent exists and three species of the placolith group, *Geophyrocapsa ericsonii*, *Gephyrocapsa* sp. A and *Coccolithus* sp. A were predominant.

Zone D. The species assemblage did not differ between the upper and middle photic layer. Five placolith species were common in this combined layer; *Emiliania huxleyi* and *Gephyrocapsa oceanica* were abundant, and *Cyclococcolithina fragilis*, *Cyclococcolithina leptoporus* and *Umbilicosphaera hulburtiana* were common. In the lower photic layer, the flora was similar to that in the northern zones.

Zone E. The species assemblage was very similar to Zone D, but *Emiliania huxleyi* was replaced by *Umbellosphaera irregularis* in the upper layer.

Zone F. The southern boundary has not been determined, but the species assemblage was quite similar to that of Zone C, perhaps being its Southern Hemisphere counterpart. Since the entire zone was not sampled, a greater number of species will probably be found in the future.

The species assemblage along traverse 5 (Fig. 4) was very similar to that for part of traverse 4 (Fig. 2) which perhaps indicates that the latitudinal distribution of oceanic coccolithophorids is fairly uniform in the North and Central Pacific. A large number of species were found between offshore Japan and 29°N along traverse 5. *Emiliania huxleyi* was dominant as in the northern half of Zone B, traverse 4. This area then seems to correspond with Zone B, traverse 4, although the species differed somewhat. Between 28°N and 1°30′N, traverse 5, which was sparsely populated,

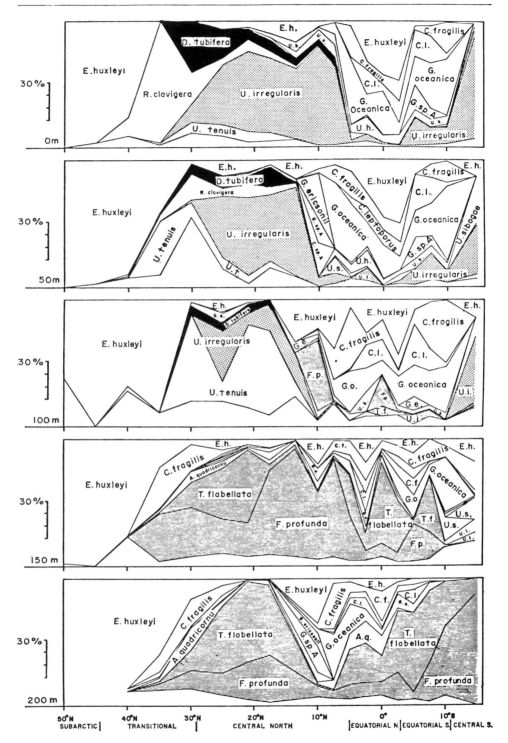

Fig. 10. Species assemblage of coccolithophorids at the surface and four subsurface levels from 50 m to −200 m, along traverse 4 (155°W). See Fig. 2 for key to abbreviations.

Umbellosphaera irregularis (90% at 21–22°N) dominated, as was also true in Zone C, traverse 4. Between 1°N and 6°S, traverse 5, *Gephyrocapsa oceanica* and *Cyclococcolithina fragilis* made up more than 90% of the samples, and between 1°N and 9°S, which crosses zones D and E, there was a transition in abundance from *G. oceanica* and *C. fragilis* to *Umbellosphaera irregularis* as on traverse 4. The latter increased at 5°S and became dominant to the south, as in Zone F, traverse 4 and at 13°S, it made up about 85% of the samples.

Similar species distributions were also observed along traverses 1 and 2. The equivalent of Zone B was widely developed from offshore Hokkaido and Honshu Islands to 25°N along 129°E and 142°E where *Emiliania huxleyi* was dominant. Species of the genus *Syracosphaera* were diverse here and in the Kuroshio Extension.

Coccolithophorids were scarce and *Umbellosphaera irregularis* dominated in the surface water between 25°N and 5°N, traverse 2 and south of 30°N, traverse 1. The surface water between 5°N and the Northern coast of New Guinea was, however, dominated by *Gephyrocapsa oceanica*, *Cyclococcolithina fragilis* and *Umbellosphaera irregularis*, as in the tropical area, traverses 4 and 5.

Species distribution through the entire water column

The coccolithophorid zonation for the surface water of the North and Central Pacific is evident in the species assemblage pattern for the 200 m water column (Fig. 11). *Emiliania huxleyi* dominated in the northern waters and it was rapidly replaced

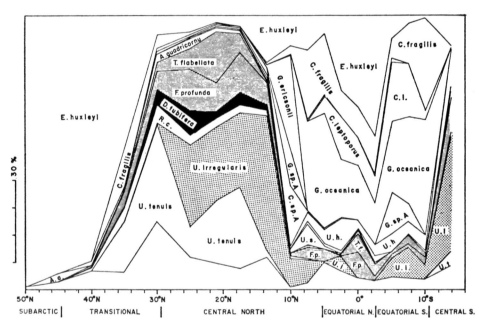

Fig. 11. Species assemblage obtained from the 200 m water column along traverse 4 (155°W). Coccolithophorids below the −200 m level are omitted. See Fig. 2 for key to abbreviations.

by a complex of species south of 30°N. It was again predominant between 2°30′N and 2°30′S. Two species of *Umbellosphaera* and two deep-water species, *Thorosphaera flabellata* and *Florisphaera profunda* made up as much as 50% and 30% of the total numbers, respectively, between 30°N and 13°30'N. *Geophyrocapsa oceanica, Emiliania huxleyi, Cyclococcolithina fragilis* and *C. leptoporus* were the major species between 7°30'N and 10°S. At 10°N, where a cold subsurface current existed, over 75% of the total numbers of cells were *Gephyrocapsa* spp. and *Coccolithus* sp. A. The percentage of *Umbellosphaera irregularis* in the total population of coccolithophorids increased rapidly south of 10°S.

A number of coccospheres were observed in the water layer deeper than 200 m along 155°W, but the concentration of coccolithophorid individuals was mostly insufficient to study the species assemblage utilizing our technique. A sample collected from 600 m at 5°N indicated that the species were quite similar to those from the 200 m water column. Half disintegrated coccospheres were, however, commonly observed. Since at this level the dissolved oxygen and pH were lowest, these are considered to be dead cells which had sunk from the upper layers to a resting boundary. The sample from 600 m at 40°N was monopolized by *Emiliania huxleyi*, as in the upper water layer (Fig. 11).

DISCUSSION

Oceanographic measurements and coccolithophorid distribution

The distribution of some coccolithophorid species was delimited by certain temperatures, with the boundaries which are essentially latitudinal and perhaps correlated with the isotherm pattern. (MCINTYRE and BÉ, 1967; MCINTYRE, BÉ and ROCHE, 1970).

Likewise a pronounced vertical thermal gradient often marked the vertical distribution of a particular species. At Sta. 3 (40°N), where the temperature decreased rapidly (0·8°C/m for 10 m) from 20 to 30 m (i.e. from 21·0–13·5°C), *Rhabdosphaera clavigera* disappeared at 30 m. A cold water channel at Sta. 10 (10°N), also had a sharp vertical thermal gradient between 30 m and 75 m (27·7–14·8°C). Three abundant species, *Discosphaera tubifera, Umbellosphaera irregularis* and *Umbilicosphaera sibogae* in the shallow water were replaced by three species of *Geophyrocapsa* and *Coccolithus* at 50 m and by *Emiliania huxleyi* at 75 m.

Umbellosphaera irregularis was dominant in the surface water between 28°N and south of 10°S along traverse 5 except in the tropics where it is equivalent to the zones D and E (Fig. 4). However, the water temperature did not differ significantly between the tropical and subtropical area (Fig. 3). Same evidence was also observed along traverse 4 (Figs. 2, 3 and 12).

The salinity in the surface waters, on the other hand, along traverses 4 and 5 changed abruptly at approximately 5°N, a change that coincided with the disappearance of *Umbellosphaera irregularis* (Figs. 2, 3, and 4). However, the variational range of salinity also did not differ markedly to the north or south as was the case for temperature. Hence, this apparent relationship does not necessarily indicate a direct relation between the two, but rather one between the boundary of the Equatorial Countercurrent and the South Equatorial Current (MARUMO, NAKAI and HASUMOTO, 1970; DIETRICH, 1963, Chart 5).

Thorosphaera flabellata and *F₁orisphaera profunda* were abundant in the deep water over a wide temperature range (10–28°C) (Figs. 7 and 12), but they seem to prefer a limited light intensity. which was generally less than 1% that at the surface although it was approximately 4% from 5°N to the equator (SATAKE, TAKAHASHI,

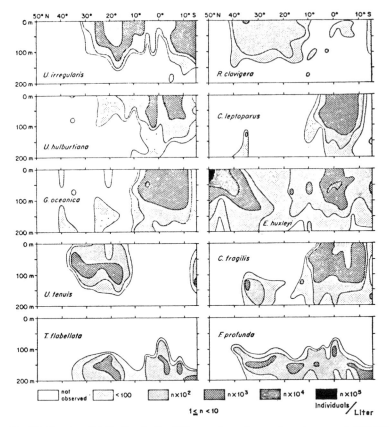

Fig. 12. Species distribution in the 200 m water column along traverse 4 (155°W).

NAKAMOTO and FUJITA, 1969). Their absence north of 40°N, and their scarcity at deeper levels at 10°N indicate that the isotherm for 10°C was a thermal barrier for these two species (Figs. 7 and 12), i.e. light intensity and water temperature together are perhaps both limiting factors.

Layered species distribution in the vertical water column

Umbellosphaera irregularis and *Rhabdosphaera clavigera*, upper photic layer species, were observed mostly in the upper 100 m layer and were quite scarce deeper (Fig. 12). *Umbilicosphaera hulburtiana* and *Cyclococcolithina leptoporus* dominated the upper 150 m and some were also observed as deep as 200 m. They can be classified as upper–middle photic species. *Emiliania huxleyi* was abundant down to 150 m over most of the area and dominated the entire 200 m layer in the northern area. *Gephyrocapsa oceanica* had a similar distribution in the tropical area. These two species then seem to prefer upper–middle euphotic layer, but they may be found over a very wide range of environmental conditions.

Umbellosphaera tenuis is distributed in the upper and middle portions of the euphotic layer, with a maximum number in the middle layer between 50 m and 100 m. *Cyclococcolithina fragilis* was observed throughout the 200 m water column. Their maximum occurred in the 125 m and 75 m levels in the Transitional and Equatorial zones, respectively, as in the Atlantic Ocean (LOHMANN, 1916) i.e. they are middle photic layer species.

Thorosphaera flabellata and *Florisphaera profunda* were mostly present in the deeper level below 100 m, with their maximum at approximately 150 m level, i.e. they are typical lower photic species.

Thus, these common species and other attendant species which are living in different depth make three vertical flora in the 200 m water column in the Central and Transitional zones (Fig. 13). However, the flora was monotonous in the Subarctic Zone, and the boundary between the upper and middle layer disappeared in the North Equatorial Zone (Fig. 9).

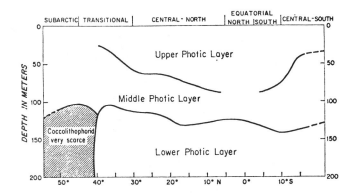

Fig. 13. Distribution of vertical coccolithophorid layers in the 200 m water column along traverse 4 (155°W).

Acknowledgements—We thank Dr. T. TSUJITA, Dr. R. MARUMO, and the scientific staff and crew members of R.V. *Oshoro Maru*, Hokkaido University and R.V. *Hakuho Maru*. Tokyo University for their co-operation during the cruises and also Dr. I. VALIELA, Marine Biological Laboratory, and Dr. W. A. BERGGREN, Woods Hole Oceanographic Institution, for critical reading of the draft manuscript and discussions. We are grateful to Dr. M. MINATO, Hokkaido University, Dr. A. McINTYRE, Lamont–Doherty Geological Observatory, and Dr. P. H. WIEBE, Woods Hole Oceanographic Institution, for valuable suggestions and discussions. Our thanks also go to Dr. J. SQUADRONI, Marine Biological Laboratory, for translation into Latin and discussion for taxonomical problems.

This investigation was supported by the National Science Foundation grant GA-30996. Use of computer facilities at the Woods Hole Oceanographic Institution was supported by National Science Foundation grant GJ-133. The early phase of the study was supported by the grant-in-aid from the Ministry of Education, Japan, A-091009, and National Science Foundation grant GA-8683.

REFERENCES

BRINTON E. (1962) The distribution of Pacific euphausiids. *Bull. Scripps Instn Oceanogr., Univ. Calif.*, **8**, 51–270.

DIETRICH G. (1963) *General oceanography, an introduction*. transl. F. Ostapoff, Interscience, 588 pp.

HASLE G. R. (1959) A quantitative study of phytoplankton from the equatorial Pacific. *Deep-Sea Res.*, **6**, 38–59.

HASLE G. R. (1960) Plankton coccolithophorids from the subantarctic and equatorial Pacific. *Nytt Mag. Bot.*, **8**, 77–88.

HONJO S. and W. A. BERGGREN (1967) Scanning electron microscope studies of planktonic Foraminifera. *Micropaleontology*, **13**, 393–406.

JOHNSON M. W. and E. BRINTON (1963) Biological species, water masses and currents. *The sea*, M. N. HILL, editor, Interscience, **2**, 381–414.

LOHMANN H. (1902) Die Coccolithophoridae, eine Monographie der Coccolithen bildenden Flagellaten, zugleich ein Kenntnis des Mittelmeerauftriebs. *Arch. Protistenk.*, **1**, 89–165.

LOHMANN H. (1916) Neue Untersuchungen über die Verteilung des Planktons im Ozean. *Sber. Ges. naturf. Freunde Berl.*, **1916**, 73–126.

MARSHALL H. G. (1970) Phytoplankton in tropical surface waters between the coast of Ecuador and the Gulf of Panama. *J. Wash. Acad. Sci.*, **60** (1), 18–21.

MARSHALL S. M. (1933) The production of microplankton in the Great Barrier Reef region. *Scient. Rep. Gt Barrier Reef Exped.* **1928–29, 2**, 112–157.

MARUMO R., editor (1970) *Preliminary report of the* Hakuho Maru *Cruise KH-69-4 (IBP Cruise)*, R. MARUMO, editor, Ocean Res. Inst., Univ. Tokyo, 68 pp.

MARUMO R., T. NAKAI and HASUMOTO (1970) Water masses, current systems and phytoplankton communities along 155°W. *Preliminary report of the* Hakuho Maru *Cruise KH-69-4 (LBP Cruise)*, R. MARUMO, editor, Ocean Res. Inst., Univ. Tokyo, 11–12.

MCINTYRE A. and A. W. H. BÉ (1967) Modern coccolithophoridae of the Atlantic Ocean—I. Placoliths and cyrtoliths. *Deep-Sea Res.*, **14**, 561–597.

MCINTYRE A., A. W. H. BÉ and M. B. ROCHE (1970) Modern Pacific Coccolithophorida: a paleontological thermometer. *Trans. N. Y. Acad. Sci.*, **32**, 720–731.

NORRIS R. E. (1961) Observations on phytoplankton organisms collected on the N.Z.O.I. Pacific Cruise, September 1958. *N. Z. J. Sci.*, **4**, 162–188.

OKADA H. (1970) Surface distribution of coccolithophores in the North and Equatorial Pacific. *Bull. geol. Soc. Japan*, **76**, 537–545.

OKADA H. and S. HONJO (1970) Coccolithophoridae distributed in Southwest Pacific. *Pacif. Geol.*, **2**, 11–21.

SATAKE K., M. TAKAHASHI, N. NAKAMOTO and Y. FUJITA (1969) *Data of light intensity at major stations for the* Hakuho-Maru *cruise KH-69-4.* (Unpublished data).

SMAYDA T. J. (1963) A quantitative analysis of the phytoplankton of the Gulf of Panama. I. Results of the regional phytoplankton surveys during July and November, 1957 and March, 1958. *Bull. Inter-Am. trop. Tuna Commn*, **7**, 191–253.

SMAYDA T. J. (1965) A quantitative analysis of the phytoplankton of the Gulf of Panama. II. On the relationship between C¹⁴ assimilation and the diatom standing crop. *Bull. Inter-Am. trop. Tuna Commn*, **9**, 467–531.

SMIRNOVA L. T. (1959) Phytoplankton in the Okhotsk Sea and Kuril Island region. (In Russian) *Trudy Inst. Okeanol., Akad. Nauk SSSR*, **30**, 3–51.

TSUJITA T., editor (1970) The *Oshoro Maru* Cruise 30 to the north and tropical Pacific Ocean and the Gulf of Carpentaria, November 1968–January 1969. *Data Rec. oceanogr. Obsns explor. Fishg*, **14**, 217–313.

WALLICH G. C. (1865) On the structure and affinities of the Polycystina. *Trans. microsc. Soc. Lond.* (2), **13**, 57–84.

APPENDIX

SYSTEMATIC DESCRIPTION FOR NEW SPECIES

Florisphaera n. gen.

Coccosphere hemiellipsoidal with shallow dome top. Coccoliths show polygonal plate shape, are all arranged in the same direction and have a concentric pattern in top view. In side view, the upper half of sphere is steplike and increases height toward center. The general view of this coccosphaere resembles a multi-petalled flower.

Holotype: HU-1 (Film), Woods Hole Oceanographic Institution. Type locality: Central Pacific.

Florisphaera nov. gen.

Coccosphaera habet diovatam formam planatumque tholum. Coccolitha polygonium laminatam formam, eadem directionemque et, in supera visione, circulos cumcentricos proponunt. In latu, sphaerae dimidia supera pars graduum facit et auget in altitudine adversus mediam partem. Flos multipetals communis aspectos coccosphaerae est.

Florisphaera profunda n. gen., n. sp.

Var. A (Plate 2, Nos. 4 and 5)

Description of coccosphere: Top diameter ranging from 3·7 to 8·5 μ, mean of 6·3 μ, and the height from 2·7 to 7·5 μ, mean 4·3 μ. The coccolith number ranges from approximately 30 to more than 100.
Description of coccolith: The most simple coccolith with general view of the quadrangular plate

has a length ranging from 0·8 to 3·1 μ, mean of 2·0 μ, and width of from 0·7 to 2·5 μ, mean of 1·5 μ. The thickness varies from approximately 0·05 to 0·25, maximum in the central part and minimum at the ends in the longitudinal direction. The side edges are straight to slightly curved and parallel to slightly tapered toward the bottom profile in its plan view. The bottom line has an irregular zigzag pattern and is almost parallel to the top profile. The top profile also has zigzag pattern and one corner is cut the angle, but some specimens show straight line.

Holotype: HU-1 (Film), Woods Hole Oceanographic Institution. Type locality: Central Pacific.

Varietas A

Descriptio coccosphaerae: Tholus planatus 3·7–8·5 μ longus (medium 6·3 μ) et 2·7–7·5 μ altus (medium 4·3 μ) est. Coccolithorum numerus prope variat ab 30 ad 100.

Descriptio coccolithi; Simplicissimus coccolithus, generalem aspetum habet quadrangularis tabulae; 0·8–3·1 μ longus (medium 2·0 μ); 0·7–2·5 μ latus (medium 1·5 μ); 0·5–0·25 μ crassus, maximum in parte centrali, minimum in extremis directione longitudinali. Margines laterales rectae, vel leviter flectae; aequidistantes vel aliquantulum cuneatae versus fundum in aspectu superiori; paululum cuneatae versus laterem dorsalem in aspectu laterali. Linea fundi habet distortum lineamentum fere parallelum lineae culminis. Margo superior structura distorta, angulo reptuso; nonnulla vero specimina rectam lineam exhibent.

Florisphaera profunda n. gen., n. sp.

Var. B (Plate 1, No. 6, Plate 2, No. 6)

Description of coccosphere: Similar to the var. A except in size. The diameter ranges from 8·3 to 13·2 μ, mean of 10·8 μ, and the height from 4·5 to 12·0 μ, mean of 8·5 μ.

Description of coccolith: The length ranges from 1·7 to 7·8 μ, mean of 4·3 μ, and the width from 1·3 to 5·0 μ. The thickness is similar to that of var. A. The side profiles are tapered toward the bottom and have a protrusion on each side of the bottom part. The protrusion is sharp and located higher on one side, but it is slightly roundish and lower on the other side. The bottom profile is similar to the var. A, with two corners off its angles. The top profile is straight and has one distinguished peak. This peak is biased and has an obvious obtuse angle, while some specimens are slightly rounded.

Varietas B

Descriptio coccosphaerae: Similis varietati A, praeter amplitudinem mensurae: 8·3–13·2 μ lata (medium 10·8 μ); 4·5–12·0 μ alta (medium 7·5 μ).

Descriptio coccolithi: 1·7–7·8 μ longus (medium 4·3 μ). Quoad crassitudinem consulas quaeso varietatem A. Lineae laterales cuneatae versus fundum, tuberositatem habens in utroque latere fundi. Tuberositas haec acuta, altiorem positionem in uno latere habens quam in altero. Linea fundi similis varietati A, uterque angulus truncatus. Latus superius rectum, unde insignis protrutio eminet. Protrutio inclinata, retusum angulum formans. Nonnula tandem specimina quamdam rotundiorem formam exhibent.

10

Reprinted from *Geol. Soc. America Bull.* **69**:121–126 (1958)

SIGNIFICANCE OF COCCOLITHOPHORIDS IN CALCIUM-CARBONATE DEPOSITION

By M. N. Bramlette

Investigation of the living coccolithophorids by marine biologists is receiving increased attention as these forms constitute an important element of the oceanic plankton. Among the more important investigations are those of Braarud and associates at Oslo, Kamptner and predecessors at Vienna, and Deflandre in Paris. Even a brief review of present knowledge is not attempted in this paper, as this has been presented recently (Bramlette and Riedel, 1954, p. 385–386) and much more thoroughly considered by others (Deflandre, 1952, p. 439–470).

Coccolithophorids are minute beflagellate Protista that derive their food supply through photosynthesis and are usually classed as algae. The skeletal material consists of a number of coccoliths (about 2–20 microns) at or near the surface of a more or less globular body of protoplasm. The coccoliths are usually not well enough articulated to remain together in the sediments as complete skeletal remains (coccospheres), though these are found in some samples. X-ray-diffraction studies indicate that they consist of calcite in the bottom sediments, and if any originally form as aragonite these must be quickly dissolved or converted to calcite.

Studies on the stratigraphic distribution of coccolithophorids include preliminary examination of more than 1000 samples and indicate much promise for these neglected fossils in correlation and age assignments. The wide distribution of these planktonic organisms is strikingly evident in the many identical forms in strata assigned on other evidence to the middle Eocene of France, Louisiana, and California. Many samples must be examined before

Contribution from the Scripps Institution of Oceanography, New Series, No. 954. This paper is based in part on results of research carried out by the University of California under contract with the Office of Naval Research.

stratigraphic ranges are reasonably well established, but it now appears that their vertical ranges are comparable with other fossils, such as Foraminifera. For example, a peculiar and easily identified form, described by Deflandre from the upper Eocene of New Zealand, has now been found in the upper Eocene only of such widely separated regions as Australia, Trinidad, Alabama, and France.

Samples from the type Danian (generally considered uppermost Cretaceous) show an assemblage very different from that of the underlying Maestrichtian, and a closely comparable assemblage is found in the Midway of Alabama (usually assigned to the Paleocene) overlying an assemblage like that of the Maestrichtian. Though some discontinuity is evident between the type Danian and underlying Cretaceous, it does not seem to mark any very great time break, and the marked difference in the assemblages may prove to be largely due to some ecological difference such as change in surface-water temperatures. The coccoliths of both the Danian and Midway are more like those of the early Tertiary than like those of the underlying Cretaceous. This and other lines of evidence remain to be considered thoroughly, however, before an age equivalence between the Danian and Midway might be suggested.

A number of distinctive coccolithophorids seem to have developed near the end of the Oligocene or in the early Miocene, as interpreted from the sequences in Italy, Trinidad, and Indonesia. Some additional evidence may thus be available on the Aquitanian and equivalents elsewhere, including the E and F part of the lettered sequence used for the Tertiary of Indonesia.

The earliest known occurrence of well-preserved and definitely recognizable remains of coccolithophorids is in the Jurassic (Deflandre, 1952, p. 459–460), but they occur in great vari-

TABLE 1.—RELATIVE IMPORTANCE OF COCCOLITHS, FORAMINIFERAL TESTS, AND OTHER CONSTITUENTS IN CALCAREOUS OOZE AND CHALK

(In per cent of total sample)

	Recent Globigerina ooze of deep Pacific — Albatross 62 (56–60 cm)	Miocene calcareous ooze of deep Pacific — Chubasco 24 (20–24 cm)	Oligocene (upper?) coccolith ooze of deep Pacific — Chubasco 38 (10–12 cm)	Oligocene (middle?) coccolith ooze of deep Pacific — Midpac 40 (130–140 cm)	Oligocene (lower?) coccolith ooze of deep Pacific — Midpac 5-1 (50–60 cm)	Eocene (middle?) coccolith ooze of deep Pacific — Midpac 25E (50–60 cm)	Cretaceous (Maestrichtian) chalk from Stevns Klint, Denmark	Cretaceous (Campanian) chalk from Travis County, Texas
Total calcium carbonate (from chemical analysis)	86	79	87	91	84	95	93	65
Coccoliths (including discoasters and related forms)	10	26	35	58	48	43	18	21
Foraminifera (including identifiable fragments)	55	25	30	5	15	32	23*	18*
Other calcareous skeletal debris	3	8	5	1	3	5	15	10
Indeterminate calcite particles	18	20	17	27	18	15	37	16
Radiolaria and diatoms	9	13	9	0	7	0	0	0
Inorganic non-calcareous residue	5	8	4	9	9	5	7	35

* In large part benthonic forms

shown in the recent sample from the Pacific (Table 1), but their dominance in the Tertiary seems to be comparable with that in the Pacific. Some cores of Tertiary age, collected from the Atlantic by Lamont Geological Observatory, were described (Ericson, Ewing, and Heezen, 1952, p. 504–507) as fine-grained chalk, and some of these samples examined by the writer contained a great abundance of coccoliths, though the samples were so small that relative percentages were not determined.

There is evidence that, during much of Tertiary time, ocean waters were warmer than at present, and oxygen-isotope data of Emiliani and Edwards (1953, p. 887) indicate that temperatures of the bottom waters may have averaged about 12° C. This would have lowered the solubility limits of calcareous deposits. The carbonate is dissolved, and little accumulates now below depths of about 4500 m in the Pacific Ocean and below somewhat greater depths in the Atlantic Ocean. Revelle has calculated (personal communication) from his graphs of the controlling variables (Revelle and Fairbridge, 1957) that the present critical depth of about 4500 m would correspond to about 6700 m for comparable solubility conditions with a Tertiary bottom-water temperature of 12°C. This would explain the occurrence of Tertiary calcareous ooze beneath "red clay" in some cores from areas where depths are too great for calcareous accumulation under present conditions. It does not seem that this solubility difference for different bottom temperatures could account for the relative proportions of coccoliths and Foraminifera in the Recent and Tertiary calcareous sediments, however, because even the smallest coccoliths are present in both, and they show little more susceptibility to dissolution than do the Foraminifera— as judged by their distribution with depth on present ocean bottoms. This is in peculiar contrast to the distinctly greater susceptibility of coccoliths to solution and recrystallization in the intrastratal waters of Cretaceous chalk formations.

The coccolithophorids thus have had a significant role in the geochemical cycle and balance of calcium carbonate in the ocean basins. An interesting and often-quoted suggestion by Twenhofel (1939, p. 327), later elaborated by Kuenen (1950, p. 392–394), was that *Globigerina* and other planktonic Foraminifera, which first appeared during the Cretaceous, have since greatly modified the major site of deposition of calcium carbonate and hence its

whole geochemical cycle in sediments. The great amount of modern *Globigerina* ooze in the deep-ocean basins represents a loss to this cycle, in contrast with the great and presumed dominance of accumulation in shallow (perhaps epicontinental) seas of pre-Cretaceous time. Much of this earlier deposition in epicontinental or marginal seas must have been incorporated again into the continental masses and returned to the cycles of erosion and deposition. The large loss to the deep seas under modern conditions was calculated by Kuenen to indicate a potential calcium-carbonate "famine" in 100–150 million years. The lesser amount postulated as lost to the deep seas in earlier times would help account for the high proportion of limestones among older rocks of the continental areas.

Kuenen (1950, p. 393) also suggests that "the present relative scarcity of calcareous deposits in shallow waters, as compared to former periods," may be similarly explained. This scarcity seems questionable in island-atoll areas, however, and in shelf-sea areas seems more probably to reflect only a greater dilution with clastic sediments from rapid erosion of unusually great topographic relief in late geological time.

It is not yet apparent that the earlier dominance of coccolithophorids was due to a greater rate of production than at present, nor that there was a subsequent (post-middle Tertiary) increase in the rate of production of the pelagic Foraminifera. The latter possibility would help the budget problem of calcium carbonate in the oceans and seems to fit the meager data suggesting a greater-than-average rate of deep-sea deposition in Quaternary time. It is not too evident, however, that the rate of supply and consequent deposition of calcium carbonate has varied greatly, and perhaps only the relative importance of the two groups of calcareous plankton has changed. At least it seems evident that the rate of supply is the critical factor for much change in the rate of carbonate deposition. At present the oceans are about saturated with calcium carbonate, and the supply has been calculated to be sufficient to replace this total amount within roughly a million years (Revelle and Fairbridge, In press).

Though the coccolithophorids seem to have been more important than the pelagic Foraminifera before late Tertiary time, their importance is not demonstrable before the Cretaceous. Unless the coccolithophorids were important earlier, however, their dominance over the pelagic Foraminifera would not alter the significant conclusion of the hypothesis on calcium-carbonate loss to the deep-sea bottoms through pelagic accumulations since Cretaceous time. Kuenen's suggestion seemed particularly inviting because both the geochemical budget during later geologic time and the excess of calcium carbonate in known pre-Cretaceous rocks present a less serious problem if the present large loss to the deep sea is abnormal and most of the carbonate was deposited on or near continental blocks during much of geological time. Attractive as this hypothesis appears, it would have no basis if such organisms as coccolithophorids or an inorganic process could have segregated much calcium carbonate in the deep-ocean bottoms before the advent of pelagic Foraminifera; both possibilities seem to need consideration.

Very fine-grained limestones in especially thin sections under a high-power microscope show a fine mosaic texture similar to aplitic—or xenomorphic granular—texture in igneous rocks. The small (commonly only a few microns) interlocking crystals indicate a recrystallization of whatever small particles were originally present. One occurrence where these were evidently coccoliths will be described. Some may equally well have been like the tiny needles of aragonite now forming on the Bahama Banks, which have been called drewite. This drewite is considered here as probably of dominantly inorganic or at least nonskeletal origin. Though this has been questioned (Lowenstam, 1955, p. 270–272), its origin is not critical to this discussion. Such material has not been found in careful examination of many Recent and Tertiary deep-sea deposits. Drewite deposition would seem more probable in warm, shallow waters, and more certainly so if it should prove to be disintegrated debris of algae such as *Halimeda*, as suggested by Lowenstam.

The fine-grained Carmel limestone of Utah might well represent a deposit comparable to the drewite of the Bahama Banks. It is comparatively thin, occurs between nonmarine formations, and thus was probably formed in relatively shallow water. Thin sections of some parts also show small pellet forms like the oölites of the Bahama Banks deposits.

Very fine-grained limestones are common, however, and rather characteristic of some geosynclinal formations that seem to represent relatively deep-water accumulations. The Tithonian and other limestones of the sub-Alpine region of France are examples.

The requirements for inorganic precipitation of much calcium carbonate include warming of the water, or decrease of carbon dioxide, or increase of salinity through evaporation. In a former time without calcareous planktonic organisms, one or more of these requirements might have been maintained to a sufficient extent near the surface of relatively deep waters, despite some inevitable mixing with water from below the surface, and inorganic precipitation might have occurred here as well as in shallow water. Conceivably the fine-grained limestones of geosynclinal accumulations could have originated in this way. Assuming inorganic precipitation under these conditions, however, would make it equally possible in the open ocean and offers no support to the hypothesis of a change to a greater proportion of deep-ocean accumulation of calcium carbonate during Cretaceous and later time.

This hypothesis is not supported by evidence of greatly increased importance of calcareous plankton either unless forms such as the coccolithophorids were, like the pelagic Foraminifera, insignificant before about this time. The coccoliths actually are found in great abundance only with the earliest occurrences of chalk in the Cretaceous, but the following considerations suggest that earlier occurrences may be represented by some pre-Cretaceous fine-grained limestones. Thus the problem would remain regarding a geochemical balance for calcium carbonate.

The original absence or paucity of planktonic Foraminifera is evident in pre-Cretaceous limestones which show comparably robust microfossils preserved, despite the evident recrystallization of the much finer matrix particles inherent in lithification to a dense limestone. Though not recognizable, the original presence of the minute coccoliths is indicated in a Cretaceous limestone of Tunisia, and this example suggests that their preservation is not expectable in other fine-grained limestones, and especially in those of greater age.

The upper Cretaceous chalk formations in central Tunisia are represented by fine-grained limestones in the thick and folded geosynclinal sequence to the north. A thin section of the Maestrichtian part of the limestones shows Foraminifera (mostly pelagic) scattered in a matrix of interlocking calcite crystals which average a few microns—just as these Foraminifera are scattered in the abundant coccoliths of the less-deformed chalk equivalent to the south.

The thick section of geosynclincal deposits of the sub-Alpine region of France includes much very fine-grained limestone of Jurassic and Cretaceous age. Thin sections show that the interlocking grains of the dense Tithonian (upper Jurassic) limestone average little larger than the sizes of coccoliths which occur in a variety of forms in the somewhat older, though much less lithified, Oxfordian marl of northern France. The only common fossil recognizable in this Tithonian is a small form classed with the planktonic tintinnids. The skeleton of this *Calpionella* is considerably larger and more robust, and thus more apt to be preserved, than any of the coccolithophorids.

This little evidence obviously does not justify any generalizations about fine-grained limestones, but the possibility that geosynclinal ones, at least, were originally like the later coccolith chalks seems to offer fewer difficulties than does an inorganic (nonskeletal) origin. This would also explain an otherwise unrelated coincidence in the occurrence of coccoliths in abundance and great variety of forms along with the earliest occurrences of the relatively unlithified calcareous deposits known as chalk. The absence of the coccoliths and their progenitors in the older fine-grained limestones would be inevitable with the recrystallization that has evidently produced the aplitic texture and obliterated whatever form the small particles originally showed.

These considerations seem to indicate, at least, that no tacit assumption that these limestones represent a fine inorganic precipitate is justified, merely because no very small fossil remains are still recognizable. Geochemical studies may help the problem for any particular limestone, as the distribution of such minor elements as barium and strontium in inorganic and skeletal calcium carbonate becomes better understood; and the carbon and oxygen isotopes may offer some evidence despite the recrystallization.

References Cited

Bramlette, M. N., and Riedel, W. R., 1954, Stratigraphic value of discoasters and some other microfossils related to Recent coccolithophores: Jour. Paleontology, v. 28, p. 385–403

Deflandre, G., 1952, Traité de Zoologie, Grassé: Paris, Masson et Cie, Tome 1, p. 439–470

Emiliani, C., and Edwards, G., 1953, Tertiary ocean bottom temperatures: Nature, v. 171, p. 887–889

Ericson, D. B., Ewing, M., and Heezen, B. C., 1952, Turbidity currents and sediments in North Atlantic: Am. Assoc. Petroleum Geologists Bull., v. 36, p. 489–511

Kamptner, E., 1954, Untersuchungen über den Feinbau der Coccolithen: Archiv. f. Protistenkunde, Bd. 100, Heft 1, p. 1–90

Kuenen, P. H., 1950, Marine geology: New York, Wiley and Sons, Inc., 568 p.

Lowenstam, H. A., 1955, Aragonite needles secreted by algae and some sedimentary implications: Jour. Sedim. Petrology, v. 25, p. 270–272

Revelle, R., and Fairbridge, R., In press, Carbon dioxide and carbonates, Chapter 10 *in* Hedgpeth, J. W. H., *Editor*, Treatise on marine ecology and paleoecology, Volume 1: Geol. Soc. America Memoir 67

Twenhofel, W. H., 1939, Principles of sedimentation: New York, McGraw-Hill, 610 p.

SCRIPPS INST. OF OCEANOGRAPHY, LA JOLLA, CALIF.
MANUSCRIPT RECEIVED BY THE SECRETARY OF THE SOCIETY, FEBRUARY 14, 1957

11

Reprinted from *Deep-Sea Research* **20**:917-921 (1973), by permission of Microforms International Marketing Corporation as exclusive copyright licensee of Pergamon Press journal back files

Deep-sea carbonates: evidence for a coccolith lysocline

WOLFGANG H. BERGER*

(*Received* 23 *April* 1973; *in revised form* 21 *May* 1973; *accepted* 4 *June* 1973)

Abstract—The most dissolution-resistant coccoliths on the Atlantic ocean floor are *Cyclococcolithina leptopora, Gephyrocapsa* sp. (*G. oceanica* and *G. caribbeanica*), and *Coccolithus pelagicus*. The relative abundance of these forms increases with depth in a fashion compatible with the foraminiferal lysocline distribution. The coccolith–foraminifera ratio may be largely controlled by redeposition processes.

INTRODUCTION

THE MAJOR facies boundary in the deep sea is the so-called calcite compensation depth (CCD), which separates calcareous ooze from carbonate-poor sediment. The boundary tends to follow depth contours in any one area. There is evidence that it is the result of a rather abrupt increase in calcite dissolution rates in the deep ocean (PETERSON, 1966). This level of increase, the 'hydrographic lysocline', apparently results in a 'foraminiferal lysocline' on the seafloor (BERGER, 1971). The purpose of this note is to present evidence that the foraminiferal lysocline is a coccolith lysocline as well.

DISSOLUTION RANKING OF COCCOLITHS

The enrichment of foraminiferal assemblages with resistant species through partial dissolution (MURRAY, 1897; SCHOTT, 1935; ARRHENIUS, 1952; PHLEGER, PARKER and PEIRSON, 1953; RUDDIMAN and HEEZEN, 1967) provides a yardstick for the assessment of the degree of dissolution that an assemblage has undergone (BERGER, 1968). On the basis of such assessment, it is possible to map a foraminiferal lysocline in the Atlantic, a zone separating well-preserved from poorly-preserved assemblages (BERGER, 1970). Over large parts of the deep Atlantic, the foraminiferal lysocline coincides with the abyssal thermocline marking the boundary between North Atlantic Deep Water and Antarctic Bottom Water. It is expected that this hydrographic level should also affect coccolith assemblages. To test this hypothesis, it is necessary to rank coccoliths according to their susceptibility to dissolution.

McINTYRE and McINTYRE (1971) recognized differential dissolution in coccoliths and identified *Emiliania huxleyi, Cyclococcolithina leptopora*, '*Gephyrocapsa oceanica*', and *Umbilicosphaera mirabilis* as forms enriched at depths where dissolution destroys other more fragile coccoliths.†

To obtain a dissolution ranking within the most common coccoliths on the deep sea floor, the data given by McINTYRE and BÉ (1967) were analyzed in the following way:

For each core, except those close to the coast, the nearest neighbor was found. Pairings in which cores were more than 3° lat. apart and where the deeper core was from less than 3500 m depth were not considered. Thus, biogeographic variations were kept to a minimum, and most of the difference in species lists in such pairs may be ascribed to dissolution effects. For each species, it was noted whether its abundance increases ('gain') or decreases ('loss') when going from shallow to deep water, or whether there is little effect ('draw'). Gains, draws, and losses were scored by the chess system (1, 1/2, 0). The resulting score card is given in Table 1. Obviously, all species are losers to some degree, an effect of presence–absence data, which were used to supplement the percentage data. The resulting rank order is interpreted as a solubility ranking.

*Scripps Institution of Oceanography, University of California, San Diego, P.O. Box 109, La Jolla, California 92037, U.S.A.

†Before Boudreaux and Hay (in HAY, MOHLER, ROTH, SCHMIDT and BOUDREAUX, 1967) erected the species *Gephyrocapsa caribbeanica*, references to '*Gephyrocapsa oceanica*' (McINTYRE and BÉ, 1967) are in fact references to both *G. oceanica* and *G. caribbeanica* combined. In their paper on coccolith distributions, McINTYRE and BÉ (1967) figure a coccosphere of *G. caribbeanica* under the label *G. oceanica*. In the present report, reference is made to *Gephyrocapsa* sp., therefore, wherever McINTYRE and BÉ (1967) use '*G. oceanica*'.

Table 1.   *Dissolution ranking of coccoliths, based on data by* McIntyre *and* Bé (1967). *The method of scoring is explained in the text.* '*Pairings*': *number of paired samples available for analysis.* '*Per cent*': *ratio of scores to pairings, in percentage of pairings.* Gephyrocapsa *sp. includes both* G. oceanica *and* G. caribbeanica (*see footnote to p.* 917).

Species	Score	Pairings	Per cent
1. *C. annula*	2	12	17
2. *C. fragilis*	3·5	20	18
3. *U. tenuis*	5	21	24
4. *D. tubifera*	1·5	6	25
5. *E. huxleyi*	7·5	26	29
6. *U. irregularis*	3·5	12	29
7. *U. mirabilis*	8	25	32
8. *R. stylifera*	6·5	20	33
9. *H. kamptneri*	9	27	33
10. *C. leptopora*	12·5	26	48
11. *Gephyrocapsa* sp.	13	27	48
12. *C. pelagicus*	13	21	62

The three most resistant forms, *C. leptopora, Gephyrocapsa* sp., and *C. pelagicus*, are clearly set apart from all other species in the present ranking. In general, when these species are dominant in a sample, the species list is short, reflecting the fact that the others have been dissolved. The sum of these three resistant species is taken as R. The distribution of R shows good agreement with respect to the foraminiferal lysocline in the Atlantic, despite the limited amount of data (Fig. 1). This agreement supports the notion that the foraminiferal lysocline is a coccolith lysocline as well. In Tertiary sediments, the percentage of discoasters, which are on the whole more resistant than the rest of the coccoliths, should prove useful in mapping the paleolysocline (cf. P. H. Roth, in Berger and von Rad, 1972, p. 802).

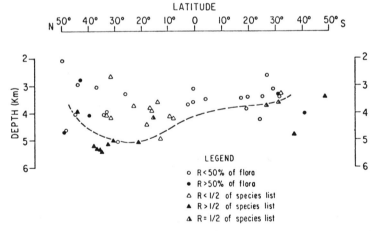

Fig. 1.   Distribution of resistant coccoliths (R = *C. leptopora* + *Gephyrocapsa* sp. + *C. pelagicus*) with respect to the Atlantic foraminiferal lysocline (dashed line; eastern branch of S. Atlantic lysocline profile omitted, since it does not apply to the data analyzed). R based on analysis of data from McIntyre and Bé (1967). Lysocline from Berger (1970).

IMPLICATIONS FOR PALEOTEMPERATURE ANALYSIS

Coccoliths have received increasing attention in recent years as paleotemperature indicators (Cohen, 1964; McIntyre, 1967; McIntyre, Bé and Roche, 1970; Bartolini, 1970; McIntyre, Ruddiman and Jantzen, 1972). The selective removal of certain coccolith species and the enrichment of sediments with resistant forms has important implications for the paleoclimatic interpretation of the assemblages (Fig. 2). Depending on the original composition, the solid cold-adapted forms are enriched through partial dissolution, or, in tropical assemblages, the robust form *Gephyrocapsa* sp. will eventually become the dominant species. A similar effect may be expected *within* certain species, which have both

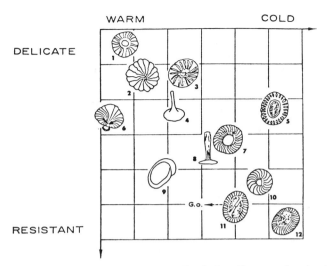

Fig. 2. Coccolith species in a temperature versus dissolution diagram, showing how differential dissolution changes the temperature aspect of a coccolith assemblage. Dissolution ranking from Table 1. Temperature ranking based on mid-points of common occurrence in McINTYRE and BÉ (1967, Fig. 16). The drawing for *Gephyrocapsa* sp. (number 11) represents *G. caribbeanica*, which is the one figured in McINTYRE and BÉ (1967), under the name *G. oceanica*. *G. oceanica s.s.* is, in fact, a tropical–subtropical form and should appear farther to the left, as indicated by the arrow labeled 'G.o.' (see footnote to p. 917).

Numbers refer to the following species:

1. *Cyclolithella annula*;   2. *Cyclococcolithina fragilis*;
3. *Umbellosphaera tenuis*;   4. *Discosphaera tubifera*;
5. *Emiliania huxleyi*;   6. *Umbellosphaera irregularis*;
7. *Umbilicosphaera mirabilis*;   8. *Rhabdosphaera stylifera*;
9. *Helicopontosphaera kamptneri*;   10. *Cyclococcolithina leptopora*;
11. *Gephyrocapsa* sp.;   12. *Coccolithus pelagicus*.

open-structured and solid varieties. Considerable morphologic variation of coccoliths of *Emiliania huxleyi* has been observed, both in the laboratory (WATABE and WILBUR, 1966) and in the field (HASLE, 1960; McINTYRE and BÉ, 1967). The variations appear to be temperature-dependent. It is noteworthy that the warm-water type is delicate and open-structured, while the cold-water type is solid and dense, both for *E. huxleyi* and for *Umbilicosphaera mirabilis* (McINTYRE and BÉ, 1967, p. 569 and p. 572). Thus, the cold-water forms of these species have a better chance of being preserved in sediments than the warm-water forms.

IMPLICATIONS FOR VARIATIONS IN THE COCCOLITH–FORAMINIFERA RATIO

McINTYRE and McINTYRE (1971) stated that, on the whole, the coccoliths show the best resistance to dissolution with depth among carbonate-secreting invertebrates (*ibid.*, p. 260), presumably mainly foraminifera. Their suggestion (first made in 1967 at the symposium on micropalaeontology of marine bottom sediments, in Cambridge, U.K.) subsequently received support from Deep Sea Drilling results, when it was found that coccoliths are relatively more abundant than foraminifera close to the CCD (Hsu and ANDREWS, 1970; HAY, 1970). However, the results here presented suggest that the dissolution behavior of coccoliths and foraminifera is rather similar, although differences in range of resistance are not excluded.

Mechanisms of separation other than differential dissolution may have to be invoked to explain the high coccolith–foraminifera ratios close to the CCD. Winnowing of deep-sea sediments, producing lag deposits, and wafting of fine material to other places, producing what may be called 'chaff' deposits, are well-documented processes (BRAMLETTE and BRADLEY, 1942; HEEZEN and HOLLISTER, 1964; EMILIANI and MILLIMAN, 1966, p. 111). Such processes could easily separate foraminifera from coccoliths on a large scale, since their settling behavior differs by orders of magnitude. Resuspension by benthonic animals should be especially helpful in initiating such separation. Results from DSDP Leg 14 are in agreement with this suggestion (BERGER and VON RAD, 1972, p. 838). Neogene chalk and marl oozes form clusters on a sand–silt–clay diagram, clusters that are entirely tied to sites, rather than to age, rate of deposition, or percentage of carbonate. Dissolution influences the clustering only when extreme. Fertility apparently varied greatly through the Neogene, as indicated by fluctuations in the abundance of siliceous fossils and by variations in the paleodepth of the lysocline. Thus, the local control of texture points to sedimentary sorting processes on the ocean floor as the first order effect on coccolith–foraminifera ratios.

CONCLUSIONS

The interpretation of coccolith assemblages on the sea floor is greatly complicated by the effects of differential dissolution. Any differences between biogeographic patterns of living coccolithophores and taphogeographic patterns of corresponding coccoliths on the sea floor cannot be ascribed to paleo-temperature fluctuations alone. Differences in the dissolution patterns of foraminifera and coccoliths should provide information on large-scale redeposition processes.

Acknowledgements—I am indebted to P. H. ROTH for informative discussions and for critical reading of the manuscript. The work was supported by the National Science Foundation, Grants GB-21259 and GA-36697.

REFERENCES

ARRHENIUS G. (1952) Sediment cores from the East Pacific. *Rep. Swed. deep sea Exped.*, 1947–1948, **5**, 1–288.

BARTOLINI C. (1970) Coccoliths from sediments of the western Mediterranean. *Micropaleontology*, **16**, 129–154.

BERGER W. H. (1968) Planktonic foraminifera: selective solution and paleoclimatic interpretation. *Deep-Sea Res.*, **15**, 31–43.

BERGER W. H. (1970) Biogenous deep-sea sediments: fractionation by deep-sea circulation. *Bull. geol. Soc. Am.*, **81**, 1385–1402.

BERGER W. H. (1971) Sedimentation of planktonic foraminifera. *Mar. Geol.*, **11**, 325–358.

BERGER W. H. and U. VON RAD (1972) Cretaceous and Cenozoic sediments from the Atlantic Ocean. *Initial Rep. Deep Sea Drilling Project*, **14**, 787–954.

BRAMLETTE M. N. and W. H. BRADLEY (1942) Geology and biology of North Atlantic deep-sea cores. Part 1. Lithology and geologic interpretations. *Prof. Pap. U.S. geol. Surv.*, **196-A**, 1–34.

COHEN C. L. D. (1964) Coccolithophorids from two Caribbean deep-sea cores. *Micropaleontology*, **10**, 231–250.

EMILIANI C. and J. D. MILLIMAN (1966) Deep-sea sediments and their geologic record. *Earth-Sci. Rev.*, **1**, 105–132.

HASLE G. R. (1960) Plankton coccolithophorids from the subantarctic and equatorial Pacific. *Nytt Mag. Bot.*, **8**, 77–88.

HAY W. W. (1970) Calcium carbonate compensation. *Initial Rep. Deep Sea Drilling Project*, **4**, 672.

HAY W. W., H. MOHLER, P. H. ROTH, R. R. SCHMIDT and J. E. BOUDREAUX (1967) Calcareous nanno-plankton zonation of the Cenozoic of the Gulf Coast and Caribbean–Antillean area, and trans-oceanic correlation. *Trans. Gulf-Cst. Ass. geol. Socs.* **17**, 428–480.

HEEZEN B. C. and C. D. HOLLISTER (1964) Deep-sea current evidence from abyssal sediments. *Mar. Geol.*, **1**, 141–174.

HSU K. J. and J. E. ANDREWS (1970) Lithology. *Initial Rep. Deep Sea Drilling Project*, **3**, 445–453.

MCINTYRE A. (1967) Coccoliths as paleoclimatic indicators of Pleistocene glaciation. *Science*, **158**, 1314–1317.

MCINTYRE A. and A. W. H. BÉ (1967) Modern Coccolithophoridae of the Atlantic Ocean—I. Placoliths and cyrtoliths. *Deep-Sea Res.*, **14**, 561–597.

MCINTYRE A. and R. MCINTYRE (1971) Coccolith concentrations and differential solution in oceanic sediments. In: *The micropalaeontology of oceans*, B. M. FUNNELL and W. R. RIEDEL, editors, Cambridge University Press, 253–261.

MCINTYRE A., A. W. H. BÉ and M. B. ROCHE (1970) Modern Pacific Coccolithophorida: a paleonto-logical thermometer. *Trans. N.Y. Acad. Sci.*, (II), 32(6), 720–731.

MCINTYRE A., W. F. RUDDIMAN and R. JANTZEN (1972) Southward penetrations of the North Atlantic Polar Front: faunal and floral evidence of large-scale surface water mass movements over the last 225,000 years. *Deep-Sea Res.*, **19**, 61–77.

MURRAY J. (1897) On the distribution of the pelagic Foraminifera at the surface and on the floor of the ocean. *Nat. Sci.*, **11**, 17–27.

PETERSON M. N. A. (1966) Calcite: rates of dissolution in a vertical profile in the central Pacific. *Science*, **154**, 1542–1544.

PHLEGER F. B, F. L. PARKER and J. F. PEIRSON (1953) North Atlantic Foraminifera. *Rep. Swed. deep sea Exped.*, **7**, 1–122.

RUDDIMAN W. F. and B. C. HEEZEN (1967) Differential solution of planktonic Foraminifera. *Deep-Sea Res.*, **14**, 801–808.

SCHOTT W. (1935) Die Foraminiferen in dem äquatorialen Teil des Atlantischen Ozeans. *Dt. Atlant. Exped. Meteor* 1925–1927, **3**, 43–134.

WATABE N. and K. M. WILBUR (1966) Effects of temperature on growth calcification, and coccolith form in *Coccolithus huxleyi* (Coccolithineae). *Limnol. Oceanogr.*, **11**, 567–575.

12

Copyright ©1975 by the Geological Society of America
Reprinted from *Geol. Soc. America Bull.* **86:**1079–1084 (1975)

Coccolith Sedimentation by Fecal Pellets: Laboratory Experiments and Field Observations

P. H. ROTH
M. M. MULLIN   *Scripps Institution of Oceanography, University of California, San Diego, La Jolla, California 92037*
W. H. BERGER

ABSTRACT

Copepods readily ingested coccospheres during feeding experiments. Fecal pellets contained coccoliths in a state of excellent preservation, although some mechanical breakage occurred, especially among delicate forms. Fecal pellets collected in deep waters of the eastern tropical Pacific contained mainly siliceous plankton remains but also some coccoliths showing good to moderate preservation. Some coccolith-bearing fecal matter apparently reached the sea floor, even well below the carbonate compensation depth. The implications of fecal transfer to paleontology and geochemistry may be of the greatest importance. *Key words: deep-sea sedimentation, fecal pellets, coccolith preservation.*

INTRODUCTION

The transfer of calcareous and siliceous skeletal particles from near-surface waters where they are produced to the sea floor where they dissolve or become buried is a process of paramount importance to the geochemistry of the oceans. The available information on this process is exceedingly sparse, however. Bramlette (1961), and many workers since, recognized two paradoxes regarding the transfer: (1) very fine particles appear to drift but little from the place of production to that of deposition, and (2) coccoliths and diatoms accumulate on the sea floor despite a presumably long settling time spent in undersaturated waters. Accelerated sinking is usually invoked to reconcile these observations (see Smayda, 1970, 1971, for reviews). Schrader (1971) presented direct evidence for such accelerated sinking by fecal transport in the sedimentation of diatoms.

Similar fecal transport must be assumed for coccoliths. In addition to the problems of drift and dissolution during settling, it is obvious that the amount of coccoliths suspended in the water column below the photic zone (Hasle, 1959; Marshall, 1968; Okada and Honjo, 1970), when combined with Stokesian settling, is insufficient to provide the flux necessary to account for observed sedimentation rates in areas where coccoliths are well preserved. The fluxes, therefore, must be provided by much larger particles. Such particles are low in concentration due to their short transit times and are rarely sampled therefore. Analogous problems arise in the transfer of organic matter (Menzel, 1967). Large particles could be provided by fecal pellets, especially copepod pellets, since copepods are the dominant component of zooplankton.

From these considerations, we decided to check whether coccoliths can survive ingestion by copepods and whether coccoliths are indeed present in fecal pellets recovered from the deep sea — in the water and on the sea floor. We believe that our results, although of a preliminary nature, will encourage a greatly intensified effort in this neglected but important field of deep-ocean sedimentation.

FEEDING EXPERIMENTS

Experimental Methods

Planktonic copepods, *Calanus pacificus* and *Rhincalanus nasutus* were captured by net off La Jolla, California, in April 1974 and held in the laboratory at 15° to 20°C until use within 24 hr later. Unialgal suspensions of coccolithophorids were made up in filtered sea water from cultures grown in an enriched sea-water medium. The clones employed were *Cricosphaera elongata* (Droop, 1955), originally isolated by I. J. Pinter of the Haskins Laboratory, New York; *Cricosphaera carterae* (Braarud and Fagerland, 1946), probably from the North Pacific Central Gyre, isolated by J. B. Jordan; and *Emiliania huxleyi* (Lohmann, 1902) from Oslo Fjord, isolated by E. Paasche. Copepods were rinsed in filtered sea water and transferred by pipette to the suspensions of coccolithophorids. After 24 to 48 hr, fecal pellets were picked from the vessels, rinsed in filtered sea water, and prepared for electron microscopy. Cells in aliquots of the cultures used to make up the suspensions were similarly prepared, so that the coccolith structure in the fecal pellets could be compared to that of the particular clone the copepods had ingested.

Ultrastructure of the Coccoliths Used in the Experiments

The two clones of *Cricosphaera* used have identical coccolith morphology and could not be distinguished in the scanning electron microscope (SEM). The coccosphere of *Cricosphaera elongata* is covered with about 200 eliptical coccoliths approximately 2 μm long, 0.5 μm wide, and 0.2 μm thick (Fig. 1A). An organic membrane covers the coccoliths and obscures most of the ultrastructural details of the coccoliths. There are some indications, especially from broken specimens in fecal pellets, that the individual coccolith is composed of two closely appressed cycles, which are about 0.1 μm wide and composed of approximately 12 nonoverlapping elements. The lower proximal cycle is only about half as thick as the distal cycle. The illustrations and descriptions of *Hymenomonas (Cricosphaera) carterae* published by Braarud and others (1952), Manton and Leedale (1969), and Outka and Williams (1971) agree quite well with the forms here either identified as *Cricosphaera elongata* or *C. carterae;* some (for example, Manton and Leedale, 1969) show forms of *Hymenomonas (= Cricosphaera) carterae* with somewhat broader cycles of elements (shields), but forms with narrower shields very similar to the ones observed in our clones have also been illustrated (Outka and Williams, 1971). Modern electron micrographs of *Cricosphaera elongata* have not been published, and therefore it is not possible to decide whether the clones originally identified as *C. elongata* (Fig. 1A) and *C. carterae* (Fig. 1B) are indeed identical, as would be suspected from the coccolith

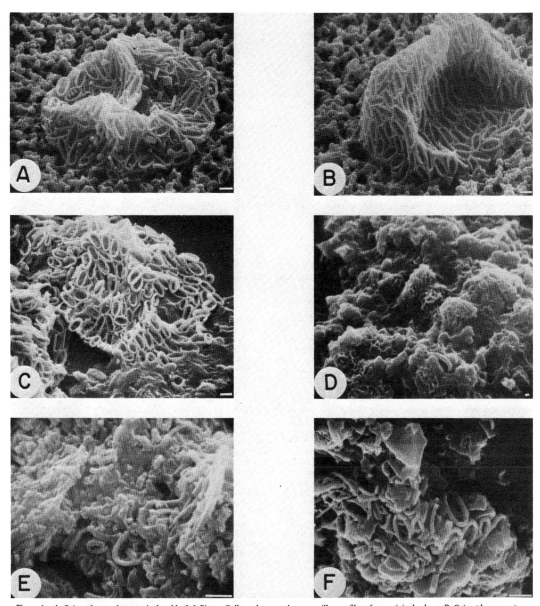

Figure 1. A. *Cricosphaera elongata*, isolated by I. J. Pinter. Collapsed coccosphere on millepore filter, from original culture. B. *Cricosphaera carterae*, clone isolated by Jordan from central North Pacific gyre. Coccosphere from original culture on millepore filter collapsed during drying. C. Coccosphere of *C. elongata* in fecal pellet. Most of coccoliths are still attached to organic membrane; others have come off. D. Fecal material from copepod with numerous coccospheres, isolated coccoliths, and fragments of *Cricosphaera elongata*, mixed in with organic debris. E. Strongly fragmented coccoliths of *Cricosphaera elongata* from fecal pellet. Elements of two cycles of coccoliths are more visible than in the original samples; perhaps part of the organic coating has been removed in digestive tract of copepods. F. Whole coccoliths and fragments of *Cricosphaera carterae* from fecal pellet of copepod. Scale bar on all pictures indicates 1 μm.

ultrastructure. The coccoliths of the clone of *Emiliania huxleyi* have the usual structure, which has been well described by many authors (for example, McIntyre and Bé, 1967). However, some unusual variations in shape of the coccoliths were observed. Some of the coccospheres are covered by elliptical and circular coccoliths; others have only elliptical or only circular coccoliths (Fig. 1A). Circular forms of *Emiliania huxleyi* have to our knowledge not been described from natural assemblages in the open ocean, and apparently they are aberrant forms due to prolonged culturing. Light microscopy of the coccoliths of *E. huxleyi* shows that they are well calcified, because the usual birefringence is observed (Fig. 2C).

Fecal Pellets from Feeding Experiments

The fecal pellets produced by the two species of copepods are rod-shaped, about 2mm long, and 0.5 mm in diameter (Fig. 2D). An organic membrane encloses the content. Slight pressure with a paint brush or needle breaks the organic membrane, and the content of the fecal pellets can then be spread on a cover slip and prepared for observation in the light microscope and SEM.

Results

The fecal pellets of copepods feeding on *Cricosphaera elongata* and *C. carterae* contained a large number of coccoliths and whole coccospheres, together with unidentifiable organic debris and coccolith fragments (Fig. 1C through 1F). It was not possible to determine whether the "coccospheres" were empty and consisted only of the organic membrane with a coccolith cover, or whether they still contained protoplasm. (R. E. Norris [personal commun., 1974] has observed coccolithophorids that still contained chromatophores and a nucleus in fecal pellets of copepods.) The number of isolated coccoliths was considerably larger in the fecal pellets than in original cultures, where only an occasional isolated coccolith was observed. About a third of the coccoliths were still attached to the coccosphere membrane, another third appeared as isolated coccoliths, and a third of the coccoliths were fragmented and made up the coccolith debris (Fig. 1E, 1F). The coccoliths of *Cricosphaera elongata* are not very sturdy, and it is to be expected that more robust coccoliths would be less damaged by the passage through the copepod guts. Some of the fecal pellets of the copepods that were fed *Emiliania huxleyi* contained coccoliths in large numbers (Fig. 2C). Others contained very few coccoliths. The coccoliths of *Emiliania huxleyi* found in fecal pellets were well preserved. Even the delicate central grille (Fig. 2B) was still intact. Some of the elements showed mechanical damage; breakage could have occurred during preparation of the sample or during passage of the coccoliths through the copepod mouthparts or guts. It is, however, noteworthy that the most delicate part of the coccoliths, the central grille, was preserved in the coccoliths from the fecal pellets. This would indicate that mechanical breakage, rather than dissolution, is responsible for the damage to coccoliths during feeding and digestion.

FIELD OBSERVATIONS

During the Benthiface Expedition to the eastern equatorial Pacific in May and June 1973, deep-water tows and biological dredges recovered fecal pellets and gut contents of planktonic and benthonic animals (see Adelseck and Berger, 1975).

In a Bongo tow taken at a depth of 3,400 m, a large fecal mass contained mainly fragments of siliceous organisms (Fig. 3A, 3B). The rare coccoliths present were moderately well preserved and showed mainly breakage. The delicate central grille of *E. huxleyi*

was preserved (Fig. 2E), whereas the elements of the shield were broken. This would indicate that mechanical damage, rather than dissolution, caused the fragmentation. Delicate forms like *Scapholithus* sp. and a fragment of *Rhabdosphera clavigera* were still preserved (Figs. 2F, 3A). Another fecal mass from the gut of an ostracod contained both well-preserved coccoliths of *E. huxleyi* (Fig. 3C) and specimens of *Umbilicosphaera sibogae* that showed signs of etching.

A fecal pellet from inside the gut of a brittle star collected at a depth of 3,600 m, several hundred metres below the regional carbonate compensation depth (CCD), contained very well preserved coccoliths, such as *Gephyrocapsa oceanica*, with a preserved central grille (Fig. 3E). Isolated shields of *C. leptopora* (Fig. 3F) and fragments of other species were also common. The dominant components of the fecal pellet from the brittle star gut, however, were diatoms, fragments, and clay minerals. The sediment within the gut of the brittle star, but not enclosed in a fecal pellet, contained mostly clay and only rare, poorly preserved coccoliths.

All the fecal pellets and gut contents studied so far were taken in the water column at great depth or from the sea floor. They might not be representative for fecal pellets produced in the photic zone. Trapping experiments have been conducted to collect such fecal pellets in the California Current, but they were unsuccessful because of loss of equipment. Future experiments are planned.

DISCUSSION

Our experiments show that coccolithophorids are readily ingested by at least some copepods, pass their guts relatively unharmed, and occur in large numbers in fecal pellets. Some mechanical breakage occurs, especially among delicate forms (*Cricosphaera*). There are indications that etching is unimportant compared to breakage during passage through the copepod guts. Thus, fecal pellet transport of coccoliths is a feasible mechanism of sedimentation. Our field observations, although spotty, suggest that this mechanism may indeed be important in the deep sea. This conclusion is in agreement with Lohmann's assessment (1902, p. 92). He wrote, "The main transportation is accomplished by fecal pellets of plankton-eating animals in which coccoliths are as crowded and numerous as on the sea floor, according to the beautiful illustrations of the Challenger reports." Subsequent observations suggesting that copepods and other organisms are highly important in providing the coccolith flux to the sea floor are widely scattered in the literature. Both Lohmann (1902) and Schiller (1930) used the gut contents of appendicularians and salps as a rich source of coccolithophorids for their systematic studies. These organisms cause even less damage to coccoliths than do copepods, since feeding is by ingestion of mucus and no grinding by molariform mouthparts occurs. Murray and Hjort (1912) reported the common occurrence of coccoliths in excrements of copepods and concluded "that the digestive juices of copepods cannot have acid reactions." They also found abundant coccoliths in salps. Esterly (1916) reported that some copepods were packed with coccoliths (mostly *Emiliania huxleyi*). Marshall (1924) studied gut contents of the copepod *Calanus finmarchicus* for a whole year and found that *Emiliania huxleyi* was present during spring and summer months in up to 17 percent of the guts. Marshall and Orr (1955) conducted feeding experiments with the same copepod species using various algae as food, among others, *Hymenomonas carterae*, *Cricosphaera elongata*, and *Emiliania huxleyi*, all of which were readily ingested by the copepods. Marshall and Orr (1956) noted that "the fecal pellets examined from nauplii (of *Calanus finmarchicus*) which had been feeding on a culture of this species (*Cricosphaera elongata*) were packed full of coccoliths which had probably come from ingested cells." Heinrich (1958) listed coccolithophorids as significant components in gut contents of most

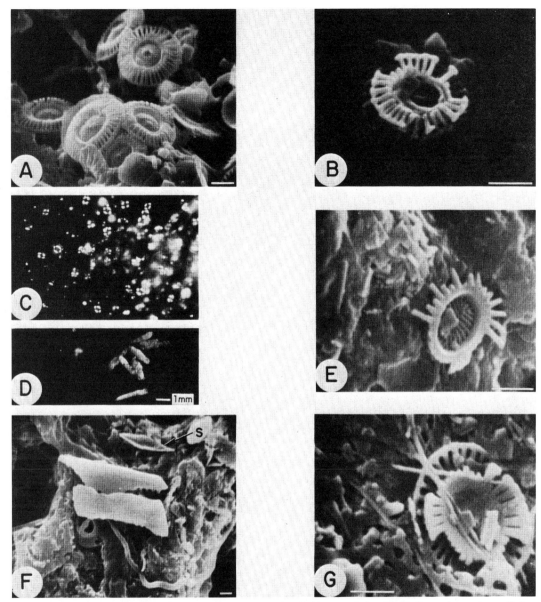

Figure 2.   A. *Emiliania huxleyi*, clone used for feeding experiments, originally isolated by E. Paasche from Oslo Fjord; parts of coccospheres, with circular coccolith in upper half of the picture. B. *Emiliania huxleyi*, from fecal pellet by copepods fed only this species. Note that central grille, partly obscured by organic matter, is preserved, whereas some elements seem broken off. C. Light micrograph of fecal material of copepods fed only *E. huxleyi*. Large number of coccoliths are present. D. Fecal pellets of copepods *Calanus pacificus* and *Rhincalanus nasutus*. E. *Emiliania huxleyi* with well-preserved central grille and broken elements, from fecal pellet collected in Bongo tow at depth of 3,400 m in eastern tropical Pacific (Benthiface Tow #4). F. Fecal material from pellet collected in same tow, with well-preserved specimens of *Gephyrocapsa oceanica* and *Scapholithus* (s), a rather delicate form, together with fragments of siliceous plankton fragments. G. *Emiliania huxleyi*, with well-preserved central area, some broken shield elements surrounded by siliceous fragments. From fecal pellet collected in Benthiface Tow #4. Scale bar on all pictures indicates 1 μm.

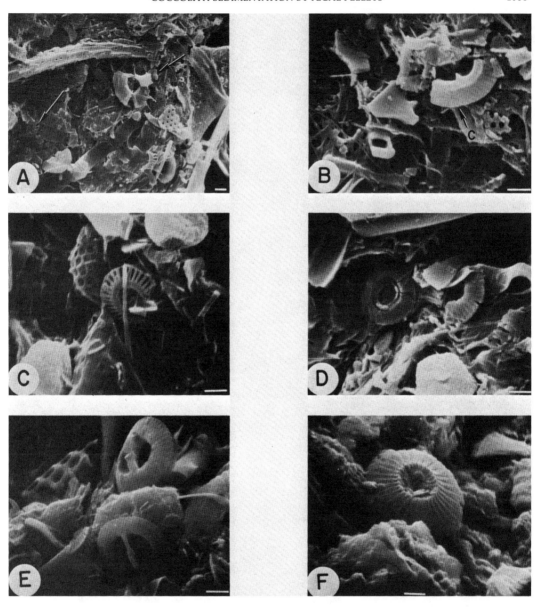

Figure 3.   A. Fecal pellet material with diatom and radiolarian and coccolith fragments. Letter r indicates fragment of rather delicate form, *Rhabdosphaera clavigera;* e, well-preserved *Emiliania huxleyi*. Two fragments of *Gephyrocapsa* also present. Fecal pellet collected in Benthiface Tow #4, depth of 3,400 m. B. Fecal material showing high degree of fragmentation of both siliceous plankton remains and coccoliths (c). Benthiface Tow #4. C. Well-preserved *Emiliania huxleyi*, with elements still connected along margin and only minor breakage. From ostracod guts, collected in Benthiface Tow #4. D. Somewhat etched and fragmented coccoliths of *Umbilicosphaera sibogae* (= *U. mirabilis*) from same ostracod guts as C. E. *Gephyrocapsa oceanica* with central grille preserved, fragment of *Emiliania huxleyi*, and fragmented siliceous plankton remains. From fecal pellet found in intestinal cavity of brittle star collected at depth of 3,600 m in eastern tropical Pacific. F. Proximal shield of *Cyclococcolithina leptopora* from same fecal pellet as E. Scale bar on all pictures indicates 1 μm.

copepod species investigated. Mullin (1966) found coc-colithophorids in copepods belonging to the families *Calanidae*, *Eucalanidae*, and *Aetididae* from the Indian Ocean. Arashkevich (1968) also reported coccolithophorids from the gut contents of five species of filter-feeding copepods from the northwestern Pacific. Harding (1974) mentioned that 11 species of deep-sea copepods out of 44 species investigated contained coccoliths in small numbers. Most recently, Honjo (1975) reported on coc-coliths in fecal pellets and emphasized the potential significance of fecal transfer.

We conclude that fecal transfer of coccoliths (as well as of other materials) is likely to be of great importance. The paleontological, sedimentological, and geochemical implications of such transfer could be profound. The probability of a living coccosphere deliver-ing coccoliths to the deep sea floor will be affected by the kind of predation experienced by the coccospheres ("preserving" or "re-cording" versus "destructive" predation; Berger, 1975). The sedimentary record, therefore, should reflect (among other factors) the effect of differential predation. It cannot be assumed that re-cording predation and destructive predation will be equally distributed with respect to major fertility and temperature regimes. Indeed, it is likely that changes in fertility regimes through geologic time will have important effects on sedimentation by causing changes in the transfer patterns. The puzzling change in coccolith/foram ratios at the end of the Tertiary Epoch (Bramlette, 1958) needs investigation under this aspect. Geochemically, fecal pellet transfer may be the major mechanism by which the typical concentration profiles for nutrients are established in the deep ocean, with low concentrations of nitrate, phosphate, and silicate, as well as decreased values of alkalinity in the uppermost waters. The significance of these profiles for deep-sea sedimentation has been discussed elsewhere (Berger, 1970).

ACKNOWLEDGMENTS

We are grateful to C. G. Adelseck and T. C. Johnson for able assistance during the field work and to J. B. Jordan for the original cultures of the coccolithophorids. We thank H. J. Schrader (Kiel) and S. Honjo (Woods Hole) for stimulating discussions on the sub-ject of fecal pellet transfer of diatoms and coccoliths.

REFERENCES CITED

Adelseck, C. G., and Berger, W. H., 1975, On the dissolution of planktonic foraminifera and associated microfossils during settling and on the sea floor, *in* Sliter, W. V., Bé, A.W.H., and Berger, W. H., eds., Dissolu-tion of deep-sea carbonates: Cushman Found. Foram. Research Spec. Pub. 13, p. 70–81.
Arashkevich, Y. G., 1968, The food and feeding of copepods in the north-western Pacific: Oceanology (Moscow), v. 9, p. 695–709.
Berger, W. H., 1970, Biogenous deep-sea sediments: Fractionation by deep-sea circulation: Geol. Soc. America Bull., v. 81, p. 1385–1402.
——1975, Biogenous deep-sea sediments: Production, preservation, and in-terpretation, *in* Riley, J. R., and Chester, R., eds., Treatise on chemical oceanography, Vol. 3: London, Academic Press (in press).
Braarud, T., and Fagerland, E., 1946, A coccolithophorid in laboratory cul-ture, *Syracosphaera carterae* n. sp.: Norske Vidensk.–Akad. Oslo Skr., Mat.–Naturv. Kl., 1946, no. 2, 10 p.
Braarud, T., Gaardner, K. R., Markali, T., and Nordli, E., 1952, Coc-colithophorids studied in the electron microscope. Observations on *Coccolithus huxleyi* and *Syracosphaera carterae*: Nytt Mag. Bot., v. 1, p. 129–134.

Bramlette, M. N., 1958, Significance of coccolithophorids in calcium-carbonate deposition: Geol. Soc. America Bull., v. 69, p. 121–126.
——1961, Pelagic sediments, *in* Sears, M., ed., Oceanography: Washing-ton, D. C., Am. Assoc. Adv. Sci. Pub. 67, p. 345–366.
Droop, M. R., 1955, Some new supra-littoral protista: Marine Biol. Assoc. United Kingdom Jour., v. 34, p. 233–245.
Esterly, C. O., 1916, The feeding habits and food of pelagic copepods and the question of nutrition by organic substances in solution in the water: Univ. Calif. Pub. 2001, v. 9, p. 253–340.
Harding, G.C.H., 1974, The food of deep-sea copepods: Marine Biol. Assoc. United Kingdom Jour., v. 54, p. 141–155.
Hasle, G. R., 1959, A quantitative study of phytoplankton from the equatorial Pacific: Deep-Sea Research, v. 6, p. 38–59.
Heinrich, A. K., 1958, On the nutrition of marine copepods in the tropical region: Akad. Nauk SSSR Doklady, v. 119, p. 1028–1031.
Honjo, S., 1975, Dissolution of suspended coccoliths in the deep-sea water column and sedimentation of coccolith ooze, *in* Sliter, W. V., Bé, A.W.H., and Berger, W. H., eds., Dissolution of deep-sea carbonates: Cushman Found. Foram. Research Spec. Pub. 13, p. 114–128.
Lohmann, H., 1902, Die Coccolithophoridae, eine Monographie der Coc-colithen bildenden Flagellaten, zugleich ein Beitrag zur Kenntnis des Mittelmeerauftriebs: Archiv Protistenkunde, v. 1, p. 89–165.
Manton, I., and Leedale, G. F., 1969, Observations on the microanatomy of *Coccolithus pelagicus* and *Cricosphaera carterae*, with special ref-erence to the origin and nature of coccoliths and scales: Marine Biol. Assoc. United Kingdom Jour., v. 49, p. 1–16.
Marshall, H. G., 1968, Coccolithophores in the northwest Sargasso Sea: Limnology and Oceanography, v. 13, p. 370–376.
Marshall, S., 1924, The food of *Calanus finmarchicus* during 1923: Marine Biol. Assoc. United Kingdom Jour., v. 13, p. 473–479.
Marshall, S. M., and Orr, A. P., 1955, On the biology of *Calanus finmarchicus*, VIII, Food uptake, assimilation and excretion in adult and Stage V *Calanus*: Marine Biol. Assoc. United Kingdom Jour., v. 34, p. 495–529.
——1956, On the biology of *Calanus finmarchicus*, IX, Feeding and diges-tion in the young stages: Marine Biol. Assoc. United Kingdom Jour., v. 35, p. 587–603.
McIntyre, A., and Bé, A.W.H., 1967, Modern Coccolithophoridae of the Atlantic Ocean — I., Placoliths and cyrtoliths: Deep-Sea Research, v. 14, p. 561–597.
Menzel, D. W., 1967, Particulate organic carbon in the deep sea: Deep-Sea Research, v. 14, p. 229–238.
Mullin, M. M., 1966, Selective feeding by calanoid copepods from the In-dian Ocean, *in* Barnes, H., ed., Some contemporary studies in marine science: London, George Allen & Unwin, Ltd., p. 313–334.
Murray, J., and Hjort, J., 1912, The depths of the ocean. A general account of the modern science of oceanography based largely on the scientific researches of the Norwegian steamer *Michael Sars* in the North Atlan-tic: London, Macmillan, 821 p.
Okada, H., and Honjo, S., 1970, Coccolithophoridae distributed in south-west Pacific: Pacific Geology, v. 2, p. 11–21.
Outka, D. E., and Williams, D. C., 1971, Sequential coccolith mor-phogenesis in *Hymenomonas carterae*: Jour. Protozoology, v. 18, p. 285–297.
Schiller, J., 1930, Coccolithineae, *in* Dr. L. Rabenhorst's Kryptogamen — Flora von Deutschland, Österreich und der Schweiz, Vol. 10, Pt. 2: Leipzig, Akademische Verlagsgesellschaft, p. 89–267.
Schrader, J.-H., 1971, Fecal pellets: Role in sedimentation of pelagic diatoms: Science, v. 174, p. 55–57.
Smayda, T. J., 1970, The suspension and sinking of phytoplankton in the sea: Oceanography and Marine Biology Ann. Rev., v. 8, p. 353–414.
——1971, Normal and accelerated sinking of phytoplankton in the sea: Marine Geology, v. 11, p. 105–122.

Manuscript Received by the Society October 21, 1974

13

COCCOLITHS: PRODUCTION, TRANSPORTATION AND SEDIMENTATION

SUSUMU HONJO

Woods Hole Oceanographic Institution, Woods Hole, Mass. (U.S.A.)

(Received and accepted January 20, 1976)

Abstract

Honjo, S., 1976. Coccoliths: production, transportation and sedimentation. Mar. Micropaleontol., 1: 65–79.

The presence of coccolith ooze on the deep-sea floor and the well preserved suspended coccoliths in the under-saturated water column is explained by accelerated and communal sinking of coccoliths and coccospheres in small zooplankton's fecal pellets. The community structure in the euphotic layer will be replicated in the underlying thanatocoenosis with high resolution without significant time-lag between the production and the deposition. The euphotic biocoenosis of coccoliths drift only a few hundred kilometers at the maximum while they descend at the rate of more than 150 m per day through 5,000 m of water column where the variance of advection is approximately 5 km a day. At an Equatorial Pacific station it was estimated that 92% of coccoliths produced in the euphotic layer were thus being transported to the deep-sea bottom. Coccolithophore production provides several grams of calcite to a square meter of sea floor per year. Coccoliths do not undergo significant change during passage through copepods and other planktonic grazers. The membrane which covers a copepod's fecal pellet is coherent particularly in cold water. It protects contents from spilling and from dissolution and accelerates sinking rates by providing a smooth surface. The membrane is biodegraded rapidly in warm water and its contents exposed. As such naked pellets sink, coccoliths are shed and suspended in aphotic water. The majority of freed coccoliths will be dissolved in the undersaturated water column before arriving at the bottom. Approximately 8% of coccoliths produced were estimated to be remineralized in the water column at the above-mentioned station if Peterson's dissolution rate of calcite is applicable.

Introduction

A coccolith is usually no more than several microns in diameter but its surface area is large for its volume because of its complex architecture. From a purely Stokesian point of view (Lerman et al., 1974), a coccolith sinks very slowly. The typical residence time estimated from laboratory sinking rate measurements is of the order of 100 years in deep sea. The sinking rates, however, varies by species. If coccoliths are transported to the deep-sea bottom by individual sinking, physical processes will be important to their motion. Eddy diffusivity is also effective in destroying the original assemblage of small particles. As a consequence, the latitudinal zonal distribution of coccolithophore species assemblage will be completely disturbed by the time the coccoliths settle on the deep-sea floor (Paradox 1).

An average single coccolith should dissolve shortly after it crosses the calcite saturation depth (CSD) if we accept the *in situ* dissolution rate of calcite measured by Peterson (1966). Consequently, no suspended coccoliths should exist in the undersaturated deep-sea water column.

The origin of the coccolith ooze on the deep-sea floor below the CSD cannot be explained by raining of individual coccoliths (Paradox 2). These problems have been known as "the paradoxes of coccolith sedimentation" (Bramlette, 1961; Smayda, 1970; Honjo, 1975a; Roth et al., 1975) and will be elaborated in this paper.

Accelerated and communal sinking of coccoliths successfully explains the field observation of the distribution of coccoliths in aphotic water column (Honjo, 1975a), its comparison to the productive community (Honjo and Okada, 1974) and physical, chemical oceanographic considerations on small carbonate particles. The main channel which transports coccoliths efficiently from the productive layer to the deep-sea floor appears to be raining fecal pellets produced by numerous small zooplankton in the upper water column. Such a possibility was first suggested by Lohmann (1902). Since then many workers reported the presence of coccoliths in the gut contents of various zooplankton (Murray and Hjort, 1912; Esterly, 1966; Mullin, 1966). Coccoliths were found in fecal pellets of zooplankton by Marshall and Orr (1956, 1962), Bernard (1963), and Roth et al. (1975). The protective role of fecal pellets has been more directly demonstrated by scanning microscopy (Schrader, 1971; Honjo, 1975a; Roth et al., 1975) as well as by results from laboratory feeding experiments of coccolithophores to common oceanic copepods, and by analyses of fecal contents collected by a sediment trap placed on the deep-sea floor.

The purpose of this article is to review the fate of coccoliths produced in the euphotic layer while they are passing through the deepwater column. It focuses on the taphonomy of coccoliths — the geographical and ecological correlation between living communities and their thanatocoenosis, and re-evaluates the utility of coccoliths as a paleo-oceanographic tool in relation to their vertical transportation mode in the deep sea. The rates of coccolith production, transportation, and sedimentation are estimated.

Production of coccoliths

Living coccolithophores are abundant in almost any environment of the euphotic layer except in the Arctic region (McIntyre and Bé, 1967; McIntyre et al., 1970). The latitudinal distribution of nannofloral zones is most closely related to the water mass distribution (Okada and Honjo, 1973) and is similar to the planktonic foraminiferal zonation (Bradshaw, 1959). The biomass distribution in the pelagic environment is not uniform but considerably less patchy than near shore or marginal sea environments (Okada and Honjo, 1975). In the Central Pacific Ocean, coccolithophores are distributed almost exclusively in the upper 50 m of water in the sub-Arctic zone. In moderate temperature zones and tropical water, they were usually most abundant in a layer between 50 m and 100 m reaching cell density of $10^4 \ l^{-1}$ in the Pacific Ocean. This layer coincided with the chlorophyll maximum layer (Honjo, 1975b), and often had a high species diversity. The Pacific Central Gyres had concentrations an order of magnitude less than the fringe area (Honjo and Okada, 1974). The cell numbers of coccolithophores in the euphotic water column (200 m deep) at the end of the summer were more than $10^{10} \ m^{-2}$ in the sub-Arctic zone, less than $10^9 \ m^{-2}$ in the Central Gyre and $4-1 \times 10^9$ in the equatorial zone in the Central Pacific (Honjo and Okada, 1974).

The production rate in the equatorial zone is more constant throughout the year than the other areas where the seasonal fluctuation is high. The growth rate of phytoplankton was estimated to be about $0.2-0.3$ doublings day^{-1} in the $70-80$ m mixed layer in the Central Gyre of the North Pacific Ocean (Eppley et al., 1973). Conservatively, this growth rate implies coccolithophore turnover rates of $4-10$ days in temperate to tropical water. The average number of coccoliths per coccolithophore in the equatorial zone is about 20. The annual production of coccoliths is thus on the order of $10^{12} \ m^{-2}$ (2×10^{12} $yr^{-1} \ m^2$ at 10°N 155°W, late summer: Honjo, 1975a). The volume of an average placolith of $5 \ \mu m \times 3 \ \mu m \ \phi$ is $22 \times 10^{-12} \ cm^3$ (according to geometric calculation based on SEM survey — an "average coccolith" is similar to an average leaf of a forest; a large margin of error is involved). A coccolith should weigh approximately 8×10^{-12} g if entirely calcite. This value is about an order of magnitude larger than Paasche's (1962) estimate from ^{14}C measurement on cultured *Emiliania huxleyi* whose volume is one of the smallest of living coccoliths. Thus, the annual mineralization rate of calcium carbonate to form coccoliths is estimated to be on the order of 10 g m^{-2} in the equatorial Pacific (13 g y^{-1} m^{-2} at 10°N 155°W: Honjo, 1975a).

The sinking of phytoplankton in the oligotrophic pelagic water enhances the prospects of survival. Sinking is advantageous for a number of reasons such as migration from the nutrient depleted micro-layer of water immediately outside of the non-motile cell (Munk and Riley, 1952; Pasciak and Gavis, 1974). Thus, phytoplankton can escape from probable death in the energy-rich/nutrient-depleted upper layers of the photic zone to less competitive, nutrient-rich deeper water. Evidence suggests that when a phytoplankton population in the photic zone approaches nutrient exhaustion sinking rates increase (Smayda, 1970).

Some coccolithophore species prefer distinct depth ranges for their biotopes. Others, such as *E. huxleyi*, *Cyclococcolithina fragilis* and *C. leptopora*, are ubiquitous in the euphotic layer. *Gephyrocapsa oceanica* and *Discosphaera tubifera* are distributed in shallower depth in the Central Pacific. The biotope of *Umbellosphaera irregularis* is limited to the upper 75 m of the Pacific Central Gyres and the lower limit coincides with the 23°C isotherm (in a late summer). *U. tenuis*, with its coccolith morphology similar to that of *U. irregularis*, is distributed directly below the latter. The number of living coccolithophores decrease sharply between 200 and 400 m. Such vertically limited biotopes of living coccolithophores, regardless of the local isopicnals, suggest that they are at very least able to control their position in the vertical column. Descending speed is most probably controlled by using coccoliths as ballast for the cell (Plate II, 2). However, there has been no data, *in vitro* or in the field (except Marshall, 1966) to prove that living cells can float by themselves (Smayda, 1970).

The grazing pressure in the pelagic euphotic layer is so high (Menzel and Ryther, 1962), that the majority of the coccosphere population is consumed by zooplankton leaving the rest as seed population (Frost, 1975). The death of coccolithophores is thus insignificant and the number of coccoliths disintegrated from dead cells will be small. The origin of suspended free coccoliths in the euphotic layer may mainly result from the sloughing off to maintain buoyancy.

Sinking mode of coccoliths (Paradox 1)

A classical model of coccoliths sedimentation — individual sinking of discrete coccoliths — is tested in the following in terms of physical (Paradox 1) and chemical (Paradox 2) processes in the deep-oceanic water column.

Descending speed of a coccolith

The rates of descent of two species of coccoliths, *Emiliania huxleyi* and *Cruciplacolithus neohelis*, from a laboratory culture, were measured at 4°, 10° and 18°C by a microscopic method similar to that of Riley (1943). The sinking rate of *E. huxleyi* coccoliths was approximately 1.6 μm sec^{-1} at 18°C, 1 μm sec^{-1} at 10°C, and less than 1 μm sec^{-1} at 4°C. *C. neohelis* coccoliths sank slightly faster. These velocities are an order of magnitude smaller than the ones computed for a calcite sphere of the same diameter from the Stokes formula. The apparent densities (Lerman et al., 1974) of these coccoliths were estimated to be less than 1.2; a significant amount of water sank with them unlike an object with a smooth surface like a sphere. Sinking speed of a hypothetical microdisc of a diameter similar to an *E. huxleyi* coccolith at 4°C ($n = 0.015$) was approximately 22 μm sec^{-1} (Lerman et al., 1974, fig. 1, 3 μm diameter; thickness 0.3 μm; apparent density 1.5; broadside settling).

Surface current

The occurrence of wind and thermally induced convection cells (such as Langmeir cell) has been shown to keep phytoplankton suspended in near surface water for a prolonged time (for a detailed review see Smayda, 1970). The seawater transport of the Gulf Stream is between 100 and 150×10^6 m^3 sec^{-1} and such a volume is maintained as far as 50°N in the North Atlantic (Worthington, 1972). The width of the stream is only approximately 100 km but the high speed results in a large flux of loads. The average surface speed is normally 90—120 cm sec^{-1} but can be faster in the middle of the current (L.V. Worthington, personal communication,1975). The current speed drops to several cm sec^{-1} at 1,000 m depth. As much as $2—5 \times 10^6$ tons of water spins off from the stream per day and mixes with adjacent water masses. According to Worthington (1972), a smaller eddy exists far off Nova Scotia and a significant mass of water moves up as far as 57°N. A coccolith produced in the Florida Strait could ·thus reach the Labrador Sea suspended within several meters of surface layer, within the same season. Other ocean-wide current systems like the Kuroshio may create similar situations.

Gulf Stream rings, cyclonic eddies of a few 100 kilometers in diameter, which are generated by offspring from the meandering main stream can be another source of disturbance of suspended small particles in the near surface water. Perhaps 15—20 eddies exist at any one time in the North Atlantic Ocean. They can move across the eastern North Atlantic within a few years until they cease or are merged with the Gulf Stream (Richardson et al., 1973). An eddy maintains the cyclonic motion and its spread reaches 100 cm sec^{-1} near the center at the surface. Such cyclonic motion of water decreases downward but is still observed down to 1,000 m. The contained energy depends upon the life history of the eddy.

Subsurface advection, eddy diffusivity, and winnowing of coccoliths

Recent research has shown low-frequency, large-scale deep-water circulation (e.g., Gould

et al., 1974). Float tracks of SOFAR buoys, free suspension buoys whose density is adjusted to keep them afloat within designated ranges of depth of the ocean (Rossby et al., 1975), and current meter vectors obtained from various depths of the deep ocean, demonstrated that the deep-sea current is more or less like a "random walk". The horizontal variance in a relatively low-energy area such as the northeastern corner of the Sargasso Sea was on the order of 5 km day^{-1} (10^4 μm sec^{-1}) throughout the water column.

Residence times of common coccoliths are estimated between less than 50 years (a large *Cyclococcolithina leptopora*) to more than 150 years (examples: a small *Gephyrocapsa ericsonii* and "eliptical discs" of *Thorosphaera flabelata*, Okada and Honjo, 1973, plate 2, no. 3) in the water column of 5,000 m. The maximum variance of a coccolith displacement in 5,000 m deep water by advection can be on the order of 10^5 km. This distance is one order larger than the linear scale of the ocean.

Dissolution of suspended coccoliths (Paradox 2)

The surface area of an "average coccolith" exposed to sea water is approximately 80 μm^2. The dissolution rate of calcite measured by Peterson (1966) at a station in the North Equatorial Pacific Ocean, was 0.03 mg cm^2 yr^{-1} (average of the measured dissolution rates between 1,700 and 3,400 m). A coccolith should dissolve within a year while sinking less than 100 m. Therefore, there should be no discrete suspended coccoliths in the calcite-undersaturated water column except in its uppermost layer. If coccoliths are supplied to the deep-sea floor by raining of individual coccoliths, there would also be no chance of forming coccolith oozes unless supplied laterally by slumping of ooze once deposited in calcite-saturated water. The calcite saturation depth (CSD) in the North Pacific Ocean is generally less than 800 m (Takahashi, 1975), indicating that it is the most calcium carbonate-undersaturated ocean known. The CSD

143

drops sharply at 20°S southward and the mean CSD is approximately 2,000 m in the South Pacific Ocean. The mean CSD of the North Atlantic is 4,000 m. The CSD rises southward and is less than 3,000 m in the southernmost area of the South Atlantic Ocean (Takahashi, 1975; conservative estimate using Ingle's apparent solubility constant — Ingle et al., 1973).

Suspended coccoliths in the deep-water column

Coccolith ooze occupies a large area of the ocean bottom (Bramlette, 1958) and is usually far deeper than the CSD, particularly in the North Pacific Ocean. Contrary to expectations, well-preserved coccoliths are found in suspension in most of the oceans even in undersaturated water deeper than the lysocline where the calcite saturation factor (CSF) was less than 0.6 of Takahashi (1975). Intact coccoliths of *Umbellosphaera irregularis*, a fragile species, were often found suspended in 5,000 m deep water in the Pacific Ocean. The nannofloral assemblage in the euphotic layer was generally reflected throughout the aphotic water column. The species diversity of suspended coccoliths generally increased toward the Equator at all aphotic depths. Diversity did not decrease with depth but often increased at depths of 3—4 km (Honjo, 1975a, fig. 5). No evidence of selective dissolution on coccolith species in suspension was found. There was a marked increase of large specimens of *Cyclococcolithina leptopora* at 5,000 m.

The species assemblage found in the surface productive layer (Honjo and Okada, 1974, fig. 10) was repeated at depth of 0.4, 1, 2, 3 and 4 km, although a significant enrichment of *Florisphaera* was reported (Honjo, 1975a). The latitudinal zonation of species assemblages at 4 km was well defined and mirrored the nannofloral boundaries in the productive layer. The zonal demarcation of the assemblage at 1 km and 2 km was more blurred than at 4 km (Honjo, 1975a, fig. 3). This indi-

cates that the expected mixing, winnowing and dissolution predicted by the regional advection and undersaturation of seawater does not exist.

The dissolution rates of the coccolith's calcite must be at least two orders less than that of the inorganic calcite of Peterson (1966) to expect coccoliths to remain in the deep-water column of the Central Pacific Ocean, if we accept that the sinking rate of a coccolith is a few microns per second. We should thus observe progressive dissolution of suspended coccoliths with depth. As reported previously, coccoliths were uniformly well preserved and no evidence of progressive dissolution was observed. The replication of the euphotic nannofloral assemblage in the deep layer inspite of advection cannot be explained by reducing the dissolution rate.

A model which explains the situation best is to increase the rate of descent of coccoliths. The resolution of the boundary between the surface nannofloral assemblage of the Central Gyre and North Equatorial zone (Okada and Honjo, 1973) at a depth of 4 km was at least better than 300 km along the 155°W meridian. Considering the distribution of the sampling stations and the fluctuation of the boundary between the water masses, the original resolution between the boundaries of those water masses was about 200 km. The observed boundary resolution of the suspended coccoliths community was 400 km at 4,000 m; coccoliths should not drift more than 200 km while descending 4 km of water column where the variance of advection is 5 km day^{-1}. In order to achieve this, the sinking speed of coccoliths should be more than 130 m day^{-1} (1,500 μm/sec), or three orders of magnitude faster than the sinking rate of an average coccolith.

Coccolith ooze on the deep-sea floor

Fair matching of coccolith species assemblages between the surface water community ("surface-water" implies less than a meter from the surface) and the thanatocoenosis in core-top samples has been reported by McIntyre and Bé (1967) and McIntyre and McIntyre (1972) from the Atlantic Ocean. However, a direct comparison of the living community of coccolithophores to the thanatocoenosis of coccoliths in core-top samples from the deep-sea floor is in many cases difficult because the selective dissolution of coccoliths involves many complex factors (Berger, 1973; Schneidermann, 1973; Roche et al., 1974). A recent study by Roth and Berger (1975) in the South Pacific Ocean also demonstrated general agreement; species adapted in warm water such as *Discosphaera tubifera* was only found on the sea floor of the tropical area, while *Coccolithus pelagicus*, a cold-water species, was found distributed in the core tops collected from south of 42°S. Eurythermal species such as *Emiliania huxleyi* (mixture of two ecophenotypes of McIntyre and Bé, 1967) and *Gephyrocapsa caribbeanica* was found from the equatorial area to 65°S. *Umbellosphaera irregularis* and *U. tenuis*, species common in the Central Gyre (Okada and Honjo, 1973), were abundant in the sediment surface underlying the South Pacific Central Gyre between 15 and 30°S.

Sediment-trap experiment

A sediment trap (Wiebe et al., 1976), at 2,200 m depth, set for two months in the Tongue of the Ocean, Bahamas, collected a large number of fecal pellets of zooplankton. Approximately 85% of the fecal pellets were less than 5×10^6 μm^3 in volume, and usually a few hundred microns long and several tens of microns wide (Plate I, 1A). Approximately 80% of the pellets were light-colored and packed with coccoliths, diatoms and clay mineral-like particles and rarely contained exoskeletal segments of small zooplankton. Although their appearance was similar to that of pellets described above, approximately 20% of the pellets were dark brown and mainly composed of amorphous organic matter such as partially digested cell material containing almost no hard tissue of phytoplankton (Plate I, 1B). The appearance was similar to fecal pellets of cultured copepods. Copepods were most abundant grazers but pellets produced by other filter feeders such as salps may also have been included.

The preservation of coccoliths in the fecal pellets was excellent (Plate I, 2—7). Fragile holococcoliths (Plate I, 6a) and fragile *U. irregularis* were often intact; coccolithophores of *Emiliania huxleyi* were commonly found (Plate I, 2a). Some coccoliths of *E. huxleyi* had incomplete rays which appeared to be in the process of dissolving. This was probably caused by incomplete growth, however, such malformation of coccoliths is often observed in cultures (Plate I, 1a). An average fecal

---

PLATE I

1. A type of pellet with greenish color and packed with hard tissue of phytoplankton (*A*). The pellet was most probably produced by a small marine copepod. Dark brown pellet (*B*), usually containing no skeletal remains. The origin is unknown. *Emiliania huxleyi* (*a*) with incomplete rays probably due to malformation and not as a result of dissolution. Portion of glass-fiber from fiber-glass substrate (*b*). All specimens illustrated in this plate were collected from Tongue of the Ocean, Bahamas, 2,200 m deep by the Wiebe sediment trap. Bar: 5 μm.   2. An intact coccosphere of *E. huxleyi* (*a*) found the artificially broken surface of a pellet. A large coccolith of *Cyclococcolithina leptopora* (*b*) showing signs of significant dissolution while other coccoliths in the same pellets are well preserved. The seawater at this depth (2,200 m) was not undersaturated in terms of calcite (Takahashi, 1975). It suggests a possibility of multiple digestion. Bar: 3 μm.   3. *Thoracosphaera* sp. (*a*) and *Helicopontosphaera kamptneri* (*b*) found on the surface of a pellet. Bar: 5 μm.   4. Well preserved *Umbilicosphaera sibogae* and *Emiliania huxleyi*. Bar: 3 μm.   5. High-resolution SEM micrograph of a coccolith of *Calciosolenia* sp. (*a*) and part of *Umbilicosphaera sibogae* (*b*), from an artificially broken surface of a pellet. Bar: 1μm.   6. A holococcolith (*a*); several species of holococcoliths were found in fecal pellets. "Skirt crystallites" of *Thorosphaera flabellata* (*b*) (see Okada and Honjo, 1973, plate 2, fig. 3). Bar: 3 μm.   7. An intact specimen of *Ceratolithus cristatus* (*a*). *Cyclococcolithina leptopora* (*b*). Uningested cell material (*c*). Bar: 5 μm.

PLATE I

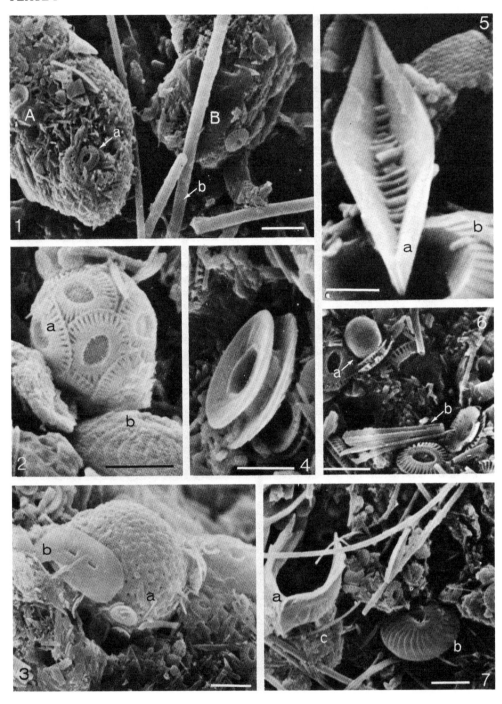

pellet of this type contains 0.8 μg of $CaCO_3$ (average of 30 fecal pellets, measured by atomic absorption spectroscopic method). This is equivalent to approximately 1×10^5 coccoliths or 5,000 average-size coccospheres.

The sinking velocity of fecal pellets collected from the sediment trap was measured in laboratory. They ranged from 50 to 225 m day^{-1}, with a mean of 160 m day^{-1} — 1,800 μm sec^{-1} at 5°C (Wiebe et al., 1976, fig. 1, B). This rate is an order of magnitude slower than the general sinking speed of planktonic foraminiferal tests (Berger and Piper, 1972; Adelseck and Berger, 1975). The estimated horizontal drift of an average fecal pellet while it sinks in the water column of 5,000 m depth, where the water advects with the average variance of 5 km day^{-1}, is approximately 175 km. Thus, the resolution of replication of the bio- and thanatocoenosis would be better than 200 km. The smallest pellets collected by the trap experiment may drift as much as 450 km but their contribution may be insignificant to the total flux.

Well-preserved coccoliths suspended in the undersaturated water column could be shed from rapidly descending fecal pellets. By being shed from fast-sinking pellets, fresh coccoliths are injected into deep water "laterally". Such "lateral-enterants" are fully

exposed to the undersaturated seawater, stalling at the depth they leave the host pellets. Dissolution proceeds along the sutures of their crystallites; disintegration into small fragments and remineralization occur efficiently. Only recently spilled coccoliths are thus collected by filtering seawater which explains the presence of undissolved coccoliths in undersaturated water. All the suspended individual coccoliths, therefore, would eventually dissolve before they arrive at the sea floor and would not disturb the bio-thanatocoenosis correspondence.

Feeding experiment of coccolithophores to zooplankton; preliminary results

In order to reproduce accelerated sinking of coccoliths in the laboratory, common marine copepods, *Centrapoges typicus* and *Acartia tonsa*, were collected near Cape Cod and maintained in large seawater tanks where they were fed fresh laboratory cultured coccolithophores and coccoliths of *Emiliania huxleyi* and *Cruciplacolithus neohelis* (isolated from the Sargasso Sea). The coccolithophore culture was kept under temperature and nutrient conditions most favorable for the production of morphologically complete coccoliths. Copepods fed well on both species

---

PLATE II

1. Coccospheres as well as coccoliths of laboratory cultured *Cruciplacolithus neohelis* were fed to a small copepod, *Acatia tonsa*. Filtering segments, a portion of the second maxiella, were examined under SEM. Smallest spines shown by *a*. The average distance between adjacent spines are over 8 μm, obviously far larger than the coccolith diameter. A mucus material (*b*) is secreted and glues materials together to form an aggregation of manageable sized small particles. (All samples in this plate were prepared by the critical-point dehydration technique and cold cathode coating by Au—Pd under Ar gas.) Bar: 5 μm.   2. Laboratory cultured coccospheres of *Emiliania huxleyi*. Full grown coccoliths (*a*) are shed during culture, probably for the purpose of ballasting the cell buoyancy. Bar: 3 μm.   3. A part of a fecal pellet shown in fig. 4. Pellicle (*a*), protective membrane, covers the surface of the fecal pellet. Pellicle (*b*) is biodegraded and partly exposes the content of the pellet. Pellicle (*c*) is completely removed and pits developed by inward biodegradation and falling-out of contents. (Photographed after leaving the pellets approximately 30 min at 20°C in natural seawater.) Bar: 10 μm.   4. A fecal pellet produced by laboratory cultured copepods *A. tonsa* which were fed mixed algal food of *E. huxleyi*, *C. neohelis* (both as coccospheres) and two species of diatoms. The pellicle covers half of the surface. Bar: 50 μm. 5. Laboratory cultured *C. neohelis*. Bar: 5 μm.   6. A part of the pellet produced by a laboratory cultured *A. tonsa* fed with the coccospheres shown in fig. 5. Many coccospheres retain their original shape and some are completely intact. The pellicle is thoroughly biodegraded and coccoliths and coccospheres are spilled out of the pellet. The pellets were left in natural seawater over 6 hours at 20°C. Bar: 5 μm.

PLATE II

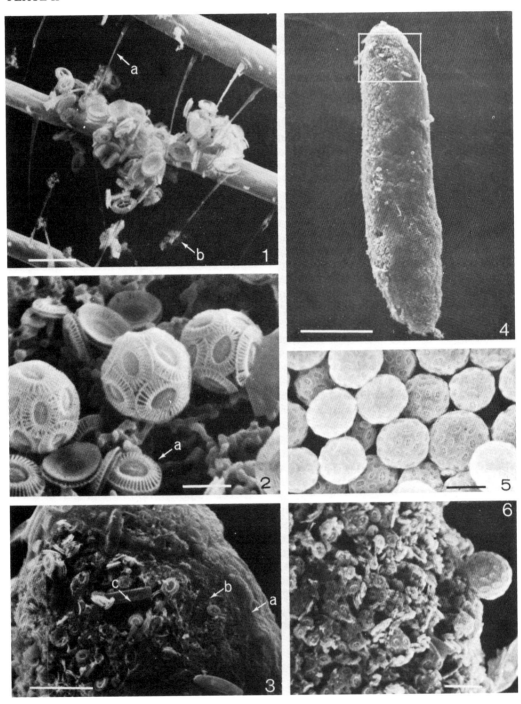

of coccolithophores. A typical feeding rate at 20°C by *A. tonsa* was 2.6 × 10⁴ day⁻¹ when they were fed in water with cell concentrations of 2.6 × 10⁵ l⁻¹ (an average natural concentration). Pure coccoliths of *E. huxleyi* and *C. neohelis* which were separated from the culture were fed in filtered seawater. Copepods fed on these pure coccoliths and produced fecal pellets which were packed by coccoliths. In either case, coccoliths were intact in the fecal pellets (Plate II, 3). No evidence was found that they were broken into small fragments. Fragmentation of large diatom frustules as the result of copepod ingestion is widely reported (see, for example, Marshall and Orr, 1956; Schrader, 1971). More efficient feeding on particles as small as coccospheres and discrete coccoliths is possible because the filter size of *A. tonsa* is significantly coarser than the diameter of coccospheres of *E. huxleyi* and *C. neohelis* (Plate II, 1).

All of these fecal pellets were covered by a thin protective membrane, or "pellicle" (Marshall and Orr, 1956). The pellicle was continuous, and usually with no breakage evident on pellets examined under SEM after dehydration by the critical-point technique (Plate II, 3a). A pellicle protects the pellet contents from disintegration and chemical attack. Thus, such protective membrane plays an important role in linking the bio- and thanatocoenosis. When a fresh fecal pellet stands in normal surface seawater, micro-organisms collect on the pellicle surface and strip it within a short time (Plate II, 3b,c). Biodegradation proceeds further on the organic remains which bind coccoliths together and gradually particles disintegrate near the surface of pellets and the pellet finally fragments. Such microbial activity is roughly proportional to the water temperature. For example, at 25°C the pellicle is degraded within minutes whereas the pellicle remains intact for days at 15°C.

Percentage of coccoliths reaching the deep bottom

The standing stock of suspended coccoliths

in the Equatorial aphotic water column from 400 to 5,000 m was on the order of 10¹⁰ m⁻² (Honjo, 1975a) and represented 10--100 mg m⁻² of calcite. Typically, the total suspended coccoliths at 10°N 155°W was approximately 4 × 10¹⁰ m⁻² or 300 mg m⁻² of calcite. The annual turn-over rate is the inverse of residence time and is roughly 3.5 yr⁻¹ using Peterson's calcite dissolution rate (0.03 mg cm⁻² yr⁻¹) in the undersaturated Pacific water. Therefore, the annual remineralization rate in the undersaturated water column with a square meter section is on the order of 100 mg. At 10°N 155°W the remineralization rate was estimated as 900 mg m⁻² yr⁻¹, or about 8% of the annual production of coccoliths in the euphotic layer; the rest of the coccoliths are assumed to reach

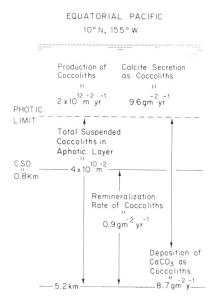

Fig. 1. An estimate of the rates of production, suspension, remineralization and deposition of coccoliths and the CaCO₃ (calcite) in form of coccoliths at 10°N 155°W. The estimation of the remineralization rate in the undersaturated water column and the rate of deposition of coccoliths are based on the dissolution rates of calcite spheres measured by Peterson (1966) in the Central Equatorial Pacific Ocean: 0.03 mg cm⁻² yr⁻¹. The sublysoclinal dissolution rates were not applied.

the bottom in zooplankton fecal pellets (Fig. 1).

The post-glacial sedimentation rates of biogenic $CaCO_3$ on the Atlantic Ocean floor is estimated as 5.5 to 15 g yr^{-1} m^{-2} from the Equatorial Atlantic Ocean (Turekian, 1965, table II, p. 102). These cores were collected from the water depth shallower than the CSD but deeper than aragonite saturation depth in the Atlantic Ocean (Berner et al., 1976), eliminating the effect of dilution by pteropod shells (Chen, 1968). The typical carbonate fraction of a core contained 20% coccoliths and 80% planktonic foraminiferal tests. Recent estimates by Früterer (1974) indicate that the ratio of coccolith to other carbonate fraction was as large as 50% in a calcareous sediment from the more productive eastern Equatorial Atlantic near Africa. The annual coccolith production estimated in the Equatorial Pacific is approximately 8 g yr^{-1} m^{-2} and it is at least the same order of magnitude as the carbonate sedimentation rate in the Atlantic. Unfortunately, the Equatorial Pacific is largely covered by calcite-undersaturated water and a direct measurement of sedimentation rates is difficult.

The density of coccoliths suspended in the upper aphotic layer were generally positively related to the total standing stock of coccoliths in the overlying euphotic water column (see Honjo, 1975a, text-fig. 2, p.117; and Honjo and Okada, 1974, text-fig. 5, p. 214). The standing stock of suspended coccoliths in the longitudinal profile of the aphotic Central Pacific Ocean along 155°W (Honjo, 1975a, lower figure in p. 117) decreases rapidly with depth while keeping species assemblage constant (Honjo, 1975a, text-fig. 3, p. 121). The distribution of coccolith density bears no relation to the CSF profile in that area (Takahashi, 1975, text-fig. 7, p. 20). Such distribution patterns can be explained by the dispersion of sinking fecal pellets. In the Transitional and Equatorial zones of the North Pacific (Okada and Honjo, 1973), the density of suspended coccoliths is high possibly because the flux of fecal pellets and consequent-

ly the number of coccoliths shed from the sinking pellets in the aphotic water column are large. The density of suspended coccoliths was one or two orders less in the aphotic water of the Central Gyre; reflecting the low productivity in the euphotic layer. The upper 50 m of the sub-Arctic zone in the North Pacific was highly productive (dominated by *Emiliania huxleyi*) in late summer (Okada and Honjo, 1973). However, the density of suspended coccoliths in the underlying aphotic water column was low (less than 100 l^{-1}). This can be explained by minimal biodegradation of pellicles at low temperatures, particularly below the thermocline (Okada and Honjo, 1973, fig. 7, p. 565). The coccoliths were better contained in fecal pellets under intact pellicles and less coccoliths were supplied to the water column by shedding.

A model of coccolith sedimentation in the water column; a summary (Fig. 2)

Production of coccoliths in the euphotic layer is variable according to season, local nutrient supply, and other hydrographic factors. In equatorial Central Pacific waters where the seasonal fluctuations are small, annual production is estimated as 7 to 22 $\times$ 10^{12} m^{-2} of coccoliths or 7 to 22 $\times$ 10^{12} g m^{-2} of calcite (typically 8 g m^{-2}). Coccospheres produced in the euphotic zone usually shed coccoliths (to adjust sinking speed) and result in numerous free coccoliths suspended in the euphotic layer.

The majority of coccolithophores are grazed by zooplankton such as copepods and aggregated into fecal pellets. Coccoliths are generally not dissolved or crushed when coccospheres are ingested by copepods. Fecal pellets are covered by pellicles so that the contents are protected from immediate chemical dissolution. However, a pellicle is often biodegraded in warm water within a short time and lose the coccoliths and other contents. Biodegradation may go further and upon digestion of the organic matter which holds the particles together they are free to

150

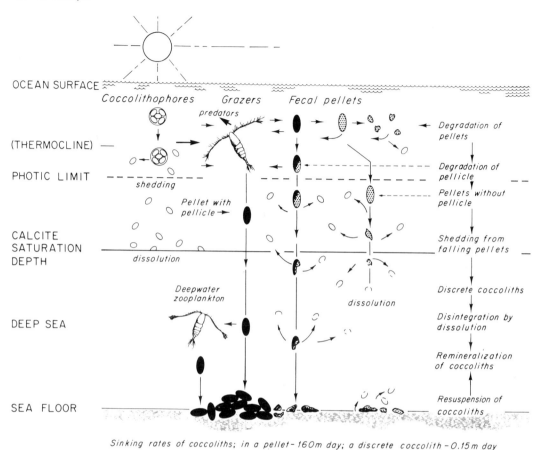

Fig. 2. A model of the relationship between the production, transportation, dissolution and deposition of coccoliths in open, deep ocean. Scales are not in proportion.

disaggregate. Some suspended coccoliths and broken pellets are reingested by grazers. A small number of individual coccoliths sink through the upper aphotic water column, across the CSD, all are quickly dissolved (i.e., in less than a year).

Deposition of coccoliths on the deep-sea floor is mainly via the sinking of fecal pellets of small but abundant zooplankton such as copepods. About 10^5 coccoliths or approximately 1 μg of $CaCO_3$ can descend with a pellet at a speed of 160 m day^{-1} and within a month or less reach the sea floor at a depth of 5,000 m. During this time dissolution of con-

tained coccoliths is negligible. An estimate in the Equatorial Pacific indicates that 92% of coccoliths produced in the euphotic layer reach the underlying deep-sea floor. The rate of sedimentation of coccoliths approximates the production rate. Such accelerated sinking allows the resolution of the bio-thanatocoenosis correspondence to within a few hundred kilometers while they sink through a deepwater column where the variance of advection is several kilometers per day. The pellets with biodegraded pellicles sink roughly 25% slower than the ones with complete pellicles. Coccoliths are constantly shed from such uncovered fecal pellets at all depths. These coccoliths will dissolve and disappear in the vicinity of where they are shed from the host pellet and will have no effect on the taphonomy of coccoliths. The estimated rate of coccolith remineralization in the undersaturated water column is approximately 8% or less in the Equatorial Pacific stations when Peterson's (1966) dissolution rate of calcite in that area is applied.

Discussion

The recycling of ingested coccoliths through the food chain, particularly in the upper euphotic layer, appears complex and warrants further study. A number of fecal pellets will be regrazed by other zooplankton, although fragments of pellets are preferred because of the more convenient size. Although such reingestion occurs within copepod food chains, the coccoliths may not suffer extensive damage. Some coccoliths may be transferred through multiple trophic levels. Coccoliths ingested by a fish or fish larvae having strongly acidic stomachs may be dissolved. Since the trophic level of fish is high, the quantity of coccoliths remineralized in their guts would be very small.

The community of coccolithophores living in cold water, below the thermocline or in high latitudes, should be better represented in the bottom thanatocoenosis. On the other hand, rapid biodegradation of pellicles and subsequent fragmentation of pellets in warm water would cause the coccolithophore community to be transferred through more complex food chains.

The budget described here is based upon Peterson's *in situ* dissolution rate of calcite. If the rate differs by factors from biogenic calcite (R.A. Berner, personal communication, 1975; Lerman et al., 1974) the basic configuration of the model would not change. However, the ratio of remineralized coccoliths in the undersaturated water column to the entire production would change significantly. Estimation of the coccolith flux in the North Atlantic Ocean remains a problem because our knowledge of calcite saturation in the water column is inconsistent (Edmond, 1974; Takahashi, 1975; and Milliman, 1975) and the productivity/standing crop of suspended coccoliths in the aphotic layer are not precisely known.

Dissolution of coccoliths (McIntyre and McIntyre, 1971; Berger, 1973; Schneidermann, 1973; Roth and Berger, 1975) or calcite cvergrowth (Bukry et al., 1971; Roth, 1973; Schlanger et al., 1973) after arrival at the bottom will be influenced by the mode of deposition (Broecker and Broecker, 1974). Microbial activity is very slow in the deep sea (Jannasch and Wirsen, 1973) and pellicles and organic matrix of pellets are biodegraded slowly or are bioturbated by mud-filtering benthos. A persistent pellicle protects coccoliths resuspended into the nepheloid layer by bottom currents. The interstitial water of fecal pellet is semi-independent from the deep-sea water. Thus, straight application of the dissolution chemistry of carbonates in seawater may not be useful in understanding the rate of dissolution of calcareous ooze.

Acknowledgements

I would like to thank Dr. P.H. Wiebe for his generous permission to use his unpublished data. I am grateful to Drs. K.O. Emery, W.A. Berggren, D.A. Ross, G. Jackson, G.D. Grice, L. Brand, and R.A. Berner for their critical reading and/or constructive discussions. I was provided valuable information on physical oceanography from Drs. V. Worthington and W.J. Schmitz. My special thanks are due to Mrs. M. Goreau for scanning microscopy and culture of coccolithophores, Mr. M. Roman for providing zooplankton culture and feeding experiments, and Miss S. Bernardo for editorial assistance.

This investigation was made possible through financial support by the Oceanography Section of the National Science Foundation under Grant GA 30996 and DES71-00509 A02. Contribution from the Woods Hole Oceanographic Institution No. 3680.

References

Adelseck, C.G. and Berger, W.H., 1975. On the dissolution of planktonic foraminifera and associated microfossils during settling and on the sea floor. In: W. Sliter, A.W.H. Bé and W.H. Berger (Editors), Dissolution of Deep-Sea Carbonates. Cushman Found. Foraminiferal Res., Spec., Publ., No. 13: 70–81.

Berger, W.H., 1973. Deep-sea carbonates: evidence for a coccolith lysocline. Deep-Sea Res., 20: 917–921.

Berger, W.H. and Piper, D.J., 1972. Planktonic foraminifera: differential settling, dissolution and redeposition. Limnol. Oceanogr., 17: 275–287.

Bernard, F., 1963. Vitesse de chute en mer des amas palmeloides de Cyclococcolithus. Ses conséquences pour le cycle vital des mers chaudes. Pelagos, 1: 1–34.

Berner, R.A., Berner, E.K. and Keir, R.S., 1976. Aragonite dissolution on Bermuda Pedestal: its depth and geochemical significance. Earth Planet. Sci. Lett., in press.

Bradshaw, J.S., 1959. Ecology of living planktonic foraminifera in the North and Equatorial Pacific Ocean. Contrib. Cush. Found. Foraminiferal Res., 10(2): 25–64.

Bramlette, M.N., 1958. Significance of coccolithophorids in calcium carbonate deposition. Geol. Soc. Am. Bull., 69: 121–126.

Bramlette, M.N., 1961. Pelagic sediments. In M. Sears (Editor), Oceanography. Am. Assoc. Adv. Sci. Publ., 67: 345–366.

Broecker, W.S. and Broecker, S., 1974. Carbonate dissolution on the western flank of the East Pacific rise. In: W.W. Hay (Editor), Studies in Paleo-Oceanography. Soc. Econ. Paleontol. Mineral., Spec. Publ., No. 20: 44–57.

Bukry, D., Douglas, R.G., Kling, S.A. and Krasheninnikov, V., 1971. Planktonic microfossil biostratigraphy of the northwestern Pacific Ocean. Deep Sea Drilling Proj. Initial Rep., 6: 1253–1300.

Chen, C., 1968. Pleistocene pteropods in pelagic sediments. Nature, 291: 1145.

Edmond, J.M., 1974. On the dissolution of carbonate and silicate in the deep ocean. Deep-Sea Res., 21: 455–480.

Eppley, R.W., Rengers, E.H., Venrick, E. and Mullin, M.M., 1973. A study of plankton dynamics and nutrient cycling in the Central Gyre of the North Pacific Ocean. Limnol. Oceanogr., 18(4): 534–551.

Esterly, C.O., 1966. The feeding habits and food of pelagic copepods and the question of nutrition by organic substances in solution in the water. Univ. Calif. Publ. 2001, 9: 253–340.

Fournier, R.O., 1968. Observations of particulate organic carbon in the Mediterranean Sea and their relevance to the deep-living Coccolithophorids Cyclococcolithus fragilis. Limnol. Oceanogr., 13: 693–694.

Frost, B.W., 1975. A threshold feeding behavior in Calanus pacificus. Limnol. Oceanogr., 20: 263–266.

Früterer, D., 1974. Plankton organisms as sediment contributors in the fine-grained fraction of surface sediments off West Africa. Symp. Marine Plankton and Sediments, and 3rd Planktonic Conference, Kiel, 1974, p. 25 (abstract).

Gould, W., Schmitz, W.J. and Wunsch, C., 1974. Preliminary field results for a mid-ocean dynamics experiment (MODE-O). Deep-Sea Research, 21: 911–931.

Harbison, G.R. and Gilmar, R.W., 1976. The feeding rates of the pelagic tunicate, Pegca confeclerata and other two salps. Limnol. Oceanogr., in press.

Harding, G.C.H., 1974. The food of deep-sea copepods. Marine Biol. Assoc. U.K. J., 54: 141–155.

Honjo, S., 1975a. Dissolution of suspended coccoliths in the deep-sea water column and sedimentation of coccolith ooze. In: W. Sliter, A.W.H. Bé and W.H. Berger (Editors), Dissolution of Deep-Sea Carbonates. Cushman Found. Foraminiferal Res., Spec. Publ., No. 13: 115–128.

Honjo, S., 1975b. Biogeography and provincialism of living coccolithophorids in the Pacific Ocean. In: A.T.S. Ramsay (Editor), Oceanic Micropaleontology. Academic Press, London, in press.

Honjo, S. and Okada, H., 1974. Community structure of coccolithophorids in the photic layer of the Mid-Pacific Ocean. Micropaleontology, 20: 209–230.

Ingle, S.E., Culberson, C.H., Hawley, J.E. and Pytkowicz, R.M., 1973. The solubility of calcite in sea water at atmospheric pressure and 35‰ salinity. Mar. Chem., 1: 295–307.

Jannasch, H.W. and Wirsen, C.O., 1973. Deep-sea microorganisms: In situ response to nutrient enrichment. Science, 180: 641–643.

Lerman, A. Lal, D. and Dacey, M.F., 1974. Stoke's settling and chemical reactivity of suspended particles in natural waters. In R.J. Gibbs (Editor), Suspended Solids in Water. Plenum Press, New York, N.Y., pp. 17–47.

Lohmann, H., 1902. Die Coccolithophoridae, eine Monographie der Coccolithen bildenden Flagellaten, zügleich ein Beitrag zur Kenntnis des Mittelmeerauftriebs. Arch. Protistenkd., 1: 89–165.

Marshall, H.G., 1966. Observations on the vertical distribution of coccolithophores in the Northwestern Sargasso Sea. Limnol. Oceanogr., 11: 432–435.

Marshall, S.M. and Orr, A.P., 1956. On the biology of Calanus finmarchicus, IX, Feeding and digestion in the young stages. Mar. Biol. Assoc. U.K. J., 35: 587–603.

Marshall, S.M. and Orr, A.P., 1962. Food and feeding in copepods. Rapp. Cons. Explor. Mer, 152: 92—98.

McIntyre, A. and Bé, A.W.H., 1967. Modern Coccolithophoridae of the Atlantic Ocean—I, placoliths and cyrtoliths. Deep-Sea Res., 14: 561—597.

McIntyre, A. and McIntyre, R., 1971. Coccolith concentrations and differential solution in oceanic sediments. In: B.M. Funnell and W.R. Riedel (Editors), The Micropaleontology of Oceans. Cambridge University Press, London, pp. 253—262.

McIntyre, A., Bé, A.W.H. and Roche, M.B., 1970. Modern Pacific Coccolithophoridae: a paleontological thermometer. Trans. N.Y. Acad. Sci., 2(32): 720—731.

Menzel, D.W. and Ryther, J.H., 1962. Zooplankton in the Sargasso Sea off Bermuda and its relation to organic production. J. Cons. Perm. Int. Explor. Mer. 26: 250—258.

Milliman, J., 1975. Dissolution of aragonite, Mg-calcite and calcite in the North Atlantic Ocean. Geology, 3: 461—462.

Mullin, M.M., 1966. Selective feeding by calanoid copepods from the Indian Ocean. In: H. Barnes (Editor), Some Contemporary Studies in Marine Science. George Allen and Unwin, London, pp. 313—334.

Munk, W.H. and Riley, G.A., 1952. Absorption of nutrients by aquatic plants. J. Mar. Res., 11: 215—240.

Murray, J. and Hjort, J., 1912. The Depths of the Ocean. A General Account of the Modern Science of Oceanography Based Largely on the Scientific Researches of the Norwegian Steamer *Michael Sars* in the North Atlantic. Macmillan, London, 821 pp.

Okada, H. and Honjo, S., 1973. The distribution of oceanic coccolithophorids in the Pacific. Deep-Sea Res., 20: 355—374.

Okada, H. and Honjo, S., 1975. Distribution of coccolithophores in marginal seas along the Western Pacific Ocean and the Red Sea. Mar. Biol., 31: 271—285.

Paasche, E., 1962. Coccolith formation. Nature, 193: 1094—1095.

Pasciak, W.J. and Gavis, J., 1974. Transport limitation of nutrient uptake in phytoplankton. Limnol. Oceanogr., 19: 881—888.

Peterson, M.N.A., 1966. Calcite: rates of dissolution in a vertical profile in the central Pacific. Science, 154: 1542—1544.

Richardson, P.L., Strong, A.E. and Knauss, J.A., 1973. Gulf Stream eddies: Recent observations in the western Sargasso Sea. J. Phys. Oceanogr., 3: 297—307.

Riley, G.A., 1943. Sinking rates of phytoplankton cells, laboratory experiments. Bull. Bingham Oceanogr. Coll., 8, Art. 4: 53 pp.

Roche, M.B., Geitzenauer, K.R. and McIntyre, A., 1974. Coccolith biogeography from N. Atlantic and Pacific surface sediments. In: Symp. "Marine Plankton and Sediments" and 3rd Planktonic Conference, Kiel, 1974, p. 61 (abstract).

Rossby, T., Dow, D. and Freeland, H., 1975. Float tracks for May—December, 1975. Mode Hot Line News, No. 76.

Roth, P.H., 1973. Calcareous nannofossils — Leg 17, Deep Sea Drilling Project. Deep Sea Drilling Proj. Initial Rep., 17: 695—795.

Roth, P.H. and Berger, W.H., 1975. Distribution and dissolution of coccoliths in the South and Central Pacific. In: W.V. Sliter, A.W.H. Bé and W.H. Berger (Editors), Dissolution of Deep-Sea Carbonates. Cushman Found. Foraminiferal Res., Spec. Publ., No. 13: 87—113.

Roth, P.H., Mullin, M.M. and Berger, W.H., 1975. Coccolith sedimentation by fecal pellets: Laboratory experiment and field observations. Geol. Soc. Am. Bull., 86: 1079—1084.

Schlanger, S.O., Douglas, R.G., Lancelot, Y., Moore, T.C. and Roth, P.H., 1973. Fossil preservation and diagenesis of pelagic carbonates from the Magellan Rise, Central North Pacific Ocean. Deep Sea Drilling Proj. Initial Rep., 17: 407—427.

Schneidermann, N., 1973. Deposition of coccoliths in the compensation zone of the Atlantic Ocean: Gulf Coast Sec., Soc. Econ. Paleontol. Mineral. Proc., Symposium on Calcareous Nannofossils, pp. 140—151.

Schrader, H.J., 1971. Fecal pellets: Role in sedimentation of pelagic diatoms. Science, 174: 55—57.

Smayda, T.J., 1970. The suspension and sinking of phytoplankton in the sea. Oceanogr. Mar. Biol. Ann. Rev., 8: 353—414.

Takahashi, T., 1975. Carbonate chemistry of sea water and the calcite compensation depth in the oceans. In: W.V. Sliter, A.W.H. Bé and W.H. Berger (Editors), Dissolution of Deep-Sea Carbonates. Cushman Found. Foraminiferal Res., Spec. Publ., No. 13: 11—26.

Turekian, K., 1965. Some aspects of the geochemistry of marine sediments. In: J.P. Riley and G. Skirrow (Editors), Chemical Oceanography, 2. Academic Press, London, pp. 81—126.

Wiebe, P.H., Boyd, S.H. and Winget, C., 1976. Sedimentational trap for use of above deep-sea floor with preliminary results of its use in the Tongue of Ocean, Bahamas. J. Mar. Res. (in press).

Worthington, L.V., 1972. Negative oceanic heat flux as a cause of water mass formation. J. Phys. Oceanogr., 2: 205—211.

14

Reprinted from *Cretaceous Research* 2:165–176 (1980)

Selective Dissolution of Late Cretaceous and Earliest Tertiary Calcareous Nannofossils: Experimental Evidence

H. R. Thierstein

Scripps Institution of Oceanography, University of California, San Diego, La Jolla, California 92093, U.S.A.

H. R. Thierstein. Selective Dissolution of Late Cretaceous and Earliest Tertiary Calcareous Nannofossils: Experimental Evidence. *Cretaceous Research* (1980) 2, 165–176. The effects of carbonate dissolution on the taxonomic composition of ancient nannofossil assemblages were investigated by deploying them on a deep mooring at depths of 1039 m–5500 m for 117 days in the Sargasso Sea. The relative proportions of only a few taxa change significantly with increasing dissolution: *Cruciplacolithus primus* is by far the most susceptible and *Micula staurophora* by far the most resistant species. The average dissolution indices calculated for the deployed assemblages closely follow the gradients of other dissolution indicators in the North Atlantic.

1. Introduction

The dissolution of calcium carbonate at depth in the oceans changes the taxonomic composition of recent planktonic microfossil assemblages (Berger, 1967; Ruddiman & Heezen, 1967; Parker & Berger, 1971; Roth & Berger, 1975), making biogeographic and paleo-ecologic reconstructions difficult (Thompson, 1976; Hutson, 1977). A comparable bias in pre-Quaternary microfossil assemblages has to be expected, but has so far received little attention. Information on the influence of selective dissolution on latest Cretaceous and earliest Tertiary calcareous nannofossil assemblages is needed to analyze changes in carbonate preservation across the Cretaceous–Tertiary boundary and to identify and separate geochemical processes from evolutionary and biogeographic changes during this crucial time (Thierstein, in press. The effects of dissolution on the taxonomic composition of comparatively well-preserved late Cretaceous and Danian nannofossil assemblages were investigated by exposing them to corrosive deep waters at various depths in the central North Atlantic. Studies on the carbonate chemistry of the deep waters in this area were conducted by Milliman (1977), Broecker & Takahashi (1978), and Honjo & Erez (1978).

2. Experimental procedures

Twenty well-preserved calcareous nannoplankton assemblages of Campanian, Maastrichtian and Danian age were initially prepared for the experiment. The taxonomic changes in 13 of them were subsequently determined (Table 1); they provide optimal coverage of stratigraphic, paleolatitudinal and biogeographic ranges

Table 1. Dissolution experiment assemblages

Assemblage	Age	Location	Water depth (m)	Sub-bottom depth (m)	(%) CaCO$_3$ (bulk)
DSDP 2-10-12-2, 43 cm	Late Campanian/Early Maastrichtian	North Atlantic	4612	389	90
DSDP 3-21-4-6, 85 cm	Late Campanian/Early Maastrichtian	South Atlantic	2102	94	86
DSDP 17-167-50-1, 122 cm	Late Campanian/Early Maastrichtian	Central Pacific	3176	760	99
DSDP 21-208-34-2, 28 cm	Late Maastrichtian	South Pacific	1545	587	93
DSDP 22-217-16-6, 140 cm	Early Danian	Indian Ocean	3020	420	82
DSDP 22-217-17-1, 20 cm	Late Maastrichtian	Indian Ocean	3020	421	94
DSDP 25-249-17-1, 53 cm	Middle Maastrichtian	Indian Ocean	2088	179	42
DSDP 36-327A-12-4, 14 cm	Late Maastrichtian	South Atlantic	2411	113	63
DSDP 43-384-13-2, 132 cm	Early Danian	North Atlantic	3920	167	92
DSDP 43-384-13-2, 145 cm	Early Danian	North Atlantic	3920	167	90
DSDP 43-384-13-3, 49 cm	Late Maastrichtian	North Atlantic	3920	168	91
DSDP 43-384-14-2, 40 cm	Middle Maastrichtian	North Atlantic	3920	176	96
A23-4 Corsicana Clay	Middle Maastrichtian	Onion Creek, Texas			14

of late Cretaceous and earliest Tertiary assemblages available (Thierstein & Haq, 1977; Thierstein, in press). The bulk carbonate contents of the studied samples range from 14–99% (Table 1). The sediment samples were disaggregated, washed through a 63 μm sieve and subsequently through a 25 μm nylon screen. Clay particles and broken nannolith elements of less than 2 μm were separated by repeated centrifugation. Each 2–25 μm fraction was dried and subsequently mounted by Dr S. Honjo (WHOI) as a 3–8 μm thick layer onto a polycarbonate film. Details of the mounting and deployment technique will be described by Honjo & Goreau (in preparation). The film was cut into pieces of 2 × 5 mm and glued onto a polyvinyl disc of 2.5 cm diameter. Ten of these discs, each containing 20 assemblages, were mounted in a perforated tube and suspended on a deep mooring in the Sargasso Sea at 31°32.9′N, 55°03.9′W from July 20, 1977 to November 14, 1977, at depths ranging from 1039 m to 5500 m. After recovery, the samples were rinsed in alcohol series and dried at 40°C for 12 hours. Smear-slides were prepared by carefully scraping off the nannoliths with a toothpick, using standard smear-slide preparation techniques. The taxonomic composition of 13 characteristic assemblages was analyzed quantitatively under the light microscope in the undeployed 2–25 μm fraction used for the experiment and from 4 deployment depths (1039 m, 2838 m, 4030 m and 5348 m). The original census data are available from the author upon request. Scanning electron microscope studies were done on two sensor discs, one which had not been used in the experiment and a second, which had been deployed at 5500 m depth.

3. Dissolution susceptibility of individual taxa

The choice of assemblages used in the study was aimed at covering most of the known biostratigraphic and geographic taxonomic variability. The dissolution susceptibility or dissolution resistance of individual taxa was assessed by computing the change in relative abundance of each taxon with increasing depth in each individual sample (Berger, 1968). Each undeployed assemblage was compared with the four deployed ones, each assemblage deployed at 1039 m with the three deeper ones and so on. A maximum of 10 comparisons for each taxon in each assemblage could thus be made. The difference in the percent abundance of each taxon in each pair was calculated and summed (added if it increased with depth, or subtracted if it decreased with depth), resulting in a preservation score for each taxon. The individual taxa are listed in order of decreasing susceptibility and increasing resistance to dissolution in Table 2. This dissolution ranking was established in the following way.

The census data were gathered by counting 300 specimens per sample. The taxonomic concepts of many of the species used have been discussed by Thierstein (1976). Additional references to taxonomic descriptions and illustrations can be found in Loeblich & Tappan (1966, 1968, 1969, 1970a, 1970b, 1971, 1973), Risatti

Table 2. Late Cretaceous and earliest Tertiary nannofossil taxa listed in order of decreasing susceptibility to dissolution, with average abundance, dissolution scores, number of pairings, and dissolution ratios (see text for further explanations)

Taxon	Average abundance (%)	Scores	Pairings	Ratios
Cruciplacolithus primus Perch-Nielsen, 1977	18.4	−446.5	30	−14.9
Biscutum constans (Górka, 1957) Black, 1967	3.8	−467.7	125	− 3.7
Prediscosphaera cretacea (Arkhangelsky, 1912) Gartner, 1968	9.3	−355.2	127	−2.8
Zygodiscus sigmoides Bramlette & Sullivan, 1961	9.7	−87.5	44	−2.0
Discorhabdus rotatorius (Bukry, 1969) Thierstein, 1973	0.7	−96.9	57	−1.7
Prediscosphaera spinosa (Bramlette & Martini, 1964) Gartner, 1968	1.6	−112.1	73	−1.5
Cribrosphaerella ehrenbergii (Arkhangelsky, 1912) Deflandre, 1952	4.0	−184.4	122	−1.5
Nephrolithus frequens Górka, 1957	5.9	−47.2	44	−1.1
Cretarhabdus surirellus (Deflandre, 1954) Reinhardt, 1970	5.1	−111.0	122	−0.9
Parhabdolithus regularis (Górka, 1957) Bukry, 1969	1.8	−88.7	104	−0.9
Vagalapilla aachena Bukry, 1969	0.5	−32.0	39	−0.8
Microrhabdulus decoratus Deflandre, 1959	2.0	−81.5	107	−0.8
Vagalapilla octoradiata (Górka, 1957) Bukry, 1969	1.8	−25.4	34	−0.7
Zygodiscus spiralis Bramlette & Martini, 1964	1.1	−85.4	120	−0.7
Chiastozygus litterarius (Górka, 1957) Manivit, 1971	1.0	−76.3	109	−0.7
Eiffellithus turriseiffeli (Deflandre, 1954) Reinhardt, 1965	2.8	−76.0	115	−0.7
Lithraphidites carniolensis Deflandre, 1963	1.3	−54.4	108	−0.5
Teichorhabdus ethmos Wind & Wise, 1977	0.9	−10.6	22	−0.5
Stephanolithion laffittei Noël, 1957	0.1	−9.4	20	−0.5
Cruciplacolithus tenuis (Stradner, 1961) Hay & Mohler, 1967	6.1	−13.4	29	−0.5
Lithraphidites quadratus Bramlette & Martini, 1964	0.7	−17.2	61	−0.3
Gatnerago obliquum (Stradner, 1963) Noël, 1970	0.6	−12.2	44	−0.3
Zygodiscus diplogrammus (Deflandre, 1954) Gartner, 1968	0.5	−17.3	66	−0.3
Cretarhabdus crenulatus Bramlette & Martini, 1964	0.4	−7.9	43	−0.2
Zygodiscus (shields)	1.2	−12.8	104	−0.1
Tetralithus gothicus Deflandre, 1959	1.0	−1.2	26	0
Tetralithus obscurus Stradner, 1963	1.9	0.5	21	0
Cylindralithus serratus Bramlette & Martini, 1964	2.0	22.1	96	0.2
Reinhardtites anthophorus (Deflandre, 1959) Perch-Nielsen, 1968	0.8	10.8	43	0.3
Podorhabdus (shields)	0.5	24.0	82	0.3
Crepidolithus neocrassus Perch-Nielsen, 1968	2.0	11.9	32	0.4
Misceomarginatus pleniporus Wind & Wise, 1977	1.1	5.3	14	0.4
Micula mura (Martini, 1961) Bukry, 1973	0.6	22.5	50	0.5
Cretarhabdus (shields)	1.1	41.3	89	0.5
Arkhangelskiella cymbiformis Vekshina, 1959	3.0	71.9	120	0.6
Broinsonia parca (Stradner, 1963) Bukry, 1969	1.1	15.6	18	0.9
Watznaueria barnesae (Black, 1959) Perch-Nielsen, 1968	16.9	103.6	117	0.9
Thoracosphaera operculata Bramlette & Martini, 1964	6.7	64.7	67	1.0
Tetralithus aculeus (Stradner, 1961) Gartner, 1968	1.9	129.5	95	1.4
Markalius astroporus (Stradner, 1963) Hay & Mohler, 1967	4.5	71.0	36	2.0
Braarudosphaera bigelowii (Gran & Braarud, 1935) Deflandre, 1947	4.4	39.4	19	2.1
Kamptnerius magnificus Deflandre, 1959	2.3	145.2	55	2.6
Lucianorhabdus cayeuxii Deflandre, 1959	3.5	185.9	53	3.5
Tetralithus trifidus (Stradner, 1961) Bukry, 1973	7.0	191.2	38	5.0
Micula staurophora (Gardet, 1955) Stradner, 1963	13.9	1390.1	124	11.2

(1973), Wise & Wind (1977), and Perch-Nielsen (1977). *Thoracosphaera operculata* Bramlette & Martini (1964), as listed here may include *T. deflandrei* Kamptner of Haq & Lohmann (1976) and *T. atlantica* Haq & Lohmann (1976). *Cretarhabdus crenulatus* Bramlette & Martini (1964), includes all *Cretarhabdus* with less than nine perforations in the central area, the size of which, however, is considerably smaller than in the early Cretaceous *Cretarhabdus angustiforatus* (Black, 1971; Bukry 1973). In each sample the relative abundances of 63 taxa, 4 overgrown morphotypes (imperforated *Arkhangelskiella cymbiformis*, *Kamptnerius magnificus*, and *Gartnerago obliquum*, and overgrown *Eiffellithus trabeculatus*, (see Thierstein, 1976)), and isolated shields of *Cretarhabdus*, *Podorhabdus*, and *Zygodiscus* were counted. Of this total of 70 categories, only 45 had maximum relative abundances of over 4% in any one sample. Relative abundance changes within the remaining 35 rare taxa and morphotypes are statistically unreliable at the given sample sizes and were discarded from further consideration. Only 6 of these discarded taxa and morphotypes show relative abundances exceeding 4% in any of the Late Cretaceous and early Tertiary

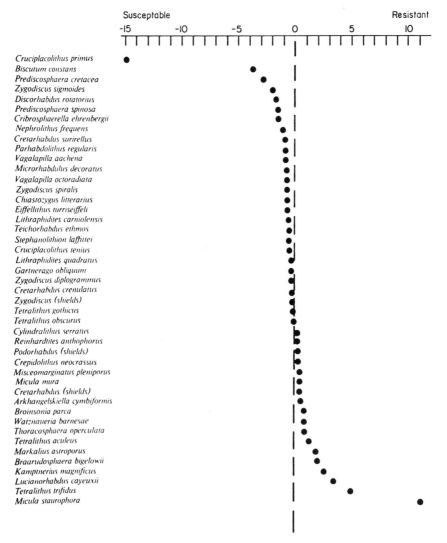

Figure 1. Dissolution ratios of Late Cretaceous and earliest Tertiary calcareous nannofossils.

assemblages studied by Thierstein (in press) and if so, only in a few samples; they are the 3 imperforated morphotypes mentioned above and *Ottavianus giannus*, *Coccolithus pelagicus*, and *Cyclagelosphaera*.

By using the difference in the relative abundance of the taxa for scoring, rather than a point system regardless of the magnitude of change (Berger, 1973), a larger range in the resulting scores is obtained, which reflects the relative extent to which dissolution affects the abundance of individual taxa. Abundant species therefore tend to accumulate higher scores, which are based on larger relative abundance changes and which are therefore of higher statistical reliability. Rare species and those showing minor and therefore statistically less reliable changes in relative abundances, tend to accumulate small absolute scores.

The first numerical column in Table 2 lists the average relative abundance of each taxon in those samples in which that taxon was encountered or should have been encountered, had it not been dissolved. The second column shows the scores resulting from the comparisons. The number of comparisons used to compute the preservation scores is given as pairings in the third column.

The magnitude of the computed score for each taxon depends on the number of comparisons made, which in turn is influenced by the biogeographic and stratigraphic provenance of the tested assemblages. To correct for this bias, the preservation scores were normalized to ratios of scores to pairings (column 4 on Table 2). The number of pairings listed is also a relative measure of the reliability of the computed dissolution ratio.

Dissolution apparently results in large changes in relative abundance of only a few taxa (Figure 1). The extreme suceptibility of *Cruciplacolithus primus*, for example, has been established in three Tertiary assemblages and resulted in a more or less proportional increase in the relative abundances of all other taxa. The sharp

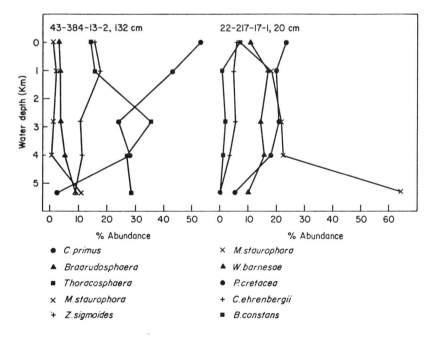

Figure 2. Relative abundance changes of dominant taxa with increasing depth of sample deployment. The earliest Tertiary assemblage 43-384-13-2, 132 cm contained about 12% reworked Cretaceous taxa (e.g. *Micula staurophora*). The Late Maastrichtian assemblage 22-217-17-1, 20 cm shows the strong abundance increase of resistant *Micula staurophora* with progressing dissolution of all other taxa.

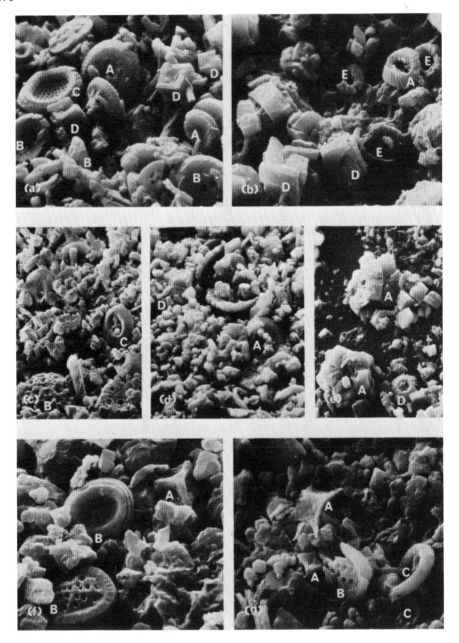

Figure 3. Effects of dissolution on Late Cretaceous and earliest Tertiary nannofossil assemblages.
(a) (c) (d) (f) are from undeployed specimen stub. (b) (e) (g) show the same assemblages after 117
days of exposure at 5500 m water depth. (a) Undeployed, well-preserved, diverse Late Maastrichtian
assemblage (DSDP 22-217-17-1, 20 cm), containing among others: 11.6% *W. barnesae* (A),
23.6% *P. cretacea* (B), 6.0% *C. ehrenbergii* (C), 7.2% *M. staurophora* (D). (b) Same late Maastrichtian assemblage as in (a) (DSDP 22-217-17-1, 20 cm) after 117 days of exposure at 5500 m. This
assemblage shows the highest dissolution index encountered at 5348 m (see also Figures 4 & 5) and
contains 10.0% *W. barnesae* (A), 5.3% *P. cretacea*, 0% *C. ehrenbergii*, 64.3% *M. staurophora*
(D), and isolated shields (E). (c) (d) Undeployed, well-preserved earliest Tertiary assemblage (DSDP 43-384-13-2, 132 cm). This undeployed assemblage shows the lowest dissolution
index encountered (see also Figures 4 & 5) and contains among others: 3.1% *B. bigelowii* (A,
partially covered by debris; note overgrowth), 14.0% *Thoracosphaera* (B), 52.9% *C. primus* (C),
and well-preserved placoliths (D). (e) The same earliest Tertiary assemblage as in (c) (d) (DSDP

increase in the proportion of *Micula staurophora* is caused by dissolution of most of the other moderately susceptible late Cretaceous nannoliths. Examples illustrating these two trends are shown in Figure 2.

The experimentally obtained dissolution ranking clearly shows that small nannoliths and those consisting of small and delicate elements and structures are dominantly dissolution-susceptible, whereas more solidly structured forms composed of larger crystallites rank among the more resistant nannofossils. This is particularly obvious in the size difference of the crystallites constituting the two end-members of the ranking list, *C. primus* and *M. staurophora*, as illustrated in Figure 3. The importance of available surface area on dissolution-precipitation processes of carbonates has also been noted in diagenetic processes and was discussed by Adelseck *et al.* (1973), Neugebauer (1974), and Schlanger & Douglas (1974). Exceptions to this rule are seen in the surprisingly high dissolution resistance of *Braarudosphaera bigelowii* and *Lucianorhabdus cayeuxii*. Pentaliths of *Braarudosphaera*, which consist of delicate, platy crystallites, are observed in Holocene sediments in box-cores from the North Atlantic at depths shallower than 3 km only (unpublished data) which suggests a high dissolution susceptibility for that taxon. This is supported by the fact that ancient deposits rich in *Braarudosphaera* characteristically occur in sediments deposited at times of lowered CCD and in regions of shallow paleodepth (Ramsay, 1971; Roth, 1974; van Andel *et al.*, 1977; Thierstein & Okada, 1979). With progressing diagenesis after burial, secondary calcite is apparently deposited preferentially on pentaliths, resulting in semi-lithified *Braarudosphaera*-chalks which consist of heavily overgrown, cone-shaped elements (Wise & Hsü, 1971; Bukry, 1978; Noël & Melguen, 1978). *Lucianorhabdus cayeuxii* is apparently a holococcolith (Wind & Wise, 1978) consisting of numerous, small (less than 1 μm) crystallites. Due to their dissolution susceptibility holococcoliths are hardly ever found in Holocene oceanic sediments, although they are known to occur commonly in the plankton (Okada & McIntyre, 1977). The apparent time-transgressiveness of *L. cayeuxii's* first appearance and its marginal distribution documented earlier (Thierstein, 1976) may therefore be due to differential preservation in shallow settings. The apparent dissolution resistance of *B. bigelowii* and *L. cayeuxii* in our experiment thus appears to be caused mainly by their susceptibility to diagenetic overgrowth, rather than by high dissolution resistance prior to burial.

4. Dissolution index

The dissolution ranking of the individual taxa can be used to characterize the preservation state of each assemblage (Berger, 1968; Roth & Berger, 1975). In this experiment this was done by computing the weighted mean of all ratios:

$$DI = \text{Sum } (P_i \times R_i)/\text{Sum } (P_i)$$

where P_i = percent abundance of taxon i, and R_i = dissolution ratio of taxon i. The change in the average dissolution index (DI) of all 13 assemblages with increasing depth of deployment as well as other indicators of carbonate dissolution are shown in Figure 4.

43-384-13-2, 132 cm) after 117 days of exposure at 5500 m. The sample exposed at 5348 m depth contained: 9.0% *B. bigelowii* (A), 28.2% *Thoracosphaera*, 2.6% *C. primus*, and strongly etched placolith shields (D). (f) Undeployed well-preserved middle Maastrichtian assemblage (Texas A 23-4) contains among others: 19.0% *M. staurophora* (A), 7.9% *A. cymbiformis* (B). (g) The same middle Maastrichtian assemblage as in (f) (Texas A 23-4) after 117 days of exposure at 5500 m. This sample shows the smallest dissolution index versus depth gradient. The assemblage exposed at 5348 m contained: 19.4% *M. staurophora* (A), 2.6% *A. cymbiformis* (B), and 16.2% isolated shields (C). Magnifications: (e) × 1200, all others approximately × 2300.

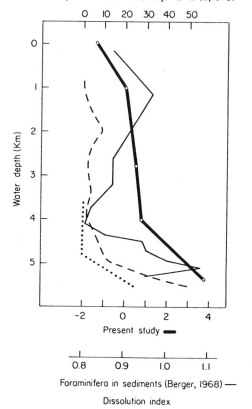

% Weight loss

Suspended Foraminifera (Milliman, 1977) – – –
Suspended Coccoliths (Honjo & Erez, 1978) ·····

Present study ━━

Foraminifera in sediments (Berger, 1968) —

Dissolution index

Figure 4. Indicators of carbonate dissolution in present-day North Atlantic. Thick solid line: average dissolution index of 13 Late Cretaceous and earliest Tertiary nannofossil assemblages deployed at 1039, 2838, 4030, and 5348 m water depth and in original fine fraction. Thin solid line: dissolution indices for equatorial Atlantic Foraminifera assemblages calculated by Berger (1968). Dashed line: percent annual weight loss of planktonic Foraminifera calcite in Sargasso Sea at Station 548 (Milliman, 1977). Dotted line: percent weight loss of bleached coccoliths of *Emiliania huxleyi* during 79 days in Sargasso Sea (Honjo & Erez, 1978).

The shape of the curve compares well with the dissolution index curve computed by Berger (1968) based on Schott's (1935) Foraminifera assemblage data from surface sediments in the equatorial Atlantic. Dissolution experiments have been carried out previously in the Sargasso Sea by Milliman (1977) and Honjo & Erez (1978). Milliman deployed planktonic Foraminifera, which were placed in cloth sacks and hung from the top of a porous container. At station 548, nearest to our experiment, he observed annual weight losses of up to 9% at depths shallower that 4 km, which increased to 46% at 5.5 km depth. Honjo & Erez (1978) reported on an additional dissolution experiment using *In-Situ Water Circulators* (ISWAC) at 3600 m, 4800 m and 5518 m. After 79 days of deployment at the Parflux station, they did not find any weight loss down to over 4800 m on bleached and unbleached coccoliths and Foraminifera and reagent calcite. Below that depth weight loss increased drastically, suggesting complete dissolution of most carbonate particles in 1–2 years. A few months later the experiment, on which this report is based, took place about 0.7 nautical miles to the SSW of Parflux station. The clearly recognizable increase in dissolution of ancient nannoliths between 1039 m and 4030 m water depth after 117 days is at variance with Honjo & Erez's (1978) earlier findings and agrees more closely with dissolution profiles given by Berger (1968) and Milliman (1977). The dissolution index at 0 km reflects the average preservation of the assemblages prior to their mounting on specimen stubs and deployment and therefore cannot be directly compared with other field data. The differences in the gradient of dissolution indicators obtained by various methods may be related to spatial and temporal variability of the carbonate saturation. The carbonate ion concentrations measured in the North Atlantic by Broecker & Takahashi (1978) varied at any particular depth over a range of about 10 μM kg^{-1}. Taking the analytical precision of their measurements into account, this range is equivalent to the gradient of the average slope over a depth of about 1.0 km, which thus would be equal to the possible variability of the depth position of the lysocline.

The dissolution indices calculated for the individual samples vary considerably as shown in Figure 5 and reflect differences in both the initial preservation state of the assemblages and their original taxonomic composition. The undeployed sample of assemblage 43-384-13-2, 132 cm has the lowest dissolution index among all of the 65 samples analyzed and this assemblage also shows the largest depth gradient in its dissolution index. In contrast, assemblage A23-4 from Texas shows a relatively high dissolution index in the undeployed sample, despite the good preservation evidenced in its ultrastructure, and a small depth gradient, despite clear evidence of dissolution in the ultra-structure of individual taxa [Figure 3(f)(g)]. This assemblage's insensitivity to dissolution most likely lies in its taxonomic composition, which is related to its biogeographic provenance and is characterized by dominance of relatively dissolution resistant taxa (e.g. *Watznaueria barnesae*, *Micula staurophora*) and absence of dissolution susceptible species.

Indexing of ancient nannofossil assemblages using changes in species frequencies

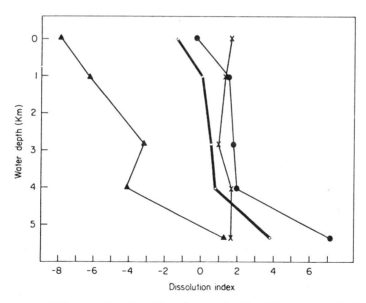

Figure 5. Dissolution indices of selected late Cretaceous and earliest Tertiary nannofossil assemblages. Sample 43-384-13-2, 132 cm has the lowest dissolution index in the undeployed assemblage and the largest range of dissolution indices among the 13 assemblages analyzed. Sample 22-217-17-1, 20 cm has the highest dissolution index encountered (at 5348 m). Compare also with Figure 2. Sample A23-4 (Corsicana clay, Texas) shows the smallest range of dissolution indices among the 13 assemblages analyzed. — Average dissolution index of 13 assemblages, ▲ DSDP 43-384-13-2, 132 cm, ● DSDP 22-217-17-1, 20 cm, × A 23-4, Corsicana clay.

to characterize their preservation state therefore does not appear to be a very sensitive tool. A similarly small and noisy nannoplankton dissolution index versus depth gradient has been found in recent coccolith assemblages in the Pacific (see Roth & Berger, 1975, Figures 20, 21.)

5. Morphologic effects of dissolution

The preservation of nannofossil assemblages as observed in the light microscope has been usefully characterized on a numbered scale following a proposal by Roth & Thierstein (1972). The utilized morphological etching features, such as development of serrate margins on coccolith shields, erosion and loss of central area structures, and isolation of proximal and distal shields, are manifest in the deployed assemblages (Figure 3). Enrichment of resistant and disappearance of susceptible taxa as determined by quantitative taxonomic evaluation of smear-slides under the light microscope can also be readily recognized under the scanning electron microscope.

6. Discussion

The major result of the experimental study is that noticeable effects of carbonate dissolution on the taxonomic compositions of Late Cretaceous and earliest Tertiary nannofossil assemblages are limited to a small number of taxa. This knowledge will greatly facilitate recognition of preservational bias in stratigraphic and biogeographic abundance patterns of nannofossil assemblages. Dominance of mainly *Cruciplacolithus primus*, and to a much lesser degree (by a factor of over 3 less) of very few other susceptible taxa, characterizes exceptionally well preserved early Tertiary assemblages. The dominance of *Cruciplacolithus primus* in earliest Danian sediments at DSDP Sites 384 and 356 in the deep North and South Atlantic, therefore, supports the postulated lowering of the CCD to abyssal depths across the Cretaceous–Tertiary boundary (Thierstein & Okada, 1979; Thierstein, 1979). The appearance of this species at the very beginning of the newly evolving Tertiary nannoplankton community in these exceptionally well-preserved assemblages from the deep sea supports the possibility that the earliest Tertiary *Markalius astroporus* Interval Zone (NP1 of Martini, 1971) is due to a widespread dissolution pulse observed in epicontinental and shallow oceanic sections (Worsley, 1974), rather than to an evolutionary sequence (Thierstein & Okada, 1979). With progressing dissolution, most of the Cretaceous taxa decrease rather evenly in relative abundance in favor of particularly resistant *Micula staurophora* and to a much lesser extent of a few other taxa. The paleogeographic distribution patterns of these few resistant taxa therefore will have to be analyzed carefully with respect to their ecological significance. *Micula staurophora*, for instance, has been reported as a major constituent of tropical and subtropical Late Cretaceous assemblages (Thierstein & Haq, 1977; Wind, 1979), but its distribution may in fact be the result of latitudinal preservation patterns possibly related to fertility.

Two alternative methods to evaluate the taxonomic bias introduced by differential preservation in Late Cretaceous and earliest Tertiary nannofossil assemblages were used for comparison with the results obtained from the experiment. Fossil assemblages from different paleo-depths were compared for taxonomic abundance changes and yielded a slightly different dissolution ranking. That result, however, was judged to be inferior because of the unknown original evolutionary and biogeographic variability between the assemblages. An attempt to rank taxa with respect to their susceptibility to overgrowth, using burial (sub-bottom) depth as a critical parameter, yielded a ranking which was intermediate, but still very similar to the other rankings. This suggests, not surprisingly, that with enhanced diagenesis overgrowth of resistant taxa proceeds simultaneously with dissolution of susceptible taxa, i.e. that small crystallites are the carbonate source for the overgrowth of larger crystallites.

Acknowledgments

Dr S. Honjo (Woods Hole Oceanographic Institution) generously mounted and deployed the samples during his deep-mooring experiment in Fall 1977. Most samples were obtained from the Deep-Sea Drilling Project, which is funded by the National Science Foundation. I thank Dr W. H. Berger & Dr P. H. Roth for their critical review of an earlier draft of the manuscript. Research supported through NSF Grant OCE76-22150.

References

Adelseck, C. G., Geehan, G. W. & Roth, P. R. 1973. Experimental evidence for the selective dissolution. and overgrowth of calcareous nannofossils during diagenesis. *Geological Society of America Bulletin* **84**, 2755–2762.

Berger, W. H. 1967. Foraminiferal ooze: solution at depths. *Science* **156**, 383–385.

Berger, W. H. 1968. Planktonic Foraminifera: selective solution and paleoclimatic interpretation. *Deep-Sea Research* **15**, 31–43.

Berger, W. H. 1973. Deep-sea carbonates—evidence for a coccolith lysocline. *Deep-Sea Research* **20**, 917–921.

Broecker, W. S. & Takahashi, T. 1978. The relationship between lysocline depth and *in situ* carbonate ion concentration. *Deep-Sea Research* **25**, 65–95.

Bukry, D. 1978. Cenozoic silicoflagellate and coccolith stratigraphy, southeastern Atlantic Ocean, Deep-Sea Drilling Project (Leg 40). In *Initial Reports of the Deep-Sea Drilling Project*, (Eds. H. M. Bolli, W. B. F. Ryan, *et al.*), Vol. 40, Washington: U.S. Government Printing Office. 635–650.

Haq, B. U. & Lohmann, G. P. 1976. Early Cenozoic calcareous nannoplankton biogeography of the Atlantic Ocean. *Marine Micropaleontology* **1**, 119–134.

Honjo, S. & Erez, J. 1978. Dissolution rates of calcium carbonate in the deep ocean; an *in situ* experiment in the North Atlantic Ocean. *Earth & Planetary Science Letters* **40**, 287–300.

Hutson, W. H. 1977. Transfer functions under no-analog conditions: Experiments with Indian Ocean planktonic Foraminifera. *Quaternary Research* **8**, 355–367.

Loeblich, A. R. & Tappan, H. 1966. Annotated index and bibliography of the calcareous nannoplankton. *Phycologia* **5**, 81–216.

Loeblich, A. R. & Tappan, H. 1968. Annotated index and bibliography of the calcareous nannoplankton II. *Journal of Paleontology* **42**, 584–598.

Loeblich, A. R. & Tappan, H. 1969. Annotated index and bibliography of the calcareous nannoplankton III. *Journal of Paleontology* **43**, 568–588.

Loeblich, A. R. & Tappan, H. 1970a. Annotated index and bibliography of the calcareous nannoplankton IV. *Journal of Paleontology* **44**, 558–574.

Loeblich, A. R. & Tappan, H. 1970b. Annotated index and bibliography of the calcareous nannoplankton V. *Phycologia* **9**, 157–174.

Loeblich, A. R. & Tappan H. 1971. Annotated index and bibliography of the calcareous nannoplankton VI. *Phycologia* **10**, 315–339.

Loeblich, A. R. & Tappan, H. 1973. Annotated index and bibliography of the calcareous nannoplankton VII. *Journal of Paleontology* **47**, 715–759.

Milliman, J. D. 1977. Dissolution of calcium carbonate in the Sargasso Sea (Northwest Atlantic). In *The Fate of Fossil Fuel CO_2 in the Oceans* (Eds. N. R. Andersen & A. Malahoff). New York: Plenum Press. 641–654.

Neugebauer, J. 1974. Some aspects of cementation in chalk. In *Pelagic Sediments: On Land and Under the Sea* (Eds K. J. Hsü & H. C. Jenkyns). Special Publication of the International Association of Sedimentologists. Vol. 1. 149–176.

Noël, D. & Melguen, M. 1978. Nannofacies of Cape Basin and Walvis Ridge sediments, Lower Cretaceous to Pliocene (Leg 40). In *Initial Reports of the Deep-Sea Drilling Project* (Eds H. M. Bolli & W. B. F. Ryan *et al.*) Vol. 40. Washington: U.S. Government Printing Office. 487–524.

Okada, H. & McIntyre, A. 1977. Modern coccolithophores of the Pacific and North Atlantic Oceans. *Micropaleontology* **23**, 1–55.

Parker, F. L. & Berger, W. H. 1971. Faunal and solution patterns of planktonic Foraminifera in surface sediments of the South Pacific. *Deep-Sea Research* **18**, 73–107.

Perch-Nielsen, K. 1977. Albian to Pleistocene calcareous nannofossils from the western South Atlantic, DSDP (Leg 39). In *Initial Reports of the Deep-Sea Drilling Project* (Eds P. R. Supko & K. Perch-Nielsen *et al.*) Vol. 39. Washington: U.S. Government Printing Office. 1099–1132.

Ramsay, A. T. S. 1971. The investigation of Lower Tertiary sediments from the North Atlantic. In *Proceedings II Planktonic Conference Rome II* (Ed. A. Farinacci). Rome: Edizioni Technoscienza. 1039–1055.

Risatti, J. B. 1973. Nannoplankton biostratigraphy of the Upper Bluffport Marl—Lower Prairie Bluff chalk interval (Upper Cretaceous) in Mississippi. In *Proceedings Symposium Calcareous Nannofossils* (Eds. L. A. Smith & J. L. Hardenbol). Houston: Society of Economic Paleontologists and Mineralogists. 8–57.

Roth, P. H. 1974. Calcareous nannofossils from the northwestern Indian Ocean, Deep Sea Drilling Project, Leg 24. In *Initial Reports of the Deep-Sea Drilling Project* (Eds. R. L. Fisher, E. T. Bunce *et al.*) Vol. 24. Washington: U.S. Government Printing Office. 969–994.

Roth, P. H. & Berger, W. H. 1975. Distribution and dissolution of coccoliths in the South and Central Pacific. In *Cushman Foundation Foraminifera Research Special Publication* (Eds W. V. *No. 13* Sliter, A. W. H. Bé, & W. H. Berger) 87–113.

Roth, P. H. & Thierstein, H. R. 1972. Calcareous nannoplankton: Leg 14 of the DSDP. In *Initial Reports of the Deep-Sea Drilling Project.* (Eds. D. E. Hayes, A. C. Pimm *et al.*) Vol. 14. Washington: U.S. Government Printing Office. 421–485.

Ruddiman, W. F. & Heezen, B. C. 1967. Differential solution of planktonic Foraminifera. *Deep-Sea Research* **14**, 801–808.

Schlanger, S. O. & Douglas, R. G. 1974. The pelagic ooze-chalk-limestone transition and its implications for marine stratigraphy. In *Pelagic Sediments: On Land and Under the Sea* (Eds K. J. Hsü & H. C. Jenkyns). Special Publication of the International Association of Sedimentologists, Vol. 1. 117–148.

Schott, W. 1935. Die Foraminiferen in dem Äquatorialen Teil des Atlantischen Ozeans. *Deutsche Atlantische Expedition Meteor 1925–1927* **3**, 43–134.

Thierstein, H. R. 1976. Mesozoic calcareous nannoplankton biostratigraphy of marine sediments. *Marine Micropaleontology* **1**, 325–362.

Thierstein, H. R. 1979. Paleoceanographic implications of organic carbon and carbonate distribution in Mesozoic deep-sea sediments. In *Deep Drilling Results in the Atlantic Ocean: Continental Margins and Paleoenvironment* (Eds. M. Talwani *et al.*). Washington: American Geophysical Union. 249–274.

Thierstein, H. R. In press. Late Cretaceous calcareous nannoplankton and the change at the Cretaceous–Tertiary boundary. In *The Deep-Sea Drilling Project: A Decade of Progress* (Eds. R. G. Douglas & E. L. Winterer). Society of Economic Paleontologists & Mineralogists Special Publication.

Thierstein, H. R. & Haq, B. U. 1977. Maestrichtian/Danian biogeographic variations in calcareous nannoplankton (abstract). *Journal of Paleontology* **51**, (supplement number 2) 28.

Thierstein, H. R. & Okada, H. 1979. The Cretaceous/Tertiary boundary event in the North Atlantic. In *Initial Reports of the Deep-Sea Drilling Project* (Eds. B. E. Tucholke, P. R. Vogt *et al.*) Vol. 43. Washington: U.S. Government Printing Office. 601–616.

Thompson, P. R. 1976. Planktonic foraminiferal dissolution and the progress towards a Pleistocene equatorial Pacific transfer function. *Journal of Foraminiferal Research* **6**, 208–227.

Van Andel, T. H., Thiede, J., Sclater, J. G. & Hay, W. W. 1977. Depositional history of the South Atlantic Ocean during the last 125 million years. *Journal of Geology* **85**, 651–698.

Wind, F. H. 1979. Maestrichtian–Campanian nannofloral provinces of the southern Atlantic and Indian Oceans. In *Deep Drilling Results in the Atlantic Ocean: Continental Margins and Paleoenvironment* (Eds M. Talwani *et al.*). Washington: American Geophysical Union. 123–137.

Wind, F. H. & Wise, S. W. 1978. Mesozoic holococcoliths. *Geology* **6**, 140–142.

Wise, S. W., Jr. & Hsü, K. J. 1971. Genesis and lithification of a deep-sea chalk. *Eclogae Geologicae Helvetiae* **64**, 157–163.

Wise, S. W., Jr. & Wind, F. H. 1977. Mesozoic and Cenozoic calcareous nannofossils recovered by DSDP Leg 36 drilling on the Falkland Plateau, southwest Atlantic sector of the Southern Ocean. In *Initial Reports of the Deep-Sea Drilling Project* (Eds. P. F. Barker, J. W. D. Dalziel *et al.*) Vol. 36. Washington: U.S. Government Printing Office. 269–491.

Worsley, T. 1974. The Cretaceous–Tertiary boundary event in the ocean. In *Studies in Paleoceanography* (Ed. W. W. Hay). Society of Economic Paleontologists and Mineralogists, Special Publication No. 20. 94–125.

15

Copyright ©1972 by the Gulf Coast Association of Geological Societies
Reprinted From *Gulf Coast Assoc. Geol. Socs. Trans.* **22**:177 (1972)

INFERRED DIAGENETIC HISTORY OF A WEAKLY SILICIFIED DEEP SEA CHALK

Sherwood W. Wise, Jr.[1], Tallahassee, Florida 32306 and Kerry R. Kelts[2], Zurich, Switzerland

ABSTRACT

At various times during the Oligocene, profuse blooms of the planktonic calcareous algae, *Braarudosphaera rosa*, contributed large numbers of braarudosphaerid pentaliths to deep ocean sediments of the South Atlantic Ocean Basin. During such intervals the carbonate compensation level for *B. rosa* was considerably depressed; nevertheless, many of the pentaliths were disaggregated into wedge-shaped segments as a result of solution. Shortly after deposition, skeletal calcite liberated by dissolution was reprecipitated as low magnesium calcite overgrowths on discoasters, coccoliths, pentalith segments, and minute particles of skeletal debris. As shown by scanning electron micrographs, extensive development of the secondary overgrowths led to the formation of low magnesium chalk laminae (up to 1 meter in thickness) within the otherwise unconsolidated Oligocene ooze sequence. Paleo-oceanographic conditions rather than absolute sediment age or depth of burial were responsible for the submarine lithification of the chalk laminae. Cementation as a result of calcite overgrowth did not reduce significantly the porosity of the chalk. It is possible that this mode of cementation during early diagenesis could protect the rock from compaction and pore closure during late diagenesis.

At one locality (Rio Grande Rise, DSDP Site 22, Sample 22/4/1), calcite cementation was followed by the deposition of silica derived from the dissolution of silicious microfossils and volcanic glass. The silica was reprecipitated as minute spherules (here named "lepispheres") of alpha-cristobalite which partially filled the interstices of the rock and are responsible for the weakly silicified condition of portions of the chalk. Clinoptilolite, a second authigenic silicate in the chalk, is readily distinguished in scanning electron micrographs from detrital quartz and mica also present in insoluble residues of the rock.

Abundant lepispheres are also present in weakly silicified chalk lenses sampled within the Eocene Horizon A radiolarian chart sequence at DSDP 29B. These results indicate that cristobalite lepispheres represent the initial stage of silicification of carbonate rock in the deep sea environment.

---

[1] Department of Geology, Florida State University
[2] Department of Geology, Swiss Federal Institute of Technology

16

Copyright ©1978 by the Société Géologique de France

Reprinted from *Soc. Géol. France Bull.* **20**:491-502 (1978)

Nannofaciès de « black shales » aptiennes et albiennes d'Atlantique sud (legs 36 et 40). Intérêt sédimentologique

par Denise NOËL * et Hélène MANIVIT **

Mots clés. — Flore, Coccolithes, Aptien, Albien, Argile, Matière organique, Sédimentation confinée, Sédimentation marine.

Océan Atlantique Sud (Leg 36, Site 330, 327, Leg 40, Site 361, 364).

Résumé. — L'analyse du nannofaciès de 65 échantillons de « black shales » aptiennes et albiennes du plateau de Falkland, du bassin du Cap et de la marge de l'Angola (Atlantique sud) apporte certaines données sur les conditions de sédimentation :
— la présence de nannoplancton témoigne d'influences marines incontestables, d'une certaine productivité des eaux de surface et de communications avec une mer ouverte ;
— le mode de fossilisation de ce nannoplancton (restes de coccosphères) indique des conditions de dépôt stagnantes, confinées, qui sont confirmées par la présence quasi constante de framboïdes de pyrite ;
— les effets de la dissolution, pouvant aller jusqu'à la disparition des corpuscules dont il ne subsiste plus qu'une empreinte, montrent que ces processus ont eu lieu dans le sédiment et non au cours de la descente dans la colonne d'eau. Il ne faut donc pas en tirer d'arguments relatifs à une grande profondeur du dépôt.

Abstract. — The nannofacies analysis of 65 samples of aptian and albian black shales from the Falkland Plateau, the Cape Basin and the Continental margin off Angola (Southern Atlantic) provides some data on depositional environment :
— the occurrence of nannoplancton shows undeniable marine conditions, some productivity of the surface water and connections with the oceanic seas ;
— the mode of fossilization (coccosphere remains) indicates stagnant and confined depositional conditions which are confirmed by occurrence of pyrite framboïds ;
— the dissolution effects which can lead to disparition of coccoliths which only remain as casts, show that dissolution processes took place in the sediment and not in the water column. Therefore, we are not allowed to deduce of them a great depth of deposition.

I. — INTRODUCTION : LE CADRE DU MATÉRIEL ÉTUDIÉ.

Les sédiments aptiens et albiens étudiés proviennent de quatre sites de forage du « *Glomar Challenger* » implantés dans l'Atlantique sud au cours des legs 36 et 40. Des études assez complètes d'ordre sédimentologique, géochimique, paléontologique, etc. ont été réalisées sur l'ensemble de ces sites de forage et ont permis, en particulier, de donner une première datation de ces faciès ainsi qu'une première description pétrographique [Initial Reports of the Deep Sea Drilling Project, vol. 36 et 40]. Des données générales sur l'extension des faciès *black shales* (ainsi que des hypothèses sur les mécanismes de cette sédimentation ont été données par Thiede et Van Andel [1977]). Dans le présent travail nous mentionnerons simplement les grandes unités rencontrées dans ces forages afin de resituer dans leur contexte les niveaux étudiés. Ce sont, en considérant les sites de forage du Sud vers le Nord :

Site 330 (leg 36) [D.S.D.P., vol. 36, p. 207-257], sur la partie orientale du plateau de Falkland, par une profondeur d'eau de 2 626 m. Le forage a traversé des séries s'échelonnant de l'Albien supérieur au socle, atteint à — 556 m (et traversé sur 19,5 m). Chronologiquement, les séries sont les suivantes :
— au-dessus du socle de gneiss granulitique, on observe un sol fossile puis des grès avec passées

---

* Lab. de géologie du Muséum, 43, rue de Buffon, 75005 Paris. E.R.A. 121, Pétrologie des roches calcaires ; R.C.P. 459, Nature et genèse des faciès confinés.

** Bureau de recherches géologiques et minières, Lab. de micropaléontologie, 45018 Orléans ; R.C.P. 459.

Note déposée le 1er février 1978, présentée le 12 décembre 1977, manuscrit définitif reçu le 5 mai 1978.

ligniteuses (16 m), sans fossiles marins, considérés [Thompson, 1976] comme des sédiments déposés en milieu marécageux subaérien ;

— puis viennent 115 m de silts et argiles terrigènes, avec débris de plantes carbonisées ; la présence de restes de Bélemmites, de restes de Pélécypodes, de rares coccolithes et Foraminifères benthiques atteste l'origine marine de ces dépôts que des analyses polliniques ont daté du Jurassique moyen à supérieur (Oxfordien) ;

— au-dessus 225 m d'argiles noires, sapropéliques, avec des passées de calcaires micritiques ont été datées du Jurassique supérieur (Oxfordien) à Aptien ;

— ces « *black shales* » sont surmontées par une centaine de mètres de boues à nannofossiles et de craie d'âge albien inférieur et moyen dans lesquelles a démarré le carottage.

Site 327 (leg 36) [D.S.D.P., vol. 36, p. 27-86], à une dizaine de kilomètres à l'Ouest du site 330, par une profondeur d'eau de 2 400 m. Le forage a en effet été implanté sur le rebord occidental de la partie surélevée du plateau de Falkland (sur le Maurice Ewing bank). Les séries traversées s'échelonnent du Pléistocène au Néocomien. Il y a donc recouvrement avec le forage précédent (du Néocomien à l'Albien), les forages 330 et 327 donnant à eux deux une section stratigraphique complète du plateau de Falkland. Pour l'intervalle qui nous intéresse dans la présente étude, les séries se présentent ainsi :

— 173 m d'argiles sapropéliques noires d'âge néocomien à aptien supérieur ;

— surmontées de 170 m d'argiles indurées et de boues à nannofossiles albiennes ;

— puis la série crétacée se poursuit avec d'importants hiatus.

Site 361 (leg 40) [D.S.D.P. vol. 40, sous presse], dans le bassin du Cap (à 180 miles à l'W-S-W de la ville du Cap), limité au Nord par la Walvis Ridge. Le forage a été effectué par une profondeur d'eau de 4 549 m. Les séries traversées s'échelonnent du Pléistocène inférieur à l'Aptien inférieur. Dans l'ordre stratigraphique on observe :

— environ 400 m d'argiles noires bitumineuses avec des intercalations silteuses noires qui représentent l'intervalle aptien inférieur-albien moyen supérieur ;

— les « *black shales* » sont surmontées de 540 m d'argiles rouges ou vertes, d'âge crétacé indifférencié.

Site 364 (leg 40) [D.S.D.P., vol. 40, sous presse], sur la marge continentale de l'Angola, au Nord de la Walvis Ridge, par une profondeur d'eau de 2 448 m. Les séries traversées s'échelonnent du Pléistocène à l'Aptien supérieur. Le forage aurait été stoppé

à quelques dizaines de mètres seulement du sel sous-jacent. Dans l'ordre stratigraphique les séries mésozoïques se présentent ainsi :

— 119 m de calcaires marneux dolomitiques avec des intercalations d'argiles sapropéliques noires finement laminées, d'âge Aptien supérieur à Albien inférieur ;

— 257 m de calcaires marneux albiens avec au sommet des récurrences de faciès « *black shales* » qui se poursuivent sporadiquement jusqu'au Santonien.

Dans ces quatre forages 65 échantillons (tab. I) des niveaux aptiens et albiens ont été étudiés en microscopie à balayage, suivant une méthode déjà décrite [Noël, 1970]. Les résultats sont exposés stratigraphiquement.

II. — L'APTIEN.

1) *Plateau de Falkland* (leg 36, sites 330 et 327).

Les « *black shales* » de ces deux sites sont des argiles généralement bien laminées, parfois fissiles, contenant en moyenne 3 à 4 % de carbone organique. Des valeurs de 6 % ont été trouvées dans les niveaux de l'Aptien supérieur. La plus grande partie de ce carbone organique consiste en un matériel amorphe, brun-jaunâtre, probablement d'origine planctonique [Thompson, 1976]. La teneur en carbonate de calcium est peu élevée (quelques % sauf au niveau des passées de calcaires micritiques). Le nannofaciès de ces « *black shales* » présente les caractères suivants :

a) *Du point de vue des restes organiques.* Les associations de nannofossiles rencontrés indiquent un âge aptien, correspondant à la zone à *Chiastozygus litterarius* pour le site 330. Pour le site 327 les zones à *Chiastozygus litterarius* et à *Parhabdolithus angustus* ont été reconnues, cette dernière zone n'ayant pas été traversée dans le site 330 à cause d'importants hiatus et d'un carottage discontinu. Les nannofossiles calcaires s'observent dans pratiquement tous les échantillons avec une densité moyenne. Seuls les échantillons 327-22-3-82-83 cm et 327-23-2-21-22 cm n'en ont pas livré à l'observation au M.E.B. Ces nannofossiles se présentent sous des aspects divers qui appellent quelques remarques.

— Ils se rencontrent souvent en *amas de coccolithes appartenant à une même espèce* (pl. I ; fig. 1), ces amas étant attribués à différentes espèces suivant les champs d'observation du sédiment. De tels amas représentent des restes de coccosphères disloquées sur place. On sait en effet qu'en fonction de la morphologie des coccolithes qui recouvrent une cellule, la coccosphère a une « cohésion » plus ou moins grande, suivant que les coccolithes voisins se chevauchent

Age	leg 36 – site 330	leg 36 – site 327	leg 40 – site 361	leg 40 – site 364
ALBIEN (zone à *Prediscosphaera cretacea)*	1-3-141-142 cm 1-5-38-39 cm 2-2-25-26 cm 2-2-45-46 cm	21-2-52-53 cm 21-3-121 cm	26-2-139-141 cm 26-6-139-141 cm 27-2-107-109 cm 28-3-98-99 cm 28-6-49-51 cm 29-4-92-94 cm 29-5-129-131 cm 30-2-57-59 cm 31-3-59-61 cm 31-3-74-75 cm 32-1-24-25 cm 32-3-119-121 cm 33-3-31-32 cm	31-1-24-26 cm 35-3-82-84 cm 36-2-38-39 cm 37-5-126-128 cm 37-4-72-73 cm 38-6-144-145 cm 38-4-137-139 cm 39-1-140-141 cm 39-6-87-88 cm 40-2-25-27 cm 40-5-49-51 cm 41-2-134-136 cm
APTIEN (zone à *Parhabdolithus angustus)*	carottage discontinu et hiatus pas d'échantillon de cette zone	22-2-56-57 cm 22-3-135-136 cm 22-3-82-83 cm 22-3-142-143 cm 23-2-21-22 cm 23-2-25-26 cm 23-2 - 105 cm	33-3-75-76 cm 33-3-76-77 cm	41-4-34-36 cm 41-2-138 cm 42-2-9-10 cm 42-6-131-132 cm 42-5-10-11 cm 43-3-100-102 cm 44-2-3-5 cm 45-2-109-111 cm
APTIEN (zone à *Chiastozygus litterarius)*	3-2-49-50 cm 3-2-70-71 cm 3-2-134-135 cm	24-1-44-65 cm 25-2-38-39 cm 25-2-88-89 cm 25-2-102-103 cm	43-3-40-41 cm 43-3-69-70 cm 48-2-142-144 cm	

FIG. 1. — Tableau des échantillons étudiés.

Pour la commodité d'expression dans le tableau, les échantillons ont été mis en colonnes juxtaposées. Mais la position sur une même ligne horizontale d'échantillons provenant de chacun des sites étudiés n'indique nullement une équivalence d'âge.

les uns les autres, ou sont seulement juxtaposés. Dans ce dernier cas, c'est la couche externe gélatineuse de la membrane cellulaire, couche externe dans laquelle sont inclus les coccolithes, qui joue en quelque sorte le rôle de ciment et assure du vivant de l'Algue la « cohésion » de la coccosphère. Mais il est évident qu'à la mort de l'Algue, de tels types de coccosphères sont dissociés et ne se retrouvent jamais intacts dans les sédiments, les coccolithes étant dispersés. Or, un mode de fossilisation courant dans les *black shales* étudiées, de même que dans de nombreux sédiments laminés [Noël et Busson, 1972 ; Noël, 1972], livre à l'observation des amas de coccolithes appartenant à une même espèce et

provenant de coccosphères disloquées sur place. Certains auteurs [Hattin, 1975 ; Honjo, 1975 ; Roth *et al.*, 1975] ont souligné le rôle joué par les Copépodes dans la formation d'agrégats de coccolithes. Ils ont démontré que ces Copépodes herbivores, broutant le nannoplancton, contribuent par leurs pelotes fécales à la sédimentation carbonatée. Mais cette interprétation ne peut s'appliquer aux amas observés dans les *black shales* sauf à imaginer des Copépodes adaptés tour à tour à une seule espèce.

Ces restes de coccosphères disloquées sur place fournissent des données intéressantes sur le milieu de sédimentation. On peut en effet en déduire : 1) l'absence de courants qui auraient vanné les sédiments et dispersé les coccolithes provenant d'une même cellule ; 2) l'absence de vie de fond qui aurait eu le même effet.

— On observe parfois également (pl. I, fig. 2) des *plages monospécifiques* qui correspondent sans doute à de véritables blooms d'une espèce et qui sont un indice de conditions restreintes.

— Outre ces coccolithes dont l'état de conservation est en général bon, le nannofaciès des « *black shales* » étudiées est également caractérisé par de fréquentes *empreintes* (pl. I, fig. 3). Ces empreintes, très nettes et très fidèles, témoignent de l'extrême finesse de la matrice argilo-bitumineuse du sédiment. De plus, ces dernières prouvent que s'il y a bien eu dissolution des nannofossiles calcaires, cette dissolution s'est produite *après* le dépôt du corpuscule et au sein du sédiment. L'existence d'une empreinte suppose en effet le jeu d'une compaction minimale et la dissolution opérée après cette compaction ne peut être imputable qu'aux eaux interstielles du sédiment. Cette dissolution plus ou moins totale (pl. I, fig. 4-5) s'est donc produite *in situ*, elle représente un phénomène diagénétique qui peut aller jusqu'à la disparition complète du coccolithe. Et toute dissolution des organismes calcaires ne doit donc pas être considérée *ipso facto* comme une dissolution dans une tranche d'eau importante, ce qui établirait une grande profondeur de dépôt.

A ces caractéristiques générales : amas de coccolithes appartenant à une même espèce, plages monospécifiques, fréquentes empreintes, il faut ajouter, à certains niveaux du site 330 la présence de *Braarudosphaera* et de *Micrantholithus* (pl. II, fig. 1). Ces pentalithes, le plus souvent dissociés en leurs éléments constitutifs, ont été observés dans des passées calcaires, gris clair, alternant avec des « *black shales* » (carotte 3, section 2, entre 70 et 49 cm). Se fondant sur l'écologie de *Braarudosphaera bigelowi* qui vit encore dans les océans actuels, certains auteurs ont voulu faire de la présence de *Braarudosphaera* dans les sédiments des indices de dessalure [Wise et Hsü, 1971 ; Bukry, 1974]. Cependant, cette

hypothèse ne rend pas compte de toutes les données de terrain, en particulier pour les niveaux oligocènes de l'Atlantique sud [Noël et Melguen, 1978].

b) *Du point de vue des minéraux.*

● *La fraction carbonatée.* Dans ces « *black shales* », la teneur en CO_3Ca est primordialement liée à la quantité de nannofossiles existant dans le sédiment.

● *La fraction argileuse*, plus ou moins importante suivant les niveaux, ne descend pas au-dessous de 50 %. Elle donne au sédiment une structure esquilleuse, à bords soulevés (pl. II ; fig. 2). Des études [R. W. Thompson, 1976] ont montré que cette fraction argileuse est dominée par l'illite et la montmorillonite, minéraux qu'il n'est pas possible de déterminer au M.E.B. où l'on ne dispose que de la morphologie des éléments. Mais sur les nannofaciès la teneur en argile du sédiment se manifeste clairement dans l'état de « propreté » variable des nannofossiles, fortement encroûtés dans les niveaux très argileux, beaucoup plus « propres » dans les lits plus calcaires (pl. II ; fig. 1).

● Autre composant habituel de ces « *black shales* », la *pyrite* observée dans la plupart des échantillons. Ce sulfure de fer — responsable de la couleur sombre des sédiments — se rencontre soit sous forme de framboïdes de cristaux octaédriques (pl. II, fig. 2), souvent nichés dans de minuscules fissures du sédiment, soit sous forme de tapissages faits de cristaux légèrement plus gros que ceux des framboïdes (pl. I ; fig. 3 et pl. III ; fig. 2). L'origine et les conditions de formation de ces framboïdes ont fait l'objet de diverses hypothèses allant de l'action de colonies de Thiobactéries [Fabricius, 1961] à de simples cristallisations inorganiques [Rust, 1935 ; Bastin, 1950]. Pour Honjo *et al.* [1965], il ne fait pas de doute que les bactéries jouent un rôle — même s'il n'est qu'indirect — en créant un milieu anaérobie et en libérant le soufre. A. Combaz [1970] a également défendu une origine directement bactérienne. R. E. Sweeney et I. R. Kaplan [1973] ont fourni

---

TEXTE-PLANCHE I. — « *Black shales* » aptiennes du plateau de Falkland.

FIG. 1. — (Leg 36 ; 327-24-1-64-65 cm). Reste de coccosphère disloquée sur place (*Corollithion* cf. *fragilis* (ROOD et BARNARD)) montrant plusieurs coccolithes de la même espèce sous des angles différents. × 5 000.

FIG. 2. — (Leg 36 ; 327-25-2-88-89 cm). Plage monospécifique de *Zygodiscus* sp. interprété comme un « bloom » de cette espèce. × 1 000.

FIG. 3. — (Leg 36 ; 330-3-2-49). Au centre, empreinte de coccolithe *(Watznaueria communis)* dissous *in situ*, en bas à droite, tapissage de cristaux de pyrite ; en haut, à gauche, une coccosphère entière bien dégagée. On remarquera la structure esquilleuse du sédiment, avec de grandes plages amorphes. × 2 000.

FIG. 4. — (Leg 36 ; 327-24-1-94-95 cm). *Watznaueria communis* montrant un premier stade de dissolution. × 11 500.

FIG. 5. — (Leg 36 ; 327-24-1-94-95 cm). Coccolithe de la même espèce, plus fortement dissous : les éléments du cycle interne du disque distal ont totalement disparu. × 13 000.

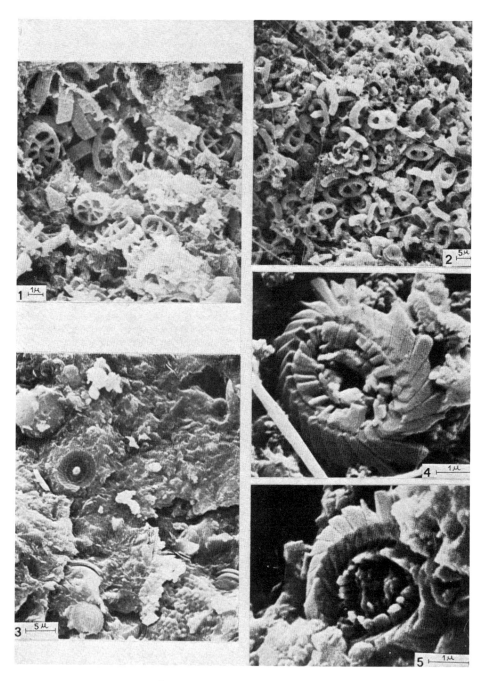

PLANCHE I *(voir légende page ci-contre)*.

des données intéressantes sur les conditions de formation des framboïdes pyriteux. Leurs études, en laboratoire et sur le terrain (bassins de Californie, de mer Rouge), les ont conduits à suggérer l'influence du ph dans les mécanismes de formation des framboïdes. L'abondance de la pyrite qui est un bon indice de confinement est importante à confirmer dans ces études de nannofaciès.

Il faut également signaler, à certains niveaux du site 330 (deuxième section de la carotte 3), l'existence de *cristoballite*. Celle-ci s'observe en quantité notable dans le niveau de calcaire induré (330-3-2-70-71 cm) intercalé au sein d'une série de « black shales » (pl. II ; fig. 3 et 4) mais elle existe également, de façon plus discrète dans les « black shales » susjacents au calcaire (330-3-2-49-50 cm). Ces lépisphères de cristoballite sont fréquentes dans de nombreux sédiments océaniques de nature lithologique variée (calcaires, argiles rouges, etc.) et sont considérées comme les témoins d'une silicification précoce de sédiment [Wise et Kelts, 1972].

Mentionnons en outre que des niveaux silicifiés sont connus dans d'autres gisements de sapropels dans des situations et avec des caractéristiques différentes, tel que par exemple le Kimméridgien du lagon d'arrière-récif d'Orbagnoux (Ain).

Les diverses caractéristiques que nous venons de voir permettent de conclure à l'existence, sur le plateau de Falkland, à l'Aptien, d'un bassin de sédimentation de type euxinique : eaux de surface en communication avec une mer ouverte, permettant la vie et le développement d'une nannoflore calcaire ; eaux de fond caractérisées par une circulation extrêmement pauvre, aboutissant à un confinement dont témoignent le type de fossilisation des coccolithes (restes de coccosphères), l'abondance de la pyrite et la teneur élevée en matière organique.

2) *Le bassin du Cap* (site 361).

L'Aptien, zone à *Chiastozygus litterarius* et à *Parhabdolithus angustus* y est représenté par des argiles noires bitumineuses intercalées de passées

gréseuses de couleur gris-verdâtre principalement au niveau des carottes 44 à 33 qui renferment également de très rares laminations calcaires d'épaisseur millimétrique.

Le nannofaciès des échantillons de « black shales » observés montre un sédiment peu structuré, avec de larges plages amorphes à contours irréguliers (pl. II ; fig. 2). On trouvera l'analyse détaillée de ces échantillons dans Noël et Melguen [1978]. Rappelons simplement les caractères essentiels des amas de coccolithes appartenant à la même espèce (restes de coccosphères disloquées sur place, pl. II ; fig. 4), pyrite en framboïdes dans les infimes fissures du sédiment (pl. III ; fig. 2) ou en tapissage. Les traces de dissolution s'observent également, soit endommageant les structures (pl. III ; fig. 3), soit ayant totalement fait disparaître les corpuscules en ne laissant plus qu'une empreinte (pl. III ; fig. 4), véritable moule externe presque parfait.

Nous avons donc, ici encore, les témoins d'un bassin de sédimentation à deux corps d'eau aux caractéristiques différentes, le corps d'eau supérieur présentant des conditions sensiblement normales, le corps d'eau inférieur avec des conditions fortement restreintes. Mais il faut noter une différence avec le plateau de Falkland. C'est, dans le bassin du Cap, l'arrivée sporadique de matériel terrigène, plus ou moins grossier, en provenance du continent. Cette origine continentale est largement attestée par la dominance des *Classopollis* dans les préparations palynologiques [McLachlan et Pieterse, 1978].

3) *La marge continentale de l'Angola* (site 364).

L'Aptien supérieur (zone à *Parhabdolithus angustus*) — dans lequel le forage a été arrêté — se présente sous un faciès de « black shales » finement laminés, alternant avec des passées de calcaires marneux ou dolomitiques de couleur plus claire. Les niveaux calcaires renferment des nannofossiles en quantité variable. Il n'y a pas été observé de pyrite, ce qui témoigne d'une oxygénation du milieu meilleure que dans les passées de « black shales ». Ces dernières

---

TEXTE-PLANCHE II

Fig. 1. — Plateau de Falkland (330-3-2-70-71 cm). Dans un niveau calcaire reste d'une coccosphère de *Corollithion fragilis* intercalée entre deux plages où abondent les pentalithes de *Micrantholithus*. Aptien × 3 000.

Fig. 2. — Plateau de Falkland (327-23-2-25-26 cm). Nannofaciès de « black shales » : framboïdes de pyrite alignés, un peu encroûtés d'argile. On remarquera la structure esquilleuse du sédiment avec particules argileuses à bords soulevés. Aptien × 2 000.

Fig. 3. — Plateau de Falkland (327-22-3-82-83 cm). « Nids » de cristoballite à la surface d'un échantillon de « black shales ». Aptien × 300.

Fig. 4. — Même échantillon, détail des lépisphères de cristoballite. × 3 000.

Fig. 5. — Bassin du Cap (361-33-3-31-32 cm). Reste de coccosphères de *Corollithion fragilis* montrant les coccolithes sous divers angles. Aptien × 5 600.

PLANCHE II *(voir légende page ci-contre).*

présentent des caractéristiques identiques à celles des sédiments étudiés dans les deux sites précédents.

III. — L'ALBIEN.

L'Aptien que nous venons de décrire, présentait donc des caractéristiques semblables dans les quatre sites étudiés, en particulier la présence de « black shales », localement intercalées de niveaux gréseux ou calcaires. Nous allons voir qu'à l'Albien, le plateau de Falkland, le bassin du Cap et la marge de l'Angola vont évoluer de façon différente.

1) Le plateau de Falkland (site 330 et 327).

Rappelons que les argiles noires aptiennes étaient intercalées de calcaires micritiques, plus ou moins argileux. Ces alternances peuvent être interprétées comme les témoins de fluctuations entre des conditions marines restreintes et plus normales. A partir de l'Albien (zone à *Prediscosphaera cretacea*) les « black shales » disparaissent totalement pour être remplacées par des marnes beige-rosé, plastiques. La nannoflore y est abondante, plus diversifiée que dans l'Aptien sous-jacent (pl. III ; fig. 5) ; elle renferme des *Nannoconus*. Les conditions sont franchement marines mais on note encore dans la carotte 1 du site 330 (330-1-5-38-39 cm ; 330-1-3-41-42 cm) la présence de pentalithes de *Braarudosphaera*. Le reste de la série crétacée est essentiellement constitué de sédiments pélagiques.

2) Le bassin du Cap (site 361).

Les faciès « black shales » observés à l'Aptien se poursuivent pendant l'Albien, toujours intercalés de passées silteuses à gréseuses, riches en fragments de plantes. Coccolithes en restes de coccosphères, framboïdes de pyrite et matière argilo-bitumineuse sont les principales caractéristiques du nannofaciès.

Il nous faut cependant signaler des niveaux présentant une lamination millimétrique, alternativement blanche et noire (361-33-3-31-32 cm). Les lamines noires montrent un nannofaciès typique des « black shales » (pl. IV ; fig. 1) ; les lamines blanches (pl. IV ; fig. 2) sont presque exclusivement constituées de *Nannoconus*. Ces nannoconus jointifs sont extrêmement bien conservés. Leurs éléments constitutifs, de forme triangulaire, sont parfaitement distincts et nullement soudés les uns aux autres comme cela s'observe fréquemment dans les calcaires du Crétacé inférieur du Sud-Est de la France par exemple, où la recristallisation est bien marquée. Notons que précisément à cause de cette absence de soudure entre les éléments constitutifs les tests de *Nannoconus* peuvent se dissocier et de ce fait n'être plus identifiables surtout à l'examen de frottis.

L'abondance de *Nannoconus (N. truiti* BRÖNNIMANN*)* dans ces laminations blanches incite à deux remarques. La première a trait aux conditions de sédimentation : les eaux de surface où proliféraient ces formes planctoniques étaient sporadiquement limpides [D. Noël, 1968], nullement troublées par des apports terrigènes. La seconde remarque concerne la distribution des *Nannoconus*. Ces nannofossiles classiques du Crétacé étaient considérés jusqu'à présent comme caractéristiques du domaine téthysien au sens le plus large. Leur présence dans certains niveaux du plateau de Falkland, leur abondance dans ces couches laminées albiennes du bassin du Cap montrent qu'il n'en est rien. Leur chemin de migration est encore mal connu et mérite des études complémentaires.

Au-dessus des « black shales » albiennes, la sédimentation crétacée dans le bassin du Cap est représentée par des argiles noires, bitumineuses, avec des intercalations sommitales gréseuses, gris-verdâtre. La teneur en carbonate de calcium est faible (5 %) et il n'y a pas été observé de nannofossiles. L'ensemble de cette série mal datée est attribué sans précision au Crétacé indifférencié.

3) La marge continentale de l'Angola.

L'Albien y est représenté par des calcaires plus ou moins marneux, généralement clairs, avec des intercalations de calcaires francs et des récurrences

---

TEXTE-PLANCHE III

FIG. 1. — Bassin du Cap (361-33-3-76-77 cm). Nannofaciès caractéristique de « black shales » avec des coccolithes bien conservés, des framboïdes de pyrite dans les minuscules fissures du sédiment. Aptien × 2 550.

FIG. 2. — Bassin du Cap (361-48-2-142-144 cm). Détail des framboïdes de pyrite constitués de grains octaédriques réguliers et, dans le coin en bas à droite, cristaux de pyrite de plus grande taille provenant d'un tapissage du sédiment. Aptien × 10 500.

FIG. 3. — Bassin du Cap (361-33-3-76-77 cm). Dans les « black shales », coccolithes partiellement dissous. Aptien × 13 800.

FIG. 4. — Bassin du Cap (361-48-2-142-144 cm). La dissolution a été totale et n'a laissé subsister que l'empreinte du corpuscule dans la matrice du sédiment. Aptien × 5 700.

FIG. 5. — Plateau de Falkland (327-21-2-52-53 cm). Niveau marneux où la nannoflore est abondante, bien diversifiée et bien conservée. Albien × 3 000.

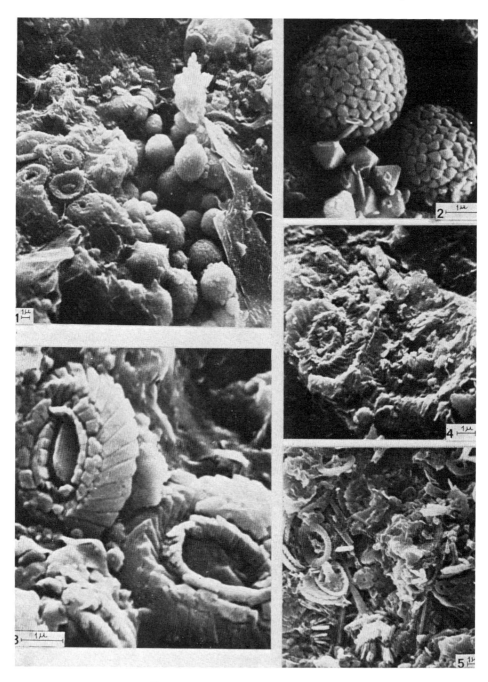

PLANCHE III *(voir légende page ci-contre)*,

TEXTE-PLANCHE IV

FIG. 1 et 2. — Bassin du Cap (361-33-3-31-32 cm). Niveau de « *black shales* » présentant des laminations millimétriques blanches. Albien ; fig. 1 : lamine noire ; nannofaciès typique de « *black shales* » : × 11 300 ; fig. 2 : lamine blanche, constituée presque exclusivement de *Nannoconus* remarquablement conservés. × 5 500.

FIG. 3. — Marge continentale de l'Angola (364-41-2-134-135 cm). Calcaire argileux où abondent les *Prediscosphaera cretacea* qui indiquent un âge albien. × 3 000.

FIG. 4. — Marge continentale de l'Angola (364-39-6-87 cm). Calcaire argileux, riche en nannofossiles, dont les aires centrales et les hampes sont bien conservées. Albien × 3 000.

mineures d'argiles noires bitumineuses. Les nannofossiles, souvent très encroûtés d'argile, sont fréquents et montrent peu de traces de dissolution : certaines structures fragiles, telles que l'aire centrale et les hampes (pl. IV ; fig. 4) sont bien conservées. La présence d'abondants *Prediscosphaera cretacea* (pl. IV ; fig. 3) nous a permis de remonter la limite Aptien-Albien, marquée par l'apparition de cette espèce.

Ce régime plus franchement marin ouvert s'est poursuivi avec des fluctuations jusqu'au Campanien, avec le dépôt de craie marneuse à nannofossiles, intercalée de récurrences de « black shales ».

IV. — Conclusions.

L'étude des nannofaciès des « black shales » aptiennes de l'ensemble des quatre forages considérés et de « black shales » albiennes de la côte occidentale d'Afrique du Sud (leg 40) présente un intérêt sédimentologique.

1) *La présence de nannoplancton*, même faiblement diversifié, témoigne d'influences marines incontestables et de communications avec une mer ouverte. Elle dénote également une certaine productivité des eaux de surface.

2) *Le mode de fossilisation* du nannoplancton (restes de coccosphères) indique des conditions de dépôt stagnantes, sans courants ni vie de fond importants susceptibles de vanner les restes de coccosphères.

3) *La « propreté » des structures des nannofossiles* est également un fait à noter et à mettre en rapport avec la teneur élevée en matière organique de ces niveaux. En effet, les structures de coccolithes bien conservées apparaissent distinctement dans le sédiment *brut* : indubitablement la matière organique — en gênant les circulations — joue en quelque sorte un rôle d'inhibiteur de la diagenèse. Mais contrairement à l'argile qui joue le même rôle d'inhibiteur mais qui incruste les structures les plus fines, la matière organique semble ne pas adhérer aux corpuscules et même les protéger de l'éventuelle fraction argileuse du sédiment.

4) *Les effets de la dissolution*, partielle ou totale, des coccolithes qui peuvent aboutir à ne laisser subsister que des empreintes très fidèles, montrent que ces processus ont eu lieu *dans* le sédiment et non au cours de la descente dans la colonne d'eau. Il ne faut donc pas en tirer d'arguments relatifs à la profondeur du dépôt.

5) *La présence quasi constante de pyrite*, en tapissages ou en framboïdes de cristaux octaédriques, fournit des indices de confinement indubitable et sans doute aussi d'activité bactérienne intense.

Ainsi donc, outre leur intérêt descriptif et analytique, de telles études de nannofaciès montrent qu'elles peuvent fournir des données intéressantes pour la reconstitution de l'histoire des bassins océaniques.

Remerciements.

Les auteurs remercient la National Science Fondation (D.S.D.P.) qui leur a permis d'obtenir les échantillons étudiés. Les sédiments du leg 40 ont été transmis par Mlle Marthe Melguen du Centre océanologique de Bretagne, à Brest, qui a participé à ce leg. Les échantillons du leg 36 ont été prélevés par l'une de nous (D.N.), après xamen des carottes correspondantes au Lamont Doherty Geological Observatory. Dans cette tâche, M. Dan Fornari a été d'une aide efficace. M. Georges Busson nous a aidé de ses conseils, tout au long de l'élaboration de ce travail, et a bien voulu critiquer notre manuscrit. Nous l'en remercions vivement. Les planches photographiques ont été réalisées par Mlle Destarac au Laboratoire de géologie du Muséum et par Mme Barthoux au Bureau de recherches géologiques et minières à Orléans.

Références bibliographiques

Barker P., Dalziel I. W. D., Dinkelman M. G., Elliott D. H., Gombos A. M., Lonardi A., Plafker G., Tarney J., Thompson R. W., Tjalsma R. J., Borch C. C. Von der et Wise S. W. (1977). — Initial Reports of the Deep Sea Drilling Project. vol. XXXVI, 1 vol. 1 079 p.

Bastin E. E. (1950). — Interpretation of ore textures. *Geol. Soc. Amer.*, Mem 45, 101 p.

Bolli H. M., Ryan W. B. F., McKnight B. K., Kagami H., Melguen M., Siesser W. G., Natland J., Longoria J. F., Proto-Decima F., Foreman J. B. et Hottman W. E. (sous presse). — Initial Reports of the Deep Sea Drilling Project, vol. XL.

Bukry D. (1954). — Coccoliths as paleosalinity indicators, evidence from Black sea. In The Black Sea-geology, chemistry and biology ; *A.A.P.G.*, Mem. 20, p. 353-363.

Busson G. et Noël D. (1972). — Sur la constitution et la genèse de divers sédiments finement feuilletés (« Laminites ») à alternance de calcaire et de matière organique ou argileuse. *C. R. Ac. Sc.*, Paris, t. 274, p. 3172-3175.

Combaz A. (1970). — Microsphérules muriformes dans les roches-mères du pétrole : hypothèse sur leur origine. *C. R. Ac. Sc.*, Paris, p. 2240-2243.

Fabricius F. (1961). — Die Strukturen des « Rogenpyrits » (Kossener Schichten, Rät) als Beitrag zum Problem der « Vererzten Bakterien ». *Geol. Rundschau*, vol. 51, p. 647-657.

Hattin D. E. (1975). — Petrology and origin of fecal pellets in upper cretaceous strata of Kansas and Saskatchewan. *J. sediment. Petrol.*, vol. 45, p. 686-696.

Honjo S. (1975). — Dissolution of suspended coccoliths in the deep-sea water column and sedimentation of coccolith ooze. *In* : Sliter, W. W., Bé, A. W. H. and Berger, W. H. (Eds.). Dissolution of deep-sea carbonates. *Cushman Found. Foram. Res. Spec. Pub.*, 13, p. 114-128.

Honjo S., Fischer A. G. et Garrison R. (1965). — Geopetal pyrite in fine grained limestones. *J. Sediment. Petrol.*, vol. 35, p. 480-488.

McLachlan I. R. et Pieterse R. E. (1978). — Preliminary palynological results site 361 Leg 40.

Noël D. (1968). — Nature et genèse des alternances de marnes et de calcaires du Barrémien supérieur d'Angles (Fosse vocontienne, Basses Alpes). *C. R. Ac. Sc.*, Paris, vol. 266, p. 1223-1225.

Noël D. (1970). — Coccolithes crétacés. La craie campanienne du Bassin de Paris. Paris (Éditions du C.N.R.S.), 1 vol.

Noël D. (1972). — Nannofossiles calcaires de sédiments jurassiques finement laminés. *Bull. Muséum national Hist. Nat.*, 3e sér., n° 75, p. 95-124, pl. 1 à 15.

Noël D. et Melguen M. (1978). — Nannofacies of Cap Basin and Walwis Ridge sediments Lower cretaceous to Pliocene (leg 40). Initial reports of the Deep Sea drilling project (leg 40), sous presse.

Roth P. H., Mullin N. M. et Berger W. H. (1975). — Coccolith sedimentation by fecal pellets : laboratory experiments and field observations. *Geol. Soc. Amer. Bull.*, vol. 86, p. 1079-1084.

Rust C. W. (1935). — Colloidal primary copper at Cornwall Mines, southeastern Missouri. *J. Geol.*, vol. 43, p. 398-426.

Sweeney R. E. et Kaplan I. R. (1973). — Pyrite framboid formation : laboratory synthesis and marine sediments. *Econ. Geol.*, vol. 68, p. 618-634.

Thiede et Van Andel T. H. (1977). — The paleoenvironment of anaerobic sediments in the late Mesozoic south Atlantic ocean. *Earth Planet. Sc. lett.*, 33, p. 300-310.

Thompson R. W. (1976). — Mesozoic sedimentation on the Eastern Falkland plateau. Initial report of the Deep Sea Drilling Project, vol. XXXVI, p. 877-891.

Wise S. W. et Hsü K. J. (1971). — Genesis and lithification of a deep sea chalk. *Eclog. Geol. Helv.*, vol. 64, p. 273-278.

Wise S. W. et Kelts K. R. (1972). — Inferred diagenetic history of a weakly silicified deep sea chalk. *Gulf Coast Assoc. Geol. Soc. Trans.*, vol. 22, p. 177-203.

179

Part III

PALEOBIOGEOGRAPHY AND PALEOENVIRONMENTAL APPLICATIONS

Editor's Comments
on Papers 17 Through 28

17   **MCINTYRE**
     Coccoliths as Paleoclimatic Indicators of Pleistocene Glaciation

18   **GEITZENAUER**
     *Coccoliths as Late Quaternary Palaeoclimatic Indicators in the
     Subantarctic Pacific Ocean*

19   **MCINTYRE, RUDDIMAN, and JANTZEN**
     Abstract from *Southward Penetrations of the North Atlantic
     Polar Front: Faunal and Floral Evidence of Large-scale Surface
     Water Mass Movements over the Last 225,000 Years*

20   **RUDDIMAN and MCINTYRE**
     Abstract from *Northeast Atlantic Paleoclimatic Changes over
     the Past 600,000 Years*

21   **HAQ and LOHMANN**
     Abstract from *Early Cenozoic Calcareous Nannoplankton
     Biogeography of the Atlantic Ocean*

22   **HAQ, PREMOLI-SILVA, and LOHMANN**
     *Calcareous Plankton Paleobiogeographic Evidence For Major
     Climatic Fluctuations in the Early Cenozoic Atlantic Ocean*

23   **HAQ**
     *Biogeographic History of Miocene Calcareous Nannoplankton
     and Paleoceanography of the Atlantic Ocean*

24   **ROTH and BOWDLER**
     Abstract from *Middle Cretaceous Calcareous Nannoplankton
     Biogeography and Oceanography of the Atlantic Ocean*

25   **BUKRY**
     Coccoliths as Paleosalinity Indicators—Evidence from Black Sea

26   **MARGOLIS et al.**
     *Oxygen and Carbon Isotopes from Calcareous Nannofossils as
     Paleoceanographic Indicators*

27   GOODNEY et al.
Abstract from *Oxygen and Carbon Isotopes of Recent Calcareous Nannofossils as Paleoceanographic Indicators*

28   DUDLEY et al.
Coccoliths in Pleistocene-Holocene Nannofossil Assemblages

The mapping of the biogeography of modern coccolithophores (see Papers 8 and 9) has shown that these distribution patterns are closely related to the thermal properties of the water masses. This makes paleobiogeography (past distribution patterns) a potentially important tool in paleoclimatic-paleoceanographic reconstructions. Papers included in this section are examples of paleobiogeographic studies of calcareous nannoplankton and their use in paleoenvironmental interpretations.

Paper 17 by McIntyre and Paper 18 by Geitzenauer were pioneering attempts to use coccoliths as paleoclimatic indicators for the Pleistocene by comparison with modern distributions. In Paper 19 McIntyre et al. were able to use the distribution patterns of coccoliths and planktonic foraminifers to document climatic shifts in the mid to high latitude North Atlantic during the late Pleistocene. They showed two significant southward shifts of the polar front and at least six northward incursions of the subtropical gyre. Their evidence indicates climatic shifts of up to 20° latitude during the last 225,000 years. A similar study of coccoliths and foraminiferal assemblages by Ruddiman and McIntyre (Paper 20) carries the analogy back to 600,000 years B. P. They identify seven climatic cycles in the North Atlantic since that time. From imput rate calculations for major sediment fractions during the last glacial and interglacial, they conclude that calcareous plankton are deposited much more rapidly during interglacial than glacial periods. These studies are excellent examples of how nannoplankton can be applied in paleoceanographic reconstructions of the relatively recent past and how analogies from modern distributions can help in interpreting such data.

Edwards (1968) attempted to reconstruct the Tertiary marine climates of New Zealand, based mainly on the relative abundance fluctuations of the cold water species *Coccolithus pelagicus* and its morphological analogs (=*Ericsonia ovalis* of Edwards). He concluded that the late Eocene was warmer than the early Eocene or the Oligocene. During the Oligocene, the southern high latitude climates deteriorated, only to improve once again in the Miocene. Edwards' results compare favorably with the climatic interpretations of subsequent studies.

The next series of papers (Papers 21 through 24) are examples of efforts to delineate pre-Pleistocene nannofossil biogeographic patterns. The abstract of the article by Haq and Lohmann (Paper 21) was the earliest attempt to map early Tertiary biogeographic patterns on an ocean-wide basis. To overcome the sparseness of data, these authors suggested a "time-slice" approach (focusing on discrete intervals of time within the available biochronologic resolution) for mapping coccolith assemblages in the Atlantic Ocean. They outlined the biogeographic patterns of twelve Paleogene time slices that led to the identification of the latitudinal (climatic) preferences of various assemblages. Their temporal shifts were then used by the authors to interpret the climatic history of the Atlantic between 65 and 25 m. y. B. P. Haq et al. (Paper 22) extended this Paleogene study by combining the coccolith and planktonic foraminiferal biogeographic data. These data suggest a series of expansions and contractions of extratropical belts, as indicated by the excursions of low and high latitude assemblages. The paleoclimatic conclusions of these studies compare well with results based on oxygen isotope studies of the Paleogene paleoclimate (see discussion in Haq, 1980). Paper 23 by Haq outlines the Miocene nannofossil biogeography of the Atlantic in synoptic time-space maps of quantitatively defined assemblages. This data also shows distinct latitudinal migrations of assemblages which suggest four climatic cycles with a time span of about 4.5 m.y. These paleoclimatic interpretations also show a correlation to the oxygen isotopic paleotemperature curves.

Another aspect of nannoplankton paleobiogeography is the dichotomy between hemipelagic and open-ocean assemblages as summarized by Bukry et al. (1971). Bybell and Gartner (1972) have documented the provincialism among mid-Eocene pentaliths. Such compositional differences between near-shore and oceanic nannofloras may be due to endemism; conversely, these patterns may be an artifact of differential preservation in shallower vs. deeper water depths.

Some aspects of Cretaceous nannofossil biogeography have been summarized by Thierstein (1976). The abstract of the paper by Roth and Bowdler (Paper 24) on the mid-Cretaceous nannofossils is the first attempt to delineate the biogeographic patterns on an ocean-wide basis for a pre-Cenozoic interval. It leads to important paleoceanographic conclusions about the early Atlantic Ocean basins.

Paper 25 by Bukry is another example of the use of coccoliths as paleoceanographic indicators. On the basis of indigenous vs. reworked assemblages and the dominance of the euryhaline species such as *Braarudosphaera bigelowi* and *Emiliania huxleyi* at certain levels, Bukry was able to draw conclusions about the onset of the last transgression that reconnected the lakelike Black Sea basin to the Mediterranean and the Holocene salinity history of the basin.

The group of papers that follow (Papers 26 through 28) are selections from recent attempts to use the coccolith carbonate for stable isotopic analyses. The analyses of the isotopic ratios of oxygen and carbon in marine calcareous plankton and benthos yield important information about the past temperature and fertility of the water masses (see discussion in Paper 23, pp. 429-431). Although coccolith isotopic variations are generally parallel to the surface planktonic Foraminifera, so far they have not been used widely for isotopic paleotemperature determinations. The problems of dissolution and the presence of noncoccolith debris in the nannofossil carbonates hamper their wider applications. Studies such as those included here, however, indicate their significant promise in paleoenvironmental work in the future.

REFERENCES

Bukry, D., R. G. Douglas, S. A. Kling, and V. Krasheninnikov, 1971, Planktonic Microfossil Biostratigraphy of the Northwestern Pacific Ocean, in *Initial Reports of the Deep Sea Drilling Project,* **6:**1253-1300.

Bybell, L., and S. Gartner, 1972, Provincialism among Mid-Eocene Calcareous Nannofossils, *Micropaleontology* **18:**319-336.

Edwards, A. R., 1968, The Calcareous Nannoplankton Evidence for New Zealand Tertiary Marine Climate, *Tuatara* **16:**26-31.

Haq, B. U., 1980, Paleogene Paleoceanography: Early Cenozoic Oceans Revisited, *Oceanol. Acta,* Special Issue: 71-82.

Thierstein, H. R., 1976, Mesozoic Calcareous Nannofossil Biostratigraphy of Marine Sediments, *Marine Micropaleontol.* **1:**325-362.

17

Reprinted from *Science* **158**:1314-1317 (1967)

COCCOLITHS AS PALEOCLIMATIC INDICATORS
OF PLEISTOCENE GLACIATION

Andrew McIntyre
Lamont Geological Observatory, Palisades, New York

Abstract. *Selected species of Coccolithophoridae from recent sediments and mid-Wisconsin glacial sediments of the North Atlantic were examined in an attempt to determine cooling effects. All species showed a definite shift southward during the glacial period. The average shift in this planktonic population was 15 degrees of latitude, with the greatest change in the eastern Atlantic. A paleoisotherm map can be drawn on the basis of the temperature boundaries of coccolithophorids. The species boundaries indicate a possible shift in position of the subtropical gyral to a glacial position roughly parallel to the 33-degree line of latitude.*

The dramatic fluctuations in Pleistocene climate are recorded in sediments in the Atlantic Ocean (*1*), but unfortunately the means of procuring these data are poorly developed. The only direct technique available at the present time is the use of oxygen isotopes (*2*). This report deals with a new approach —plotting the migration of biogeographic boundaries for temperature-restricted species of Coccolithophoridae due to Pleistocene glaciation.

Among all the microorganisms that leave fossil records in oceanic sediments, the Coccolithophoridae probably have the greatest potential as paleoclimatic indicators. In addition to their wide geographic distribution and stable mineral skeleton (calcite), these marine algae inhabit the upper euphotic zone (*3–5*) and consequently are under direct climatic control. In living species it is possible to correlate biogeographic boundaries with surface water isotherms (*4*), and this is the basis of my report.

The method of attack, being biogeographic, requires the widest possible geographical distribution of core material. This is not easily obtained, for, although the North Atlantic has been the site of intensive sampling, there remain large gaps in the core distribution. A limiting factor is that large areas of the North Atlantic basin are

below the carbonate compensation level, with a consequent lack of coccolith flora. Thus the 23 cores sampled (Table 1) are restricted to three linear belts. Two cover the shelf, slope, and

rise of both North America and Europe-Africa; the third, the Mid-Atlantic Ridge.

Choice of the particular species to be examined requires that two separate cri-

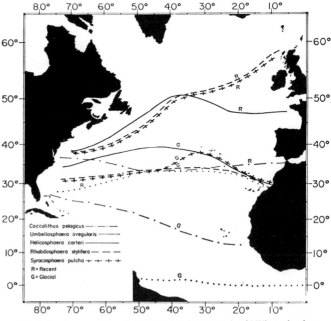

Fig. 1. Species population boundaries for Recent and mid-Wisconsin time.

teria be met. They must have relatively limited temperature ranges and a good paleontologic record. This is made difficult by the differential preservation of coccolith structural types. Some of the best temperature indicators are found among the delicate caneoliths where fossil recording is minimal.

Eight species, *Coccolithus pelagicus*, *Cyclococcolithus leptoporus*, *Helicosphaera carteri*, *Rhabdosphaera stylifera*, *Syracosphaera pulchra*, *Umbellosphaera irregularis*, *Umbellosphaera tenuis*, and *Umbilicosphaera mirabilis*, were selected on the basis of preservation and temperature range and their present distribution in the North Atlantic, compared with that of the mid-Wisconsin (last glacial period). Counts and identification of species were done with the electron microscope.

Synchronous samples from a finite time in the Wisconsin glaciation can be obtained by C^{14} dates or, where these are inadequate, by use of sedimentation rates established by radioactive dating. The rates for each core, expressed in centimeters per 1000 years, can be multiplied by a unit time in years ($24 = 24,000$ years of the glacial period) that would represent the mid-Wisconsin. This gives a depth in centimeters for each core from which synchronous samples are taken. While this technique is regularly used in geological oceanography, it is open to criticism since one must often assume that the sedimentation rate established for the upper portion of the glacial section in each core remains constant for the entire section. It is necessary to bear the limitations in mind since there is as yet no other means of obtaining the samples needed. The midglacial position of these samples was verified by planktonic foraminiferal stratigraphy.

The data (Table 2) from eight species, when plotted on distributional maps (Figs. 1 and 2), indicate a marked change in biogeographic distribution between recent and glacial sediments. The ranges of each species are delineated in Figs. 1 and 2 by two lines marking the boundaries of the recent and glacial populations. These lines are drawn on the extreme northern or southern appearances for warm and cold species, respectively.

Coccolithus pelagicus, the only species limited to cold water presently recognized in the North Atlantic (4), is restricted to subarctic and transitional waters. Its maximum southern limit coincides with the 14°C isotherm. At present this species extends to 35°N

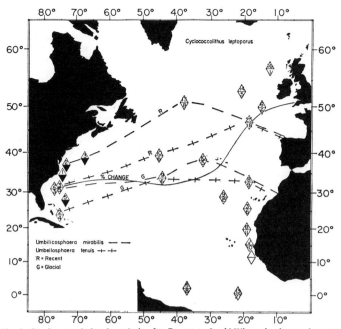

Fig. 2. Species population boundaries for Recent and mid-Wisconsin time and percent values for *Cyclococcolithus leptoporus* in diamonds with the upper recent and the lower glacial values.

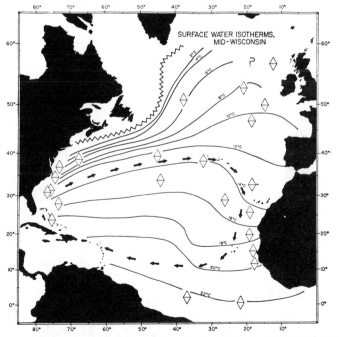

Fig. 3. Paleoisotherm map of the mid-Wisconsin North Atlantic erected with the use of coccolithophorid temperature data. The dark arrows indicate presumed position of the subtropical gyral based on coccolith boundaries of subtropical forms. The jagged line represents pack ice [after Flint].

latitude in the cold Labrador water found between the Gulf Stream and the coast of North America. In the mid-Wisconsin it ranged as far south as 13°N latitude along the African coast (Fig. 1). The average latitude shift between recent and glacial is approximately 15° latitude and even greater between living material and glacial (4).

At the opposite extreme in temperature tolerance is *Umbellosphaera irregularis* (Fig. 1), a tropical species. The minimum shift between recent and glacial boundaries for this form is approximately 20° latitude. The accuracy to which such shifts can be determined reflects the spacing of core samples. Between the glacial presence (core Nos. V16-200 and A180-72) and absence (core No. A180-56) there is a gap of 10° latitude (see Table 1).

The temperature ranges of *Helicosphaera carteri*, *Rhabdosphaera stylifera*, and *Syracosphaera pulchra* are not as limited as those of the two preceding species (Fig. 1). They are presently found in subtropical to transitional waters. All show a much more limited distribution during mid-Wisconsin time. *Helicosphaera carteri* has an average distributional difference of 10° latitude while *Rhabdosphaera stylifera* and *Syracosphaera pulchra* both have an approximate shift of 17° latitude. Note that in these three species the recent maximum distributional lines follow the northern border of the subtropical gyral (Gulf Stream) across the Atlantic Ocean. This agrees with plankton data from the North Atlantic, where the boundary between subtropical and subarctic species approximates the northern border of the Gulf Stream (4).

Umbellosphaera tenuis and *Umbilicosphaera mirabilis*, while having a satisfactory preservation record, do not show as distinctive a difference in their recent and glacial distributions as the preceding species.

Umbellosphaera tenuis is found at higher latitudes and is limited today by the 16°C isotherm; it is typical of subtropical waters. Its glacial to recent shift is from 10° to 15° latitude (Fig. 2). It is not as abundant in glacial as in recent sediments. This may be due to the fragility of its macrococcoliths.

The coccoliths of *Umbilicosphaera mirabilis*, like *Coccolithus huxleyi*, have temperature-dependent structural variations (4, 6). While these changes are gradational in *C. huxleyi*, *U. mirabilis* appears to have separate cold- and warm-water forms. The distribution of *U. mirabilis* is today bounded by the 18°C isotherm; this may be too high a value since the colder-stage coccoliths are present at higher latitudes in sediment samples from this study than was observed with either living or surface sediment material in the survey of modern forms. It is possible that the colder stage of *U. mirabilis* has a distribution similar to that of *R. stylifera* and *S. pulchra*. Nevertheless, the species shows a definite shift between recent and glacial sediments which averages 15° latitude (Fig. 2). Further evidence indicative of a cooling of the Atlantic in glacial times is the preponderance of cold-form coccoliths in glacial-age sediments from the mid-North Atlantic, a situation reversed in recent surface sediments.

Cyclococcolithus leptoporus (Fig. 2) is among the most eurythermal of the Coccolithophoridae. Today it ranges from the equator to Arctic waters. There is no apparent difference in maximum range between recent and glacial, although this is probably due to the lack of core material further north than the line Sp 10-1 through R 10-2 (see Table 1). There is a marked change in percentage distribution between recent and glacial populations, with the higher percentage (which today occurs in this species in transitional waters) indicative of the optimum range. Thus the line representing percentage change may be a rough indicator of the subtropical boundary in the Wisconsin. This type of subtle biogeographic change will require further work before definite conclusions can be drawn.

A number of other species showed some degree of biogeographic change by a shift in maximum boundaries; however, they are not plotted because of relative rarity in the core material. In the case of the ubiquitous eurythermal species (*Coccolithus huxleyi* and *Gephyrocapsa oceanica*), their distribution is similar to that of *Cyclococcolithus leptoporus*.

One interesting change, presently not usable for paleoclimatic work but important in systematics, is the reversal in dominance of *Coccolithus huxleyi* and *Gephyrocapsa oceanica* from glacial to recent (Table 2).

In today's ocean *Coccolithus huxleyi* usually constitutes over 50 percent of the flora, but in the mid-Wisconsin it shared and in some latitudes was dominated by *Gephyrocapsa oceanica*. I consider that *Coccolithus huxleyi* is a relatively recent form, none being found before the Pleistocene, and that it evolved from the *Gephyrocapsa oceanica* complex during the late Pleistocene. Similarities in form and ecology, combined with the finding of intermediate forms in Pleistocene core samples that I am now investigating, support this theory.

A comparison of the latitudinal change in flora from northern cold waters to southern warm waters in recent and glacial sediments indicates the dominance of cooler water forms in lower latitudes during the mid-Wisconsin. *Umbilicosphaera mirabilis* and *Syracosphaera pulchra* constitute a much larger percentage of the flora in glacial times in this area, while *Cyclococcolithus fragilis* and *Umbellosphaera irregularis*, subtropical to tropical forms, are nearly absent during glacial time from the North Atlantic.

The biogeographic boundaries of Coccolithophoridae species in today's seas can be correlated with surface water isotherms (4). If we assume that the present temperature ranges of the species held for the last glacial period, then the paleogeographic boundaries of species can also be assigned temperature values. If all these boundaries are

Table 1. Core locations in the North Atlantic and the depth of the glacial sample in each core.

Core No.	Location		Depth of glacial sample (cm)
	Latitude	Longitude	
A153-146	33°43'N	44°45'W	80
A156-4	34°49'N	74°41'W	844
A156-5	37°07'N	73°37'W	95
A164-59	38°42'N	67°52'W	245
A167-13	31°39'N	75°21'W	300
A167-14	31°28'N	76°28'W	300
A179-13	23°56'N	75°45'W	97
A179-17	28°00'N	73°47'W	280
A180-9	39°27'N	45°57'W	115
A180-16	38°21'N	32°29'W	140
A180-32	29°07'N	26°15'W	59
A180-48	15°19'N	18°06'W	488
A180-56	12°15'N	17°46'W	207
A180-72	00°36'N	21°47'W	120
R5-36	46°55'N	18°35'W	162
R5-54	25°52'N	19°03'W	35
R5-57	19°40'N	19°06'W	270
R10-2	56°59'N	12°28'W	100
SP8-4	32°50'N	18°32'W	65
SP9-3	53°53'N	21°06'W	220
SP9-4	50°02'N	14°46'W	200
SP10-1	51°23'N	38°04'W	150
V16-200	01°58'N	37°04'W	120

Table 2. Species composition of recent and glacial samples for each core (expressed as percentages). Since only eight species are reported from the total flora, the percentage values do not equal 100. Abbreviations: R, recent; G, glacial; x represents values that are less than 1 but greater than 0.5 percent.

Core No.	Coccolithus pelagicus		Cyclococco-lithus leptoporus		Helico-sphaera carteri		Rhabdo-sphaera stylifera		Syraco-sphaera pulchra		Umbello-sphaera irregularis		Umbello-sphaera tenuis		Umbilico-sphaera mirabilis	
	R	G	R	G	R	G	R	G	R	G	R	G	R	G	R	G
A153-146	x	5	9	6	1	2	1	x	1	0.5	x		1	1	4	3
A156-4		x	1	x				x	1							
A156-5	x	x	2	x												
A164-59	x	x	3	x	6			x	2				2		4	
A167-13		1	3	5			1	1	1							x
A167-14		x	2	3	2	4	1	1					x			1
A179-13			2	4	2	3	2	6	2	4	2			2		7
A179-17		x	5	x	1	x	1			3	3		4		6	x
A180-9	18	7	10	6	2	1	x		1				x		3	
A180-16	x	1	10	5	1	2	x		1	2.5			2	1	3	1
A180-32		1	9	15	2	3	2	1	1	x			2	1	4	8
A180-48		3	x	10	4	1			1						1	2
A180-56								2		2						6
A180-72			13	27	2	2	1		1	1	2	1	x		7	11
R5-36	x	6	25	18	1			x	1				1		x	
R5-54		1	3	6	2	4	2	2	3	1	4		4	1	9	5
R5-57	x	1	8	11	3	1	1	1	9	1					2	1
R10-2	4	2	16	1			2		4							
SP8-4		0.5	8	9	1	0.5	1	x	1	1	x		2	2	3	1
SP9-3	1	1	14	5					2							
SP9-4	1	5	6	6				x		x						
SP10-1	30	3	13	8	4			x	6				2			
V16-200			5	8	2.5	3	2	x	x	2	4	1	3	1	14	22

plotted on a map and each boundary is given the present value of the maximum temperature (isotherm), it should be possible to draw a paleoisotherm map for the glacial period. Having laid out the temperature lines for the species and interpolating between overlapping ranges, I found it possible to draw a tentative paleoisotherm map (Fig. 3). Although insufficient core coverage makes any fine adjustment of these lines impossible (a gap remains in the northwest portion of the Atlantic), two lines, the 14° and the 22°C isotherms, are established on the basis of a number of overlapping species. The addition of more core material and the mapping of other species boundaries should result in a paleoisotherm map that will be an accurate representation of the average temperature of surface water of the glacial North Atlantic.

In seven species the amount of latitudinal shift between glacial and recent is greatest along the eastern side of the Atlantic. In the three species with subtropical to transitional ranges it is a factor of 2 to 3. This distributional difference is presumably the result of the main current system in existence today and in the Wisconsin. At present the distribution of *R. stylifera, H. carteri, U. tenuis,* and *S. pulchra* is encompassed by the northern boundary of the subtropical gyral (Gulf Stream).

This boundary rises from approximately 40°N latitude off North America to over 55°N latitude off Europe. This is also true for other subtropical coccolith species not included in this report. In the mid-Wisconsin the line of species presence is relatively horizontal, running roughly parallel to the 30° latitude line with a slight southern turn along the eastern edge of the Atlantic (Fig. 3).

Admittedly the core density is somewhat low; nevertheless, the core coverage is sufficient to allow no more than a 5° fluctuation in latitude without a major change in azimuth of this line since it is bracketed by cores. If one compares data (Figs. 1 and 2), it is possible to say that the northern border of the subtropical gyral during mid-Wisconsin time flowed along or near the 33° latitude line.

From this first report, based on information gained from modern species of Coccolithophoridae, it appears that coccoliths can be used for paleoecologic studies and that the application of these studies to the problem in this report has led to the following conclusions about the effect of cooling on the North Atlantic during the Wisconsin glaciation: (i) That the maximum cooling in the mid-Wisconsin resulted in a southward shift of planktonic populations of approximately 15° latitude, with the greatest shift occurring in the eastern Atlantic; (ii) that it may be possible to erect paleoisotherm maps of surface water with the use of population boundaries of Coccolithophoridae species of known temperature range as isotherms, particularly if greater core coverage can be combined with data on additional species; and (iii) that the northern boundary of the subtropical gyral, from a present position of approximately 40° latitude off North America to over 55° latitude off Europe, was displaced to a position extending from approximately 30°N latitude off North America to approximately 38° off Europe.

References and Notes

1. D. B. Ericson, M. Ewing, G. Wollin, B. C. Heezen, *Geol. Soc. Amer.* **72**(2) 193 (1961).
2. C. Emiliani, *J. Geol.* **63**, 538 (1955).
3. G. A. Riley, *Limnol. Oceanogr.* **2**. 252 (1957); E. M. Hulburt, *ibid.* **7**, 307 (1962).
4. A. McIntyre and A. W. H. Bé, *Deep-Sea Res.,* in press.
5. H. G. Marshall, *Limnol. Oceanogr.* **11**, 432 (1966).
6. A. W. H. Bé and A. McIntyre, *Spec. Paper Geol. Soc. Amer.* **82**, 8 (abstr.) (1964); N. Watabe and K. M. Wilbur, *Limnol. Oceanogr.* **11**, 567 (1966).
8. I thank A. W. H. Bé for helpful discussion and J. Imbrie, W. Broecker, A. Gordon, and C. Drake for criticism of the manuscript. Supported by National Science Foundation grant GP 4768. Lamont Geological Observatory contribution No. 1131.

6 September 1967

18

Copyright ©1969 by Macmillan Journals Ltd

Reprinted from *Nature* **223**:170–172 (1969)

COCCOLITHS AS LATE QUATERNARY PALAEOCLIMATIC INDICATORS IN THE SUBANTARCTIC PACIFIC OCEAN

Kurt R. Geitzenauer

Department of Geology, Florida State University, Tallahassee, Florida 32306

CLIMATIC fluctuations during the Pleistocene have been revealed by isotope and palaeontological studies of deep-sea cores from the Atlantic and Indian Oceans[1-4] and the Pacific Ocean[1-6]. This article indicates the utility of two species of Coccolithophoridae as palaeoclimatic guides in high latitude areas. The eight deep-sea carbonate cores studied were collected from the National Science Research Vessel Eltanin in the South Pacific Ocean (Table 1).

Table 1. LOCATION OF EIGHT CORES COLLECTED IN THE SOUTH PACIFIC OCEAN

Core	Lat.	Long.	Depth (m)	Length (cm)
E11–1	54° 54′ S	114° 42′ W	3,475	625
E11–3	56° 54′ S	115° 15′ W	4,023	1,140
E15–16	56° 03′ S	119° 55′ W	3,054	1,217
E20–18	44° 33′ S	111° 20′ W	2,868	473
E21–24	49° 02′ S	120° 05′ W	3,319	448
E21–15	52° 01′ S	120° 03′ W	2,999	472
E21–16	54° 05′ S	119° 57′ W	2,981	564
E21–17	55° 29′ S	119° 56′ W	2,816	1,006

The abundant Coccolithophoridae species found in the cores I used are: *Umbilicosphaera leptopora* (Murray and Blackman); *Coccolithus pelagicus* (Wallich); *Gephyrocapsa oceanica* (Kamptner sensu lato); *Emiliania hyxleyi* (Lohmann) and *Helicopontosphaera kamptneri* (Hay and Mohler).

Syracosphaera pulchra (Lohmann); *Umbilicosphaera mirabilis* (Lohmann); *Rhabdosphaera stylifera* (Lohmann); *Discosphaera tubifera* (Murray and Blackman) and *Cyclolithella annula* (Cohen) occur spasmodically in the cores and never exceed 1 per cent of the total Coccolithophoridae present. With the exception of *C. pelagicus* all of these species are still living in the South Pacific[7].

Cores E11–1 and E11–3 were studied because of previous work on their radiolarian faunas[6]. Curves obtained by plotting the percentage of *U. leptopora* in the total coccolith assemblage correlate rather closely with curves given by Hays[6] based on percentage of warm water radiolarians in the total radiolarian assemblage (Fig. 1). Complete agreement between coccolith and radiolarian trends is not expected, however, because these cores show clear evidence of carbonate solution throughout much of their length, particularly at certain horizons (Fig. 1), and the quantitative characteristics of foraminiferal assemblages can be greatly affected by selective solution[8]. In this study coccoliths also indicate the effects of selective solution. The coccoliths of *U. leptopora* are larger than those of the more abundant *G. oceanica* and probably are

damaged less by solution; this leads to a high percentage of *U. leptopora* in core E11–1 at 600–625 cm and in core E11–3 at 475–525 cm where solution of carbonate has been extensive.

The percentages of *C. pelagicus* and *U. leptopora* in the total coccolith assemblage are plotted for six additional carbonate cores from a wider range of latitude (Fig. 2). An abundance peak occurs (I) at or near the top of each core immediately above a short interval of very low abundance (II). Below there are a series of peaks (III, IV and V). The scarcity of long carbonate cores has made correlation difficult in the lower portions of the longest cores. The data, however, indicate that below peak V there is a relatively long interval characterized by few *U. leptopora* (VI). In core E21–17 this long interval (containing a possible minor peak) occurs above a broad fluctuating peak (VII) that is distinguished by a general dominance of *C. pelagicus* over *U. leptopora*. Below peak VII in core E15–16 there is 7 m of core that contains consistently low percentages of both *C. pelagicus* and *U. leptopora*.

The high frequency peaks of *U. leptopora* are believed to reflect relatively warm intervals, because they correspond with warmer water radiolarian faunas, and also because rare warm water coccolithophorid species (*Syracosphaera pulchra, Discosphaera tubifera, Rhabdosphaera stylifera* and *Umbilicosphaera mirabilis*[7,9] occur only at these intervals (Fig. 2).

If the peaks represent relatively warm intervals and the intervening troughs relatively cool intervals, it should be possible to correlate them with the glacial and interglacial stages distinguished in deep-sea cores from lower latitude areas. But because high and low latitude faunas are too distinct to be correlated directly, it is necessary to establish independent chronological criteria for the cores. Palaeomagnetic data[10] are available only for the two longer cores (E15–16 and E21–17) and in these the faunal data and palaeomagnetic events do not show sufficiently consistent relationships to offer assistance in correlation. Radiometric dating is possible, however, by the excess ^{230}Th dating method using gamma-ray spectroscopy. In each of the six cores a level is determined below which essentially no excess ^{230}Th is present (Fig. 2). These horizons represent a 300,000 yr BP isochron with a possible error of ± 25 per cent[11]. Although this estimated error may represent as much as ± 75,000 yr, at worst they should provide a rough guide

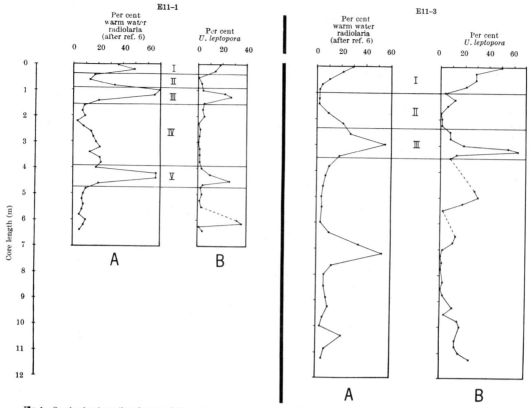

Fig. 1. Graphs showing rather close correlation, in two cores, of curves formed by the percentage of *Umbilicosphaera leptopora* in the total coccolith flora (*B*) and the percentage of warm water radiolaria, determined by Hays' (*A*). The dashed portions of the coccolith curves indicate levels at which coccoliths are essentially absent due to strong solution effects. Successive climatic events described in this article are designated by Roman numerals.

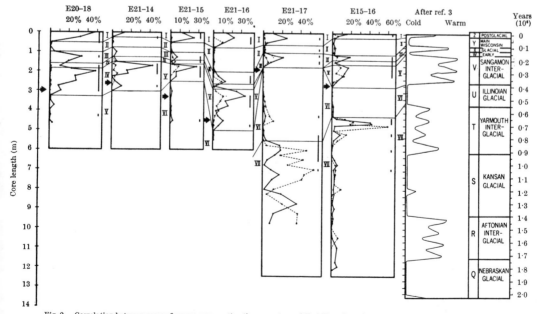

Fig. 2. Correlation between cores of curves representing the percentage of *Umbilicosphaera leptopora* (——) and *Coccolithus pelagicus* (- - -) in the total coccolith flora and with the generalized Quaternary climatic scheme of Ericson and Wollin[3] shown to the right. Successive climatic events described in this paper are designated by Roman numerals. The solid vertical lines at the right in each column indicate the occurrence of warm water coccoliths. Arrows to the left of each column indicate that portion of the core determined by the ^{230}Th method as having an age of 300,000 ($\pm$ 75,000) yr.

as to the value of *U. leptopora* peaks as palaeoclimatic and correlative tools. The results do show, in fact, that the [230]Th isochron occurs with reasonable consistency at the beginning or slightly preceding peak V in all cores.

The curves based on the frequency of *U. leptopora* compare favourably with the climatic curve established for the Atlantic and Caribbean by Ericson and Wollin[3] based on foraminiferal criteria. These authors established a chronology based on magnetic reversals.

If frequency peak V in the South Pacific Ocean cores represents a relatively warm interval and the established 300,000 yr isochron occurs at the correct horizon, then a good correlation exists between the palaeoclimatic curves I have described and those of Ericson and Wollin[3]. These authors place the beginning of the Sangamon Interglacial Stage at about 400,000 yr BP based on palaeomagnetic reversals. Ku and Broecker[12] applied the [230]Th-dating technique to a deep-sea core from the Caribbean and arrived at a date of 320,000 ($\pm$ 32,000) yr BP for the beginning of the Sangamon, which is relatively close to the age of the beginning of frequency peak V. Exact correlation of glacial and interglacial stages between high and low latitudes would not be expected, however, because it is almost certain the floral and faunal changes are not isochronous in response to major world wide climatic changes. A consistent coccolith peak (III) occurs in five of the six cores and appears to correlate well with a short interglacial during the Wisconsin Glacial Stage[3] (Fig. 2).

Although the coccolith data compare favourably with the general climatic scheme of Ericson and Wollin, it must be emphasized that this is a preliminary treatment and that oceanographic variables, other than temperature, may control the distribution and abundance of coccolith species in subantarctic latitudes of the Pacific Ocean. The coccolith species *U. leptopora* and *C. pelagicus* are, however, useful palaeoclimatic indicators in Late Quaternary sediments of the South Pacific Ocean.

I thank J. P. Kennett, H. G. Goodell, A. Niedoroda, J. K. Osmond and F. H. Cramer for reviewing the manuscript. This work was supported by grants from the US National Science Foundation to H. G. Goodell.

Received March 24; revised April 30,'1969.

[1] Ericson, D., and Wollin, G., *Micropaleontology*, 2, 257 (1956). Emiliani, C. *Geol. Soc. Amer. Bull.*, 75, 129 (1964). Lidz, L., *Science*, 154, 1448 (1966). Conolly, J. R., *Nature*, 214, 873 (1967).

[2] Ericson, D., Ewing, M., and Wollin, G., *Science*, 146, 723 (1964).

[3] Ericson, D., and Wollin, G., *Science*, 162, 1227 (1968).

[4] Glass, B., Ericson, D. B., Heezen, B. C., Opdyke, N. D., and Glass, J. A. *Nature*, 216, 437 (1967).

[5] Blackman, A., and Somayajulu, B. L. K., *Science*, 154, 886 (1966).

[6] Hays, J. D., *Prog. Oceanog.*, 4, 117 (1967).

[7] McIntyre, A., and Be, A. W. H., *Deep-Sea Res.*, 14, 561 (1967).

[8] Kennett, J. P., *NZ Dept. Sci. Indust. Res. Bull.*, 186 (1968). Berger, W. H *Science*, 156, 383 (1967). Ruddiman, W. F., and Heezen, B. C., *Deep Sea Res.*, 14, 801 (1967).

[9] Hasle, G. R., *Nytt Mag. Bot.*, 8, 77 (1960).

[10] Goodell, H. G., and Watkins, N. D., *Deep-Sea Res.*, 15, 89 (1968).

[11] Osmond, J. K., and Pollard, L. D., *Earth Planet. Sci. Lett.*, 3, 476 (1967)

[12] Ku, T. L., and Broecker, W. S., *Science*, 151, 448 (1966).

19

Reprinted from *Deep-Sea Research* **19**:61 (1972), by permission of Microforms International Marketing Corporation as exclusive copyright licensee of Pergamon Press journal back files

Southward penetrations of the North Atlantic Polar Front: faunal and floral evidence of large-scale surface water mass movements over the last 225,000 years

A. McIntyre,* W. F. Ruddiman † and R. Jantzen*

(*Received* 29 *March* 1971; *in revised form* 8 *July* 1971; *accepted* 22 *July* 1971)

Abstract—Eight cores taken on the eastern flank of the Mid-Atlantic Ridge from 30°N to 62°N document climatic shifts over the last 225,000 years. Southward penetration of polar water eliminated the coccolith carbonate, heightened the amount of ice-rafted mineral detritus, and drastically altered the foraminiferal assemblages. Only a single polar foraminiferan, left-coiled *Globigerina pachyderma*, remained to dominate the cold waters, whereas diverse subpolar species and abundant Coccolithophorida flourished in warmer times. The major glacial pulses occurred from roughly 30,000–15,000 Y.B.P. (the late Wisconsin maximum) and from 160,000–135,000 Y.B.P. Shorter and less severe cold periods are present on at least four occasions in the last 225,000 years. Conversely, marked northward incursions of species typical of the present subtropical gyre happened at least six times, the three strongest being: slightly post Terminations I (8000 Y.B.P.) and II (125,000 Y.B.P.) and the oldest approximately 175,000 Y.B.P. The continuing similarity through late Pleistocene time for foraminiferal and coccolithophorid assemblages in the six northernmost cores suggests there has existed a uniform water mass with minimal variations from 45°N to above 55°N. The southern margin (42–45°N) of this faunally-defined unit in the eastern Atlantic coincides with the southward limit of late Pleistocene polar water penetration.

*Lamont–Doherty Geological Observatory of Columbia University and Dept. Earth and Environmental Science, Queens College of C.U.N.Y.
†United States Naval Oceanographic Office, Chesapeake Beach, Md.

20

Reprinted from *Geol. Soc. America Mem. 145,* 1976, pp. 111–112

Northeast Atlantic Paleoclimatic Changes over the Past 600,000 Years

W. F. Ruddiman

AND

A. McIntyre

ABSTRACT

In the subpolar Atlantic Ocean during the Quaternary Period, water-mass environments have migrated across more than 20° of latitude, which is equivalent to temperature oscillations of the ocean surface of at least 12°C. The migrations have occurred along a northwest-trending axis at mean rates of approximately 100 m/yr sustained over intervals of several centuries. During peak glaciations, polar water moved south to lat 42°N, where an abrupt frontal system separated the cyclonic subpolar gyre from the anticyclonic subtropical gyre.

Seven complete climatic cycles have occurred in the past 600,000 yr, within which at least 11 separate major southward advances of polar water have occurred. Both in number and shape, these cycles are correlative to oxygen isotopic cycles in the western equatorial Pacific Ocean and to palynologic cycles determined from a core from Macedonia. The northeast Atlantic cycle geometries are not so uniformly saw-toothed in form as isotopic curves from the equatorial Atlantic Ocean and Caribbean Sea because of interruptions by short but severe cold climatic pulses lasting for intervals as short as a few thousand years. One such pulse, which lasted only 7,000 yr, retained at least 90% of its original peak intensity despite vertical mixing.

Quantitative determination of the absolute input rates of the major sediment fractions over the glacial and interglacial portions of the last major climatic cycle shows that coccoliths and foraminifera were deposited two to three times more rapidly during interglaciations than glaciations; in converse proportions, coarse and fine terrigenous detritus was preferentially rafted into the northeast Atlantic Ocean during glaciations. The absence of coccoliths in polar water accounts for the existence of glacial coccolith-barren zones.

At the scale of local sediment redistribution (related to siting factors), fine coccolith carbonate is most easily redistributed. The absolute abundance of all coarse and fine components increases at higher net sedimentation rates, but fine carbonate increases most rapidly.

21

EARLY CENOZOIC CALCAREOUS NANNOPLANKTON BIOGEOGRAPHY OF THE ATLANTIC OCEAN

BILAL U. HAQ and G.P. LOHMANN

Woods Hole Oceanographic Institution, Woods Hole, Mass. (U.S.A.)

(Received and accepted January 20, 1976)

Abstract

Haq, B.U. and Lohmann, G.P., 1976. Early Cenozoic calcareous nannoplankton biogeography of the Atlantic Ocean. Mar. Micropaleontol., 1: 119—194.

Biogeographic patterns of Early Cenozoic calcareous nannoplankton assemblages are delineated for the North and South Atlantic Ocean, Caribbean Sea, and Gulf of Mexico. Nannoplankton assemblages are defined by Q-mode Varimax Factor and Oblique Factor Analyses of census data on 44 taxa from 113 deep-sea and land-based samples. Examination of their latitudinal distribution through time allows recognition of those assemblages which can be used as environmental indicators. Comparison of the distributions of contemporaneous nannoplankton assemblages with the distribution of the appropriate environmental indicator assemblage permits their classification as either low-, mid-, or high-latitude nannoflora.

Early Paleocene is characterized by a high-latitude Thoracosphaerid—*Markalius astroporus* Assemblage and a mid- to low-latitude Braarudosphaerid Assemblage.

Eight Middle Paleocene—Early Eocene nannoplankton assemblages are identified and grouped according to their relative environmental distribution:

(1) Low-latitude nannoflora: the *Toweius craticulus*—*Coccolithus pelagicus* Assemblage, the Discoaster—*Cyclococcolithus formosus* Assemblage, the *Toweius craticulus*—*Ericsonia subpertusa*—Discoaster Assemblage, and the Fasciculith—Discoaster Assemblage.

(2) Mid-latitude nannoflora: the *Ericsonia subpertusa* Assemblage and the *Coccolithus pelagicus* Assemblage.

(3) High-latitude nannoflora: the *Prinsius martinii* Assemblage and the *P. bisulcus* Assemblage.

These groupings are indicated by comparison of the distribution of our Paleocene—Early Eocene environmental indicator, the high-latitude *Prinsius martinii* Assemblage, with the distributions of contemporaneous assemblages.

Seven Eocene nannoplankton assemblages are identified:

(1) Low-latitude nannoflora: the Reticulofenestrid Assemblage and the *Cyclococcolithus formosus*—Sphenolith Assemblage.

(2) Mid-latitude nannoflora: the Discoaster Assemblage, the *Reticulofenestra umbilica*—*R. bisecta*—*Coccolithus pelagicus* Assemblage, and the *Cribrocentrum reticulatum* Assemblage.

(3) High-latitude nannoflora: the *Toweius craticulus*—*Coccolithus pelagicus* Assemblage, and the *C. pelagicus*—*Cyclococcolithus formosus*—*C.* aff. *gammation* Assemblage.

Our Eocene environmental indicator is the low-latitude Reticulofenestrid Assemblage.

Five Oligocene nannoplankton assemblages are identified:

(1) Low-latitude nannoflora: the Sphenolith—Discoaster Assemblage.

(2) Mid-latitude nannoflora: the *Cyclococcolithus neogammation* Assemblage and the *Dictyococcites hesslandii* Assemblage.

(3) High-latitude nannoflora: the Reticulofenestrid *R. bisecta* Assemblage and the *Coccolithus pelagicus* Assemblages.

Our Oligocene environmental indicator is the low-latitude Sphenolith—Discoaster Assemblage.

If it is assumed (1) that the latitudinal differentiation of calcareous nannoplankton assemblages we observe in the Early Cenozoic is related to a latitudinal temperature gradient, and (2) that the ecological preferences of these asssemblages remain stable through time, then the latitudinal nannofloral migrations we recognize delineate paleotemperature changes: The maximum equatorward migration (cooling) of high- and mid-latitude nannofloras in the Paleocene occurs at about 58 m.y. B.P. This is followed by their poleward migration, the disappearance of the high-latitude nannoflora, and the appearance of a new low-latitude nannoflora. A major poleward migration (warming) occurs at about 49 m.y. B.P. The Middle Eocene is characterized by the return of high-latitude nannofloras into mid-latitudes, with the maximum equatorward migration (cooling) occurring at 48—43 m.y. B.P. Low-latitude nannoflora again gradually invade high latitudes through the Late Eocene, indicating a second major Eocene warming by at least 38 m.y. B.P. There is a well-defined migration of high-latitude nannoflora into the mid-latitudes during the Middle Oligocene, with the maximum indicated cooling between 32 and 27 m.y. B.P.; an earlier, though minor, Oligocene cooling may have occurred at about 36 m.y. B.P. During the Late Oligocene, low-latitude nannoflora migrate into high latitudes, indicating a warming by 26 m.y. B.P. These inferred paleo-temperature changes are similar to those delineated by some workers on the basis of terrestrial flora.

Nannoflora characterized by the cool-water coccolith *Coccolithus pelagicus* predominate at the equator throughout most of the Paleocene, are confined mostly to high latitudes in the Eocene, and generally remain above mid-latitudes during the Oligocene. Although this migration could indicate that the Paleocene was the coolest epoch in the Early Cenozoic, and the Oligocene the warmest, we suggest instead that this shift of *C. pelagicus* to higher latitudes is a result of evolution in its ecology.

The implications of paleobiogeography to high-latitude biostratigraphy are discussed and the *acme horizon* as a time-stratigraphic concept is informally introduced.

The following new taxa are described: *Fasciculithus rotundus, Neochiastozygus imbriei* and *Thoracosphaera atlantica.*

The following taxa are recombined: *Cyclococcolithus protoannulus. Cyclolithella bramlettei Dictyococcites hesslandii,* and *Nannotetrina alata.*

196

22

Reprinted from pages 3861–3865, 3866–3867, 3869, 3871, and 3873–3876 of *Jour. Geophys. Research* **82**:3861–3876 (1977)

Calcareous Plankton Paleobiogeographic Evidence for Major Climatic Fluctuations in the Early Cenozoic Atlantic Ocean

Bilal U. Haq

Woods Hole Oceanographic Institution, Woods Hole, Massachusetts 02543

Isabella Premoli-Silva

Institute of Paleontology, Milan University, Milan, Italy

G. P. Lohmann

Woods Hole Oceanographic Institution, Woods Hole, Massachusetts 02543

It is only since the advent of the Deep Sea Drilling Project (DSDP) and the wider availability of older Cenozoic deep-sea core material that large-scale pre-Quaternary paleoenvironmental studies using marine microplankton could be attempted. The present study is one of the first attempts at mapping the early Cenozoic (65–24 m.y. B.P.) spatial and temporal distributions of calcareous plankton of the Atlantic Ocean using the DSDP cores and reconstructing the paleoclimatic history of the region from the resultant patterns. The early Cenozoic biogeographic patterns of calcareous nannoplankton (coccoliths and discoasters) and planktonic foraminifera have been delineated on the basis of quantitatively defined assemblages. These patterns show (1) that latitudinal differentiation among the calcareous planktonic groups existed during most of the early Cenozoic, with the exception of the earliest Paleocene (65–64 m.y. B.P.), when planktonic foraminiferal assemblages were essentially homogeneous through all latitudes, showing little provinciality, and when the floral gradients exhibited by nannoplankton were more closely related to near-shore and open ocean conditions and (2) that there have been major changes in the distributional patterns of both groups: those related to major latitudinal shifts with time and those due to evolution and disappearance of various forms. As is true in the present-day ocean, the early Cenozoic patterns are considered to be controlled mainly by the latitudinal thermal gradient (climate). Thus the temporal oscillations in the assemblages are interpreted as being caused by major climatic fluctuations. Four marked cooling episodes are recorded within the early Cenozoic Atlantic Ocean: those during the middle Paleocene (60–58 m.y. B.P.), the middle Eocene (46–43 m.y. B.P.), the earliest Oligocene (37–35 m.y. B.P.), and the middle Oligocene (32–28 m.y. B.P.). A particularly marked warming episode occurred during the late Paleocene–early Eocene (54–51 m.y. B.P.), and a second, less prominent warming trend began in the latest Oligocene (28 m.y. B.P.) and continued into the early Miocene.

Introduction

The study of paleobiogeography (spatial and temporal distribution) of microfossils can be of great value in the interpretation of past climatic and ecologic conditions. When environmentally controlled faunal or floral parameters can be recognized, changes in these parameters can be effectively used in interpreting changes in the environment. The potential of marine microfossils in reconstructing paleoclimatic and paleoceanographic conditions became obvious when it was first demonstrated that the biogeographic patterns of some planktonic foraminifera showed a strong correlation with the surface temperature patterns of the ocean [*Schott*, 1935]. Later studies of the distributional patterns of extant planktonic foraminifera [*Bradshaw*, 1959; *Parker*, 1960; *Phleger*, 1960] and calcareous nannoplankton [*McIntyre and Bé*, 1967; *Okada and Honjo*, 1973] clearly indicated that the geographic ranges of taxa were related to the thermal structure of the water masses and the latitudinal climatic gradient.

These distributional characteristics have been exploited as a powerful tool in the downcore studies of Quaternary climates. Numerous quantitative methods of analysis are now available for the reconstruction of Quaternary oceanographic conditions. These methods range from simple ratio manipulations of assemblages [*Lidz*, 1966] to more sophisticated transfer function analyses which quantify the correlations between the present distributions of certain temperature-sensitive Quat-

ernary assemblages and present sea surface temperatures. The results are used to derive climatic information about the past [e.g., *Imbrie and Kipp*, 1971; *Climap Project Members*, 1976]. Such paleoclimatic inferences based on comparison with modern distributional patterns can, however, only be made as far back as the late Pleistocene, when assemblages were composed of essentially the same species as are observed today. In earlier ages we deal with assemblages that are today mostly extinct. This limitation has discouraged the use of microfossils for paleoclimatic interpretations further back in time.

To resolve this problem, the geography of fossil assemblages must first be mapped; once their spatial and temporal distributions are known, assemblages that show relative restriction to either high or low latitudes can then be used as paleoclimatic indicators. By comparison with these, all other assemblages can be ordered along a relative latitudinal (climatic) gradient. Fluctuations in the biogeographic patterns of assemblages through time can then be interpreted in terms of the dynamic changes in the climatic/oceanographic system in the geologic past.

In this study we summarize Paleogene (65–23 m.y. B.P.) biogeographic data on calcareous plankton from the North and South Atlantic Ocean. The nannoplankton and planktonic foraminiferal assemblages are quantitatively defined, and their latitudinal distribution patterns are delineated in order to recognize those assemblages that can be used as paleoclimatic indicators. Major fluctuations in the distributional patterns of these assemblages are then used to decipher the early Cenozoic climatic history of the Atlantic Ocean.

There are some basic differences between the two calcareous plankton groups (nannoplankton and planktonic foraminifera) studied here; for example, the nannoplankton are photoautotrophic, living in the euphotic zone of the water column and thus reflecting surface water conditions; the planktonic foraminifera, on the other hand, are more vertically differentiated, occupying both shallow and deeper waters, and the assemblages therefore reflect both surface and deeper water conditions representing the habitat ranges of the component taxa. In spite of these differences and the differences in size, these two calcareous plankton groups behave in a sedimentologically similar manner with respect to dissolution, diagenetic, and preservational processes. These groups are also related in the evolutionary sense by the similarity in their rates of evolution and diversity patterns [*Berggren*, 1969; *Haq*, 1973] and their biogeographic responses to environmental factors [*McIntyre and Bé*, 1967; *Okada and Honjo*, 1973; *Parker*, 1960; *Phleger*, 1960].

MATERIAL AND SCOPE OF STUDY

Most of our biogeographic data have been gathered from the samples taken by the Deep Sea Drilling Project (DSDP). Many sites both in the North and in the South Atlantic have yielded early Cenozoic calcareous plankton, although the best sample coverage is in the North and Central Atlantic, the Caribbean, and the Gulf of Mexico. Some samples from land-based marine sections, piston cores, and drilling wells have been added for those time intervals for which DSDP materials are inadequate.

The calcareous nannoplankton biogeography of the Atlantic has been reported elsewhere [*Haq and Lohmann*, 1976] for materials from Atlantic DSDP legs 1–4, 10–12, 14, and 15. In the present study we have integrated nannoplankton data from more recent DSDP legs (36, 38, 39, and 43), which is important because the additional data are from areas where little or no information was available previously. The calcareous nannofossil biogeography has also been described in the initial reports for leg 36 [*Haq et al.*, 1977a], leg 38 [*Haq and Lohmann*, 1977], leg 39 [*Haq et al.*, 1977b], and leg 43 [*Haq et al.*, 1977c]. This synthesis does not include foraminiferal data from leg 38, for which planktonic foraminifera are very rare, or from leg 43, the material of which was not available for foraminiferal analysis. Additional data for both microfossil groups have also been included from earlier Atlantic DSDP legs (1–4, 10–12, 14, and 15).

We selected the Atlantic basin for our study for the following reasons. Our objective was to study the paleobiogeography of calcareous plankton within the framework of plate tectonics and sea floor spreading in an attempt to understand the dynamic interrelationships between paleobiogeography, paleoceanography, and the evolution of the Atlantic basin; the paleogeographic and geologic history of the Atlantic Ocean is better known than that of other oceans; moreover, the early Cenozoic DSDP sites and land-based marine sections around the Atlantic are widely distributed and ensure a reasonably good coverage for most of our time intervals. Since 1972, when this study was initiated, many more DSDP sites have been added, and our data base has thus increased considerably.

APPROACH

Our approach consists of five major steps: (1) data census, (2) data reduction, (3) delineation of biogeographic patterns,

(4) identification of climatic indicator assemblages, and (5) paleoclimatic interpretation.

Our raw data are a census of fossil calcareous nannoplankton and planktonic foraminifera. In the case of nannoplankton, approximately 300 specimens are counted in random traverses of smear slides made from raw samples (see *Haq and Lohmann* [1976, pp. 125–126] for a discussion of procedural problems concerning nannofossils). In the case of foraminifera, washed samples larger than 63 μm have been split into three size fractions: >250 μm, 250–149 μm, and <149 μm. About 300 specimens are counted in random traverses of each size fraction unless the number of specimens is less than 300, in which case the total number are counted. For both nannofossils and foraminifera, many individual species have been grouped within supraspecific categories. This is necessary because individual taxa often have short vertical ranges and plots of their temporal distributions would be meaningless for long-term paleoclimatic interpretations. The groupings are based on evolutionary affinities. Thus, for example, of the 206 identified foraminiferal taxa, 38 general categories are recognized, mainly on the basis of evolutionary affinities. These broader taxonomic groups are more likely than individual taxa to remain stable in response to a possible evolution in their ecologic requirements. Such lumping thus reduces the possible effect of this factor on the biogeographic patterns (see *Haq and Lohmann* [1976] for a disucssion of this aspect).

The raw census data (number of nannofossils and foraminifera) have been used to determine plankton assemblages by Q mode varimax factor analysis. Assemblages defined in this way account for most of the observed variation among our samples and approximate natural taxonomic associations.

Once the major plankton assemblages have been recognized, their relative abundances can be plotted on time versus paleolatitude plots, thereby delineating their distributions in space and time. Examination of these plots allows us to identify relatively high latitude (cool) and low latitude (warm) assemblages.

Changes in the latitudinal (climatic) distribution of the assemblages can then be used to infer changes in the climate, i.e., shifts of low-latitude (warm) assemblages toward high latitudes, reflecting warming trends, and shifts of high-latitude (cool) assemblages toward low latitudes, reflecting cooling trends.

CHRONOLOGIC FRAMEWORK

Originally, our objective was to produce a series of synoptic paleobiogeographic maps for selected and relatively equispaced time intervals. Our choice focused on those intervals for which more widely distributed sample material was available. This resulted in a number of sets of near-contemporaneous samples ranging within discrete 'slices' of time. This 'time slice' approach was a useful working concept in constructing the initial framework of biogeographic patterns. Thus our earlier study [*Haq and Lohmann*, 1976] focused on 12 selected time slices within the early Cenozoic. After having outlined the overall picture with the time slice approach we have now attempted to study more continuous stratigraphic sections in order to refine the details, and toward this goal we have considered some of the available continuously cored sections from the more recent DSDP legs. In addition, we have also attempted to fill the gaps between time slices from the earlier legs as far as possible.

Our biostratigraphic-chronologic criteria for age determination of samples are summarized in Figure 1. This summary is based on a correlation of planktonic foraminiferal and nanno-plankton biostratigraphic zones which has been chronologically calibrated by *Berggren* [1972]. All samples were reexamined to estimate an **age** based on foraminiferal and/or nannofossil biostratigraphy. In numerous cases, dates were provided (or refined) by interpolation using known or inferred sediment accumulation rates.

EARLY CENOZOIC CALCAREOUS PLANKTON BIOGEOGRAPHY

In this section we summarize the biogeographic patterns of the major nannofossil and planktonic foraminiferal assemblages differentiated by the *Q* mode varimax factor analysis (see *Haq and Lohmann* [1976] for details of the method).

Major evolutionary changes in the floral components of the nannofossil assemblages during the early Cenozoic require separate analysis of the Paleocene–early Eocene, Eocene, and Oligocene intervals. Similarly, major foraminiferal changes within the Paleogene warrant separate consideration of the Paleocene–Eocene and late middle Eocene–Oligocene intervals.

The biogeographic patterns of only the major assemblages are presented here on time-paleolatitude plots (Figures 2–6); relative abundances are indicated by the sizes of the circles drawn around the samples. The assemblages, as defined by the factor analysis, may consist of one or more (or groups of taxa); however, here they are named after the most dominant taxon (or group of taxa) present in them. (For the composition of assemblages, see figures and captions.)

CALCAREOUS NANNOPLANKTON ASSEMBLAGES

Paleocene–Early Eocene Interval

Analysis of census data from 101 samples from the Paleocene–early Eocene interval (see Figure 7 for the paleolatitudinal distribution of samples) allows the differentiation of three major nannoplankton assemblages having distinct spatial and temporal patterns (Figure 2).

Starting in the early Paleocene at about 64–63 m.y. B.P., the *Ericsonia subpertusa* assemblage was most prominent (Figure 2a). It first appeared in the lower latitudes but later extended its range to mid-latitudes, where it persisted as an important biogeographic element until 53 m.y. B.P. (latest Paleocene), disappearing after 50 m.y. B.P., within the early Eocene.

A second major coccolith assemblage of this interval, the *Prinsius martinii* assemblage (Figure 2b), first became important in the early Paleocene at about 62 m.y. B.P. in the high latitudes; it migrated toward mid-latitude regions between 61 and 59 m.y. B.P. and then withdrew back into higher latitudes. It occurred in relatively minor amounts in the middle and low latitudes until the late Paleocene, disappearing completely after 50 m.y. B.P., within the early Eocene.

The last major coccolith assemblage of the Paleocene–early Eocene interval was the *Toweius craticulus* assemblage (Figure 2c). It first appeared in small numbers in the mid-latitudes in the early Paleocene, around 63 m.y. B.P., achieving prominence at about 60 m.y. B.P. and then extending its range into low latitudes around 59 m.y. B.P. It migrated into higher latitudes in the early Eocene at about 52 m.y. B.P.

Eocene Interval

Census data from 64 samples were analyzed to identify five major nannoplankton assemblages in the Eocene (Figure 3). Sixteen early Eocene samples in this analysis overlap those analyzed separately for the Paleocene–early Eocene interval

AGE		CALCAREOUS NANNOPLANKTON ZONES		PLANKTONIC FORAMINIFERAL ZONES
25 —	LATE OLIGOCENE	NN 1	T carinatus	G. kugleri
		NP 25	S. ciperoensis	G angulisuturalis
		NP 24	S distentus	G. opima opima
30 —	EARLY OLIGOCENE	NP 23	S. predistentus	G. ampliapertura
35 —		NP 22	H reticulata	Cassigerinella chipolensis/ Pseudohastigerina micra
		NP 21	E subdisticha	
40 —	LATE EOCENE	NP 20	S pseudoradians	G. cerroazulensis
		NP 19	L recurvus	
		NP 18	C oamaruensis	G. seminvoluta
	MIDDLE EOCENE	NP 17	D saipanensis	T. rohri
45 —		NP 16	D tani nodifer	O. beckmanni
		NP 15	N alatus	G. lehneri
				G. subconglobata
				H. aragonensis
50 —	EARLY EOCENE	NP 14	D sublodoensis	G. palmerae
		NP 13	D lodoensis	G. aragonensis
		NP 12	T orthostylus	G. formosa
		NP 11	D binodosus	G. subbotinae
		NP 10	M contortus	G. edgari
55 —	LATE PALEOCENE	NP 9	D multiradiatus	G. velascoensis
		NP 8	H riedeli	G. pseudomenardii
		NP 7	D mohleri	
		NP 6	H kleinpelli	
		NP 5	F tympaniformis	G. pusilla pusilla
60 —		NP 4	E macellus	G. angulata
				G. uncinata
	EARLY PALEOCENE	NP 3	C danicus	G. trinidadensis
		NP 2	C tenuis	G. pseudobulloides
65 —		NP 1	M astroporous	G. eugubina

Fig. 1. Summary of biochronology used for age determination of samples. Samples were assigned a minimum and a maximum age in million years before present (m.y. B.P. or Ma) on the basis of characteristic zonal nannoplankton and/or foraminifera present in them (nannoplankton zonation after *Martini* [1971]; planktonic foraminiferal zonation after *Bolli and Premoli-Silva* [1973] and *Premoli-Silva and Bolli* [1973]; correlation of biostratigraphic zones with absolute time scale after *Berggren* [1972]).

(see Figure 8 for the paleolatitudinal distribution of samples analyzed for the Eocene interval).

In the earliest Eocene the Paleocene *Toweius craticulus* assemblage (Figure 3a) continued to be prominent in the low to middle latitudes; it migrated into higher latitudes between 52 and 50 m.y. B.P. After 49 m.y. B.P. it lost prominence.

In the early Eocene a new assemblage, composed of small-to medium-sized coccoliths having reticulate central areas, the reticulofenestrid assemblage (Figure 3b), appeared, replacing the *Toweius craticulus* assemblage in the mid-latitudes around 50 m.y. B.P. (late early Eocene). It shifted slightly toward lower latitudes at about 45–44 m.y. B.P., in the late Eocene, but reestablished itself once again in the Oligocene high latitudes.

The discoaster assemblage (Figure 3c) also achieved quantitative prominence in the early Eocene around 51 m.y. B.P. It was prominent mostly in the mid-latitude regions, shifting into lower latitudes between 46 and 45 m.y. B.P. (middle Eocene) and disappearing after 38 m.y. B.P. (late Eocene).

Two other major Eocene assemblages were the *Reticulofenestra umbilica* assemblage (Figure 3d) and the *Coccolithus pelagicus* (s. ampl.) assemblage (Figure 3e). Both of

199

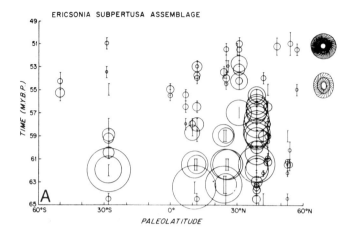

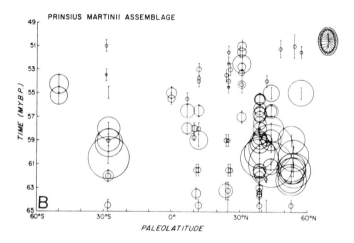

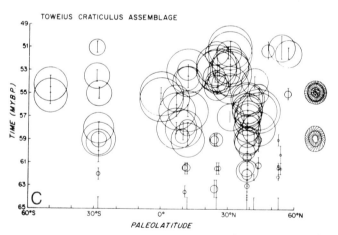

Fig. 2.   Time and paleolatitudinal distribution of the major Paleocene-early Eocene nannoplankton assemblages. Sample positions are indicated by vertical line segments, the lengths of which represent the uncertainty of the estimated age. The sizes of the circles drawn around the samples indicate the relative abundances of assemblages in the samples (circles with diameters spanning 20° of paleolatitude indicate 100% abundance). Major and minor components of the assemblages (as defined by the Q mode factor analysis) are drawn in the top right corners. (a) *Ericsonia subpertusa* assemblage (dominated by *E. subpertusa* and containing minor amounts of *Coccolithus pelagicus* (s. ampl.)). (b) *Prinsius martinii* assemblage (also contains very small amounts of chiasmoliths). (c) *Toweius craticulus* assemblage (*T. craticulus* being the major component and *Coccolithus pelagicus* the minor component).

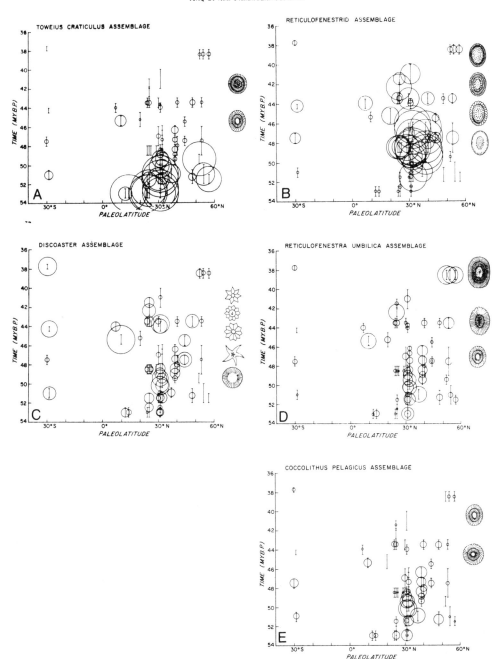

Fig. 3.   Time and paleolatitudinal distribution of the major Eocene nannoplankton assemblages (see Figure 2 caption for plotting format). (*a*) *Toweius craticulus* assemblage (*T. craticulus* and *Coccolithus pelagicus* being the major and minor components, respectively). (*b*) Reticulofenestrid assemblage (composed of small- to medium-sized reticulofenestrids). (*c*) Discoaster assemblage. (*d*) *Reticulofenestra umbilica* assemblage (dominated by this large reticulofenestrid and containing lesser amounts of *Dictyococcites bisectus* and *Coccolithus pelagicus*). (*e*) *Coccolithus pelagicus* assemblage (also contains minor amounts of *Cyclococcolithus formosus*).

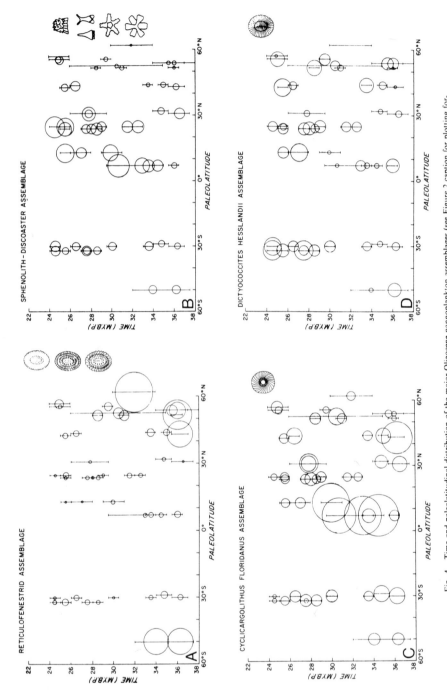

Fig. 4. Time and paleolatitudinal distribution of the major Oligocene nannoplankton assemblages (see Figure 2 caption for plotting format). (*a*) Reticulofenestrid assemblage (containing minor amounts of *Dictyococcites bisectus*). (*b*) Sphenolith-discoaster assemblage (containing sphenoliths as the dominant component and discoasters as the less dominant component. (*c*) *Cyclicargolithus floridanus* assemblage. (*d*) *Dictyococcites hesslandii* assemblage.

these had somewhat similar distributional patterns, prominent mostly in mid-latitudes. However, the former showed a shift into lower latitudes between 46 and 43 m.y. B.P. (middle Eocene) and prominence in higher latitudes at about 38 m.y. B.P. (late Eocene).

Oligocene Interval

Nannoplankton data from the 49 Oligocene samples (see Figure 9 for the paleolatitudinal distribution of samples) were factor-analyzed to define four major assemblages in this interval (Figure 4).

The Eocene reticulofenestrid assemblage appeared once again in the early Oligocene (Figure 4a), mainly in the high latitudes between 37 and 33 m.y. B.P.

Among the other major assemblages of the Oligocene the sphenolith-discoaster assemblage (Figure 4b) dominated the low latitudes between 35 and 24 m.y. B.P., the *Cyclicargolithus floridanus* (Figure 4c) and the *Dictyococcites hesslandii* (Figure 4d) assemblages were both cosmopolitan assemblages occurring throughout the Oligocene; however, the former showed relative prominence in low latitudes between 35 and 29 m.y. B.P. (early Oligocene). These cosmopolitan assemblages characterized the Oligocene biogeographic scene.

Paleoenvironmental Significance of Nannoplankton Assemblages

The following nannofossil assemblages best meet the requirements as paleoclimatic indicators, because they show the highest frequencies at either high latitudes or low latitudes.

1. In the Paleocene–early Eocene the *Prinsius martinii* assemblage shows a clear preference for the high latitudes, making only one incursion into the lower latitudes of a relatively short duration (see Figure 2b). We select this assemblage as our high-latitude paleoclimatic indicator for this interval.

2. For the Eocene the reticulofenestrid assemblage is selected as an indicator of mid-latitudes, where it is most prominent through most of the Eocene (see Figure 3b).

3. The sphenolith-discoaster assemblage shows the clearest patterns, remaining in low latitudes throughout the Oligocene, and is therefore the obvious choice as the Oligocene low-latitude indicator (see Figure 4b).

By comparing the abundances of other assemblages with the abundances of these paleoclimatic indicators the relative latitudinal positions of the major nannoplankton assemblages are indicated (Table 1).

[*Editor's Note:* The description of foraminiferal assemblages has been omitted at this point.]

TABLE 1. Relative Latitudinal Positions of the Major Nannoplankton Assemblages

Latitudinal Preference	Assemblage	Time Interval When Dominant
Paleocene–early Eocene		
Low	*Toweius craticulus*	middle Paleocene–early Eocene
Middle	*Ericsonia subpertusa*	early Paleocene–late Paleocene
High	*Prinsius martinii*	early Paleocene–late Paleocene
Eocene		
Low to middle*	reticulofenestrid	early Eocene–late Eocene
	discoaster	early Eocene–middle Eocene
Middle	*Reticulofenestra umbilica*	early Eocene–early late Eocene
Middle to high*	*Coccolithus pelagicus*	early Eocene–middle Eocene
	Toweius craticulus	early Eocene–early middle Eocene
Oligocene		
Low	sphenolith–discoaster	early Oligocene–late Oligocene
Cosmopolitan	*Cyclicargolithus floridanus*	early Oligocene–late Oligocene
Cosmopolitan	*Dictyococcites hesslandii*	early Oligocene–late Oligocene
High	reticulofenestrid	early Oligocene

*Part of range emphasized.

203

MAJOR MIGRATIONARY PATTERNS AND PALEOCLIMATIC HISTORY OF THE ATLANTIC OCEAN

In this section we use the latitudinally ranked calcareous plankton assemblages and their major spatial shifts through time to interpret the early Cenozoic history of the Atlantic Ocean. Since most of our data are from the Central and North Atlantic, our interpretations are mainly valid for latitudes of 0°–60°N. However, as is clear from many of the patterns, the assemblage characteristics and distributions are often symmetrical on both sides of the equator, and at times we have made use of the southern latitude data to confirm or integrate our results from the northern latitudes.

In the case of calcareous nannoplankton we have combined major assemblages (described here) and minor assemblages (not described here) having similar latitudinal preferences into general low-, mid-, or high-latitudinal categories (see Figure 12) to enable us to visualize the major migrationary patterns more distinctly (see *Haq and Lohmann* [1976] for a description of minor assemblages).

The summary of major migrationary patterns of both nannoplankton and foraminiferal assemblages is given in Figure 12. These patterns suggest the following sequence of paleoclimatic events in the Atlantic Ocean during the early Cenozoic.

1. The earliest Paleocene (65–63 m.y. B.P.) was characterized by generally cool climates, as is indicated by the high- to low-latitudinal spread of the intermediate low-spired subbotinid foraminiferal assemblage. This was followed by a period of relatively warm climates between 63 and 60 m.y. B.P., when the low-spired subbotinids became restricted to the higher latitudes, and at the same time a new, relatively low latitude morozovellid assemblage appeared in the low latitudes.

2. This trend was reversed around 60 m.y. B.P. (beginning of the late Paleocene), when the low-spired subbotinid assemblage once again invaded the low latitudes and the morozovellid assemblage became restricted to the lowest latitudes. The spread of the high-latitude nannofloral *Prinsius martinii* assemblage into low latitudes confirms this cooling episode, which had a peak near 59 m.y. B.P.

3. Beginning at about 57 m.y. B.P., all evidence suggests a marked warming trend in the late Paleocene and early Eocene. This is indicated by (1) the migration of the intermediate low-spired subbotinid assemblage to higher latitudes, (2) the major shift of the low-latitude morozovellid assemblage toward mid-latitudes, (3) the reversal in the migration of the high-latitude *P. martinii* assemblage from low latitudes toward high latitudes and its eventual disappearance around 54 m.y. B.P., and (4) the major latitudinal shifts of both low- and mid-latitude nannofloral assemblages (*Toweius craticulus* and *Ericsonia subpertusa*) toward higher latitudes. Peak warming occurred between about 53 and 50 m.y. B.P., within the early Eocene, when the low-spired subbotinid assemblage had a maximum high-latitudinal distribution (up to 60°N paleolatitude). This

warm trend lasted until about 47 m.y. B.P., and climates remained relatively stable. However, foraminiferal patterns show at least one minor fluctuation (cooling) within this warm interval. This occurred around 54–53 m.y. B.P., when the low-spired subbotinid assemblage reappeared in larger numbers in lower latitudes. Three assemblage events are noticeable within this interval: (1) the low prominence of low-spired subbotinid and morozovellid assemblages around 47 m.y. B.P. (typical morozovellids become extinct at that level), (2) the appearance of a new *Globigerinatheka* assemblage and the prominence of the biserial heterohelicid assemblage, and (3) the disappearance of mid- and high-latitude nannofloral assemblages (*Ericsonia subpertusa* and *Prinsius martinii*) between 53 and 50 m.y. B.P.

4. Relatively cooler climates returned between 46 and 43 m.y. B.P. (late middle Eocene). This is indicated by the spread of the intermediate biserial heterohelicid assemblage into low latitudes, followed by the appearance at about 45 m.y. B.P. of the high-latitude foraminiferal globigerinid assemblage and its extension into mid-latitudes at about 44 m.y. B.P. This evidence is corroborated by the migrations of high-latitude nannofloral assemblages (*Toweius craticulus* and *Coccolithus pelagicus*) into low latitudes at this time.

5. The middle Eocene cooling episode was followed by relatively stable and generally warmer climates in the late Eocene, lasting until 38 m.y. B.P. The high-latitude nannofloral assemblages withdrew back to the higher latitudes, and the reticulofenestrids extended their range up to 30°N. High-latitude foraminiferal assemblages also withdrew to higher latitudes during this time. A major change in the foraminiferal fauna took place in the late middle Eocene: the formerly prominent low- to mid-latitude acarininid assemblage, as well as the few lingering remnants of morozovellids, became extinct around 43 m.y. B.P.

6. In the early Oligocene, between 37 and 35 m.y. B.P., a marked cooling episode is indicated, peak cooling occurring around 36 m.y. B.P., when a new, relatively high latitude foraminiferal globigerinid assemblage invaded the lower latitudes and the Oligocene high-latitude nannofloral assemblages (reticulofenestrid and *Coccolithus pelagicus*) extended their range equatorward.

7. After an initial reversal of this cooling trend between 35 and 32 m.y. B.P., indicated by the withdrawal of high-latitude nannofloral and foraminiferal assemblages to higher latitudes, a second Oligocene cooling event occurred between 32 and 28 m.y. B.P. During this time the high-latitude assemblages once again invaded the low latitudes, and the low-latitude nannofloral sphenolith-discoaster assemblage was restricted to the low latitudes. A new, relatively low latitude foraminiferal globoquadrinid assemblage became prominent around 35 m.y. B.P. and remained within the low and middle latitudes during the cool episode.

8. The last important Paleogene climatic event was a warming trend in the latest Oligocene (beginning after 28 m.y.

B.P.). This is demonstrated by the return of the high-latitude foraminiferal assemblages to their original habitat and a poleward shift of the low-latitude globoquadrinid assemblage. The low-latitude sphenolith-discoaster nannofloral assemblage also extended its range into the mid-latitudes, and the high-latitude nannofloral assemblages withdrew from the low latitudes.

DISCUSSION

From the preceding account of the biogeography we have demonstrated two things.

1. The latitudinal differentiation (provincialism) among both calcareous planktonic groups has existed throughout most of the early Cenozoic. An exception to this was the earliest Paleocene (65–64 m.y. B.P.). During this interval, which followed a period of massive extinction of marine plankton [*Bramlette, 1965*] in the latest Cretaceous, only a few older but ecologically robust remnants, which had wide latitudinal distributions, survived. Thus owing to the wide geographic ranges of taxa and the low total diversities, the earliest Paleocene shows little or no latitudinal provinciality. The foraminiferal assemblages were essentially homogeneous through all latitudes, and the floral gradients exhibited by nannoplankton were determined more by their proximity to continental margins; i.e., the hemipelagic braarudosphaerid pentaliths were dominant near-shore elements but were rare or absent in the open ocean, where the other nannofloral element of this period, the thoracosphaerids, dominated [see *Haq et al.,*

1977c]. There was no latitudinal differentiation of nannoplankton assemblages.

2. The latitudinal provinciality changed with time, and there were major changes in the distribution patterns of both planktonic foraminifera and nannoplankton. These changes involved both major shifts of assemblages through space and time and major evolutionary changes, involving appearances of new dominant groups and disappearances of old ones, that altered the nature of the assemblages.

It was the major migrations of assemblages through the latitudes that we were able to apply as a tool to reconstruct the paleoclimatic history of the Atlantic Ocean. We inferred four marked cooling episodes within the early Cenozoic (middle Paleocene, middle Eocene, earliest Oligocene, and middle Oligocene), a marked warming during the late Paleocene and early Eocene, and a less pronounced warming event in the late Oligocene. How does our interpretation of these events compare with paleoclimatic evidence from other independent sources?

When they were compared with the paleoclimatic record on land, three of the major cooling events indicated by our data (middle Paleocene, middle Eocene, and middle Oligocene) were also shown by the analysis of the character of the leaf margins from western North America by *Wolfe and Hopkins* [1967]. High percentages of entire-margined leaves are characteristic of the subtropical modern floras, and low percentages are characteristic of cool to temperate climate floras. These authors also inferred a gradual warming climate in the late

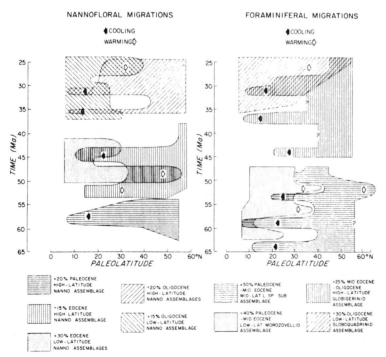

Fig. 12. A summary of the major nannofloral and foraminiferal migrationary patterns through the early Cenozoic. Migrations toward higher latitudes are interpreted as being caused by climatic warming, and those toward lower latitudes, by climatic cooling. The patterns delineated enclose all samples which contain abundances greater than those indicated in the legend. Arrows in these areas indicate the directions of the major shifts of assemblages. Major and minor nannofloral assemblages having similar latitudinal preferences have been combined to obtain composite patterns in some cases (see text for details).

Paleocene and early Eocene from their data. However, they could not show a sharp drop in temperature in the earliest Oligocene owing to lack of data points for this interval.

Although there are no clear-cut correlations, the sequence of paleoclimatic events inferred by us shows a general agreement with the paleotemperature curves obtained through $^{18}O/^{16}O$ analysis of calcareous microfossils. With the exception of the middle Paleocene cooling, most Cenozoic oxygen isotope paleotemperature curves show a sequence of climatic events similar to ours: a marked warming in the late Paleocene and early Eocene [*Douglas and Savin*, 1975; *Savin et al.*, 1975; *Margolis et al.*, 1975; *Shackleton and Kennett*, 1975], a drop in temperature in the middle Eocene [*Margolis et al.*, 1975; *Shackleton and Kennett*, 1975], a marked cooling in the earliest Oligocene [*Dorman*, 1966; *Devereux*, 1967; *Douglas and Savin*, 1975; *Savin et al.*, 1975; *Margolis et al.*, 1975; *Shackleton and Kennett*, 1975], a middle Oligocene cooling [*Douglas and Savin*, 1973; *Savin et al.*, 1975, DSDP site 167], and a late Oligocene warming trend, which continued into the early Miocene [*Dorman*, 1966; *Devereux*, 1967; *Douglas and Savin*, 1973, 1975; *Savin et al.*, 1975]. The middle Paleocene cooling event inferred by us has not been reported from the oxygen isotopic record. This may be due to the lack of well-preserved samples from the Paleocene, which has discouraged detailed isotopic analyses of this interval. Hopefully, with more extensive and better preserved samplings of this part of the record the middle Paleocene climatic event can be resolved.

Acknowledgments. This paper was presented at the Joint Oceanographic Assembly Symposium on Paleoceanography, convened by T. van Andel and held in Edinburgh, Scotland, in September 1976. The research was supported by National Science Foundation grants GA 21983 and GA 21274. The paper was reviewed by W. A. Berggren, J. P. Kennett, A. McIntyre, and R. C. Tjalsma. Woods Hole Oceanographic Institution contribution 3872.

REFERENCES

Berggren, W. A., Rates of evolution in some Cenozoic planktonic foraminifera, *Micropaleontology, 15,* 351–365, 1969.

Berggren, W. A., A Cenozoic time-scale—Some implications for regional geology and paleobiogeography, *Lethaia, 5,* 195–215, 1972.

Bolli, H. M., and I. Premoli-Silva, Oligocene to Recent planktonic foraminifera and stratigraphy of leg 15 sites in the Caribbean Sea, in *Initial Reports of the Deep Sea Drilling Project, Leg 15,* pp. 475–498, U.S. Government Printing Office, Washington, D. C., 1973.

Bradshaw, J. S., Ecology of living planktonic foraminifera in the north and equatorial Pacific Ocean, *Contrib. Cushman Found. Foraminiferal Res., 10,* 25–64, 1959.

Bramlette, M. N., Massive extinctions in the biota at the end of Mesozoic time, *Science, 148,* 1696–1699, 1965.

Climap Project Members, The surface of the Ice-Age earth, *Science, 191,* 1131–1137, 1976.

Devereux, I., Oxygen isotope paleotemperature measurements on New Zealand Tertiary fossils, *N. Z. J. Sci., 10,* 988–1011, 1967.

Dorman, F. H., Australian Tertiary paleotemperatures, *J. Geol., 74,* 49–61, 1966.

Douglas, R. G., and S. M. Savin, Oxygen and carbon isotope analyses of Cretaceous and Tertiary foraminifera from the central North Pacific, in *Initial Reports of the Deep Sea Drilling Project, Leg 17,* pp. 591–605, U.S. Government Printing Office, Washington, D. C., 1973.

Douglas, R. G., and S. M. Savin, Oxygen and carbon isotope analyses of Tertiary and Cretaceous microfossils from Shatsky Rise and other sites in the North Pacific Ocean, in *Initial Reports of the Deep Sea Drilling Project, Leg 32,* pp. 509–520, U.S. Government Printing Office, Washington, D. C., 1975.

Haq, B. U., Transgressions, climatic change, and diversity of calcareous nannoplankton, *Mar. Geol., 15,* 25–30, 1973.

Haq, B. U., and G. P. Lohmann, Early Cenozoic calcareous nannoplankton biogeography of the Atlantic Ocean, *Mar. Micropaleontol., 1,* 119–197, 1976.

Haq, B. U., and G. P. Lohmann, Remarks on the Oligocene calcareous nannoplankton biogeography of the Norwegian Sea (D.S.D.P. leg 38), in *Initial Reports of the Deep Sea Drilling Project, Leg 38,* U.S. Government Printing Office, Washington, D. C., in press, 1977.

Haq, B. U., G. P. Lohmann, and S. W. Wise, Calcareous nannoplankton biogeography and its paleoclimatic implications—Cenozoic of the Falkland Plateau (D.S.D.P. leg 36) and Miocene of the Atlantic Ocean, in *Initial Reports of the Deep Sea Drilling Project, Leg 36,* U.S. Government Printing Office, Washington, D. C., in press, 1977a.

Haq, B. U., K. Perch-Nielsen, and G. P. Lohmann, Contribution to the Paleocene calcareous nannofossil biogeography of the central and SW Atlantic (Ceara Rise and Sao Paulo Plateau, D.S.D.P. leg 39), in *Initial Reports of the Deep Sea Drilling Project, Leg 39,* U.S. Government Printing Office, Washington, D. C., in press, 1977b.

Haq, B. U., H. Okada, and G. P. Lohmann, Paleobiogeography of the Paleocene-Eocene calcareous nannoplankton from the North Atlantic Ocean, in *Initial Reports of the Deep Sea Drilling Project, Leg 43,* U.S. Government Printing Office, Washington, D. C., in press, 1977c.

Imbrie, J., and N. Kipp, A new micropaleontological method for quantitative paleoclimatology application to a Late Pleistocene Caribbean core, in *The Late Cenozoic Glacial Ages,* edited by K. K. Turekian, chap. 5, pp. 71–181, Yale University Press, New Haven, Conn., 1971.

Lidz, L., Deep-sea Pleistocene biostratigraphy, *Science, 154,* 1448–1452, 1966.

Margolis, S. V., P. M. Kroopnick, D. E. Goodney, W. C. Dudley, and M. E. Mahoney, Oxygen and carbon isotopes from calcareous nannofossils as paleoceanographic indicators, *Science, 189,* 555–557, 1975.

Martini, E., Standard Tertiary and Quaternary calcareous nannoplankton zonation, in *Proceedings of the 2nd Planktonic Conference, Rome (1970),* pp. 739–785, Edizioni Technoscienza, Rome, 1971.

McIntyre, A., and A. H. W. Bé, Modern Coccolithophoridae of the Atlantic Ocean, I, Placoliths and crytoliths, *Deep Sea Res., 14,* 561–597, 1967.

Okada, H., and S. Honjo, The distribution of oceanic coccolithophorids in the Pacific, *Deep Sea Res., 20,* 355–374, 1973.

Parker, F. L., Living planktonic foraminifera from the equatorial and S.E. Pacific, *Sci. Rep. Tohoku Univ., Ser. 2, 4,* 71–82, 1960.

Phleger, F. B., *Ecology and Distribution of Recent Planktonic Foraminifera,* 297 pp., Johns Hopkins Press, Baltimore, Md., 1960.

Premoli-Silva, I., and H. M. Bolli, Late Cretaceous to Eocene planktonic foraminifera and stratigraphy of leg 15 sites in the Caribbean Sea, in *Initial Reports of the Deep Sea Drilling Project, Leg 15,* pp. 499–528, U.S. Government Printing Office, Washington, D. C., 1973.

Savin, S. M., R. G. Douglas, and F. G. Stehli, Tertiary marine paleotemperatures, *Geol. Soc. Amer. Bull., 86,* 1499–1510, 1975.

Schott, W., Die Foraminiferen in dem äquatorialen Teil des Atlantischen Ozeans, *Wiss. Ergeb., Deut. Atlantischen Exped. Vermess. Forschungsschiff Meteor 1925–1927, III(3), Sect. B,* 43–134, 1935.

Shackleton, N. J., and J. P. Kennett, Paleotemperature history of the Cenozoic and the initiation of antarctic glaciation: Oxygen and carbon isotope analyses in D.S.D.P. sites 277, 279, and 281, in *Initial Reports of the Deep Sea Drilling Project, Leg 29,* pp. 743–755, U.S. Government Printing Office, Washington, D. C., 1975.

Wolfe, J. A., and D. M. Hopkins, Climatic changes recorded by Tertiary land floras in northwestern North America, in *Tertiary Correlations and Climatic Changes in the Pacific, 11th Pacific Science Congress, Tokyo,* pp. 67–76, 1967.

(Received November 1, 1976;
revised March 15, 1977;
accepted March 16, 1977.)

206

23

Reprinted from pages 414–432 and 442–443 of *Micropaleontology* **26**:414–443 (1980)

Biogeographic history of Miocene calcareous nannoplankton and paleoceanography of the Atlantic Ocean

Bilal U. Haq
Woods Hole Oceanographic Institution
Woods Hole, Massachusetts 02543

ABSTRACT

Biogeographic patterns of Miocene calcareous nannoplankton in the North and South Atlantic Ocean, the Caribbean Sea and the Gulf of Mexico are apparent from a Q-mode factor analysis of census data from 49 DSDP sites. In a total of 444 relatively well-preserved Miocene samples, 5 major and 1 minor (but relatively important) assemblages were recognized, most of which show distinct spatial and temporal distribution patterns. The major assemblages (identified by the dominant taxa) are the *Dictyococcites minutus, Cyclicargolithus floridanus, Coccolithus pelagicus, Reticulofenestra pseudoumbilica-R. haqii,* and *Discoaster-Sphenolithus* assemblages. Four of the major assemblages show distinct shifts through latitudes that are interpreted as a response to changing climate. Four warming and cooling cycles of 4 to 4.5 m.y. duration are identified in the Miocene. Both the paleobiogeographic and available isotopic data show the warming episode between 17 and 15.5 Ma and the cooling trend that followed, as well as the warming episode between 9 and 7 Ma, to have been particularly extreme in both Hemispheres, and probably of global extent. The sharp cooling centered at 15 Ma and the biogeographic changes at this time suggest this event to have been associated with the development of an extensive ice-cap on Antarctica.

INTRODUCTION

This study is a part of a continuing effort to delineate the spatial and temporal distribution patterns of Cenozoic calcareous nannoplankton in the Atlantic Ocean. Among the world's oceans, the Atlantic has the maximum latitudinal extent, ranging from subpolar to equatorial, in both Northern and Southern Hemispheres. This ocean is thus ideal for the study of the temporal evolution of the latitudinally differentiated phytoplankton such as calcareous nannoplankton, whose distribution is closely related to the surface circulation patterns. The availability of relatively well-distributed drill cores in the Atlantic facilitate a comparison between the patterns in the two Hemispheres, the glacio-climatic histories of which has been different at least since the middle Cenozoic time.

Paleogeographically, the Miocene Atlantic Ocean was much as it is today, the major exception being the existence of the Central American connection between the Pacific and the central Atlantic until the late Pliocene. Seafloor spreading subsequent to Magnetic Anomaly 6 time (about 21 Ma) has only slightly changed the position of the plates surrounding the Atlantic (Sclater et al., 1977). General circulation patterns in the North Atlantic became established during the Early Miocene when water from the Norwegian Sea began to spill over in significant quantities into the northern North Atlantic (Talwani and Udinstev, 1976). A number of extra-Atlantic paleogeographic events, however, have modified the circulation and climatic patterns in the Atlantic Ocean since that time. During the Early Miocene (about 18 Ma) the collision of Africa and Eurasia finally separated the eastern Tethys from its western part and stopped the trans-Tethyan circulation westward. The closure of the Tethyan connection significantly modified the flow of the eastern limb of the ancestral Gulf Stream by deflecting it southward (Berggren and Hollister, 1974). As the Gulf Stream evolved into a self-contained gyral system it considerably modified the climates in the northern Atlantic. For example, numerous incursions of tropical faunal elements have been recorded in the Middle and Late Miocene North Atlantic, some 10° north of their present limit (Cifelli, 1976), indicating large-scale water mass migrations. A warm branch of the Gulf Stream apparently continued to flow along the Newfoundland Coast in Late Miocene time (Berggren and Hollister, 1974).

In Messinian (Late Miocene) time the remnant western Tethys (Mediterranean) suffered the well-documented "Salinity Crisis" (Hsü et al., 1973; Adams et al., 1977). The crisis began at about 6.2 Ma when a tectonically restricted connection between the Mediterranean and the Atlantic was severed due to a drastic drop in global sea-level (Ryan et al., 1974; Hsü et al., 1978). The elimination of Mediterranean recirculation into the open Atlantic apparently had little or no effect on the surface circulation patterns in the North Atlantic (Berggren and Hollister, 1974). It was not until the transgression in the earliest Pliocene (5 Ma) that the connection with the open Atlantic through the Straits of Gibraltar was re-established, the Mediterranean basins were once again submerged under deep and open marine water (Hsü et al., 1978), and the circulation patterns developed as we know them today.

The late Miocene sea level drop that led to the final isolation of the Mediterranean and the Messinian Salinity Crisis is considered to be global by some workers (Van Couvering et al., 1976; Adams et al., 1977) and may have been caused by an increase in the volume of the Antarctic ice-cap at that time (Shackleton and Kennett, 1974). An extensive ice-cap on Antarctica is commonly believed to have developed during the middle Miocene between 15 and 14 Ma (Shackleton and Kennett, 1975; Savin et al., 1975; Kennett, 1977), but may have existed as early as the Early Miocene around 19.5 Ma (Stump et al., 1980). The development of the ice-cap must have had important significance for the global climate. Schnitker (1980) suggested that significant glaciation on Antarctica developed after the final submergence of Iceland-Faeroe Ridge (13 Ma). According to this model the higher salinity North Atlantic surface water (which had mixed with saltier Mediterranean water) flowed into the Norwegian Sea, and was converted to NADW (North Atlantic deep water) that flowed out over the Iceland-Faeroe Ridge. After its traverse southward across the Atlantic, NADW upwelled off Antarctica and provided the excess moisture needed for an extensive snow-ice accumulation. The formation of an extensive ice-cap on Antarctica and the intensification of the circum-Antarctic current resulted in steepening of thermal gradients between southern high and equatorial latitudes and between surface and bottom water temperatures in the late Early Miocene (Savin et al., 1975). This change in the thermal structure of the oceanic water masses coincides with a major change in the latitudinal provinciality of surface-dwelling microplankton (see Discussion).

The purpose of the present study is to delineate the Miocene biogeographic patterns of calcareous nannoplankton and to evaluate these with respect to the known Miocene paleogeographic history of the Atlantic Ocean. The history of the Miocene climatic variations in this ocean, as interpreted from the faunal and isotopic records is still sketchy. Temporal migration patterns of nannoplankton should provide a more complete picture of climatic variations and the differences between the Northern and Southern Hemisphere climates.

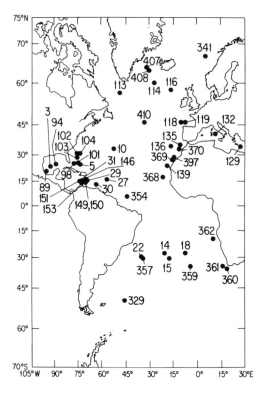

FIGURE 1

Atlantic Ocean DSDP sites from which core samples were used to delineate Miocene calcareous nannoplankton biogeography.

STUDY MATERIAL

With the exception of 2 sites from the Mediterranean, all the material studied comes from the Miocene of the Atlantic Ocean DSDP (Deep Sea Drilling Project) sites (fig. 1). The sites were cored on various DSDP legs ranging between Legs 1 and 47. The sites are well distributed in the Caribbean Sea and the Gulf of Mexico, and there is also good coverage in mid latitudes of both North and South Atlantic basins. There is adequate coverage in the northern high latitudes; however, there is a dearth of drill-sites in the South Atlantic high latitudes, and only 1 site (329) could be included in the present study (see Appendix I for a list of DSDP sites used in this study and their location).

Initially, samples were selected from the Miocene sections based on the published biostratigraphic information in the DSDP Initial Reports. Samples were then examined to verify (or correct) the biostratigraphic ages assigned to them by shipboard micropaleontologists. In addition, each sample was carefully examined to observe the degree of calcite dissolution and over-

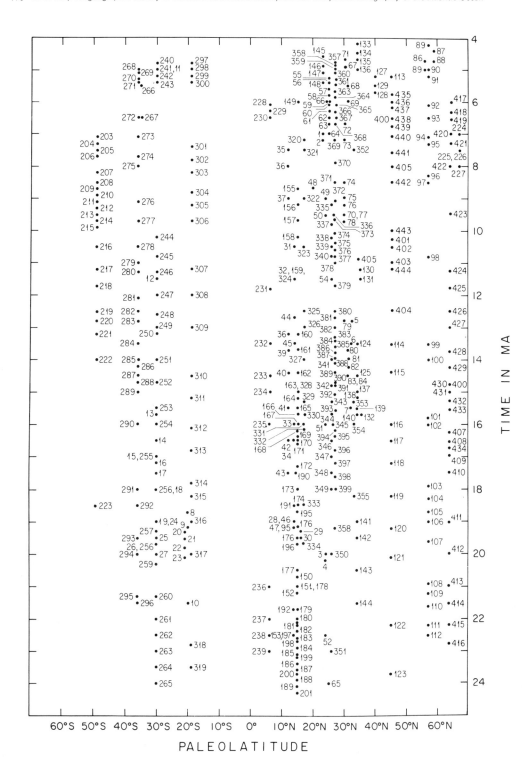

PALEOLATITUDE

TIME IN MA

growth on nannofossils. By including only those samples containing both robust and delicate (and thus more solution susceptible) forms for final data census, the potential problem of differential dissolution affecting species composition (Berger, 1973; Roth and Berger, 1975) was minimized. Of more than 2800 samples originally selected, only 444 were adequately preserved for data census (for final list of samples used in this study, see Appendix II). Figure 2 shows the position of these samples in a time-paleolatitude grid.

BIOSTRATIGRAPHY AND TIME-SCALE

All samples included in the nannofloral analysis were placed within a uniform biostratigraphic framework. The Miocene nannofossil zonation used here is after Bukry (1975) and the ages assigned to the zonal boundaries are based on the Berggren and Van Couvering (1974) and Ryan et al. (1974) time-scales for the Neogene with some modifications to the ages of the latest Miocene zones in view of more recent biochronologic data (Haq et al., in press). After the biostratigraphic assignment, each sample was assigned an absolute age in Ma (millions of years before present) based on the position of the sample within the zonal scheme and the rate of sediment accumulation in that part of the section. These age assignments were then used in the paleolatitude-time grid (see fig. 2) and for plotting the biogeographic patterns presented in figures 3 through 8.

METHODOLOGY
Data census

After age assignments to the selected samples the nannofossils in each sample are counted under the optical light microscope. Approximately 300 individuals are counted in random traverses of smear slides made from raw samples. Thus the basic data consists of census of fossil nannoplankton. These data are expected to differ from the original distribution and composition of taxa due to a number of biologic, stratigraphic, and preservational factors. These factors and the minimization of their effects have been discussed in some detail by Haq and Lohmann (1976). In spite of the suspected differences between the bio- and thanatocoenoses, meaningful reconstructions of biogeographic patterns and the changes in these patterns with time, can be made from the census data.

In this study a total of 25 taxonomic categories were counted (all species were counted, but some were grouped into larger categories for various reasons—see section on Taxonomic Concepts). The general statistics of the first 12 taxonomic categories that were included in the factor analysis are given in table 1. Of the 25 taxonomic units counted (see Appendix II) only the first 17 units occurred consistently in space and time, and of these only the first 12 were sufficiently important numerically to be included in the transformation of raw census data into assemblages by means of a Q-mode Factor Analysis. (See Appendix III.)

TABLE 1
General statistics of first 12 variables from raw data.

Variables	Average Percent	Standard Deviation	Minimum Value	Maximum Value
1 *Discoaster* spp.	9.15	12.29	0.0	69.80
2 *Sphenolithus* spp.	6.07	9.08	0.0	57.00
3 *C. floridanus*	13.21	22.28	0.0	90.00
4 *C. pelagicus*	10.30	12.85	0.0	75.20
5 *D. hesslandii*	0.73	2.7	0.0	24.80
6 *D. antarcticus*	1.80	8.32	0.0	89.50
7 *D. minutus*	25.11	24.25	0.0	87.50
8 *R. pseudoumbilica*	10.97	14.55	0.0	72.50
9 *R. haqi*	9.23	10.32	0.0	48.10
10 *Reticulofenstra* spp.	0.70	2.16	0.0	15.10
11 *U. jafari*	3.1	7.14	0.0	45.90
12 *D. abisectus*	1.80	3.40	0.0	32.90

Data reduction

The large data matrix of relative abundance of taxa in 399 samples was transformed into assemblages by means of Q-mode Varimax Factor Analysis (data from the last 45 samples was not available at the time Factor Analysis was performed; see Appendix I). This technique is designed to portray relationships among samples, as characterized by their taxonomic compositions. The technique has been described in detail by Jöreskog et al. (1976); and its application to nannofossil biogeographic data was discussed and illustrated by Haq and Lohmann (1976). The Varimax factor assemblages resulting from the Q-mode factor analysis are defined in such a way that the constituent taxa contribute most to the samples considered, approximating natural taxonomic associations. In this study, however, the last step of the analysis performed by Haq and Lohmann (1976), namely, Oblique Factor solutions, was not considered necessary and the analysis was ended after orthogonal Varimax rotations. The principal difference in the orthogonally vs. obliquely rotated solutions is that in the former the assemblages remain uncorrelated

FIGURE 2
Time and paleolatitudinal distribution of 444 samples used for delineating Miocene calcareous nannoplankton biogeographic patterns. Same time and latitudinal grid has been used to construct contour and perspective plots of assemblages illustrated in figures 3–8. For the key of sample numbers and their location in DSDP site/core/section see Appendix II. For geographic location of sites see Appendix I.

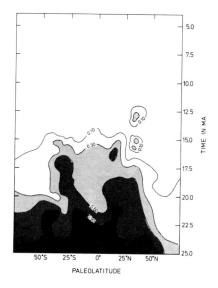

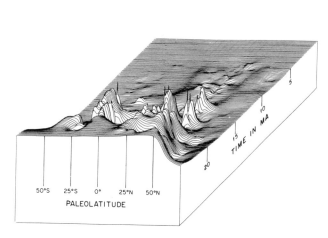

FIGURE 3

Spatial and temporal distribution patterns of *Cyclicargolithus floridanus* assemblage. Left: Planar isoline plot of assemblage dominance; contour values and heaviness of shaded areas is proportional to relative abundance (factor loadings) of assemblage. Some intermediate contour lines have been eliminated for clarity. Right: semi-perspective block-diagram of assemblage dominance. Two horizontal scales represent paleolatitude and time. Height of peaks (vertical scale) is proportional to relative abundance of assemblage.

with each other, while in the latter they become cor-related (Cooley and Lohnes, 1971). In the present study the initial application of Oblique Factor solutions did not yield results significantly different from those obtained by orthogonal rotation and were thus not used in the final application of the technique used here.

The mathematical analysis was performed on an IBM 370/165 system computer at the DATACENTRALEN in Stockholm, through the University of Stockholm in Sweden. The Klovan and Imbrie (1971) CABFAC pro-gram was used for the Q-mode factor analysis; an updated version includes procedures for both Varimax rotation and the optional, Oblique Factor solutions.

Data presentation: The Varimax Factor assemblages have been presented in planar isoline maps on a paleo-latitude vs. time grid, and on computer-simulated, semi-perspective, block diagrams which allow an easily discernable, three-dimensional representation of bio-geographic data (Flodén and Haq, 1980). Hand gen-erated three-dimensional models have sometimes been used to show biogeographic patterns (e.g. Honjo and Okada, 1974); however, when the data base is large, such plots become time consuming and impractical without computer help.

In the isoline and semi-perspective maps presented here the input data consist of an X-coordinate (paleo-

latitude), a Y-coordinate (time in Ma) and a Z-coordi-nate (a vertical parameter; factor-loadings in this case) for each data point. The plotting program can accept a practically unlimited number of Z-values, distributed in any form within the X-Y grid. The plotter scales and isoline density in planar maps, as well as scales, vertical exaggeration and line density in semi-perspective dia-grams can be chosen to suit the requirements of each illustration in this versatile plotting technique (see Flodén and Haq, 1980). The special appeal of semi-perspective diagrams is that the distributional charac-teristics of the assemblages and the fluctuations in their relative geographic dominance with time can be seen at a glance. The disadvantage of these diagrams is that actual abundance values of the assemblages cannot be discerned; planar isoline maps are necessary for this.

TAXONOMIC CONCEPTS

Initially I attempted to count separately every species represented in each sample (see Haq and Berggren, 1978, and Backman, 1978, for our species concepts). However, it quickly became evident that such an effort would be extremely time consuming and, in some cases, meaningless, when the light microscope was used to distinguish taxa. Many taxa, even with a minimal overgrowth of calcite, are commonly indis-

211

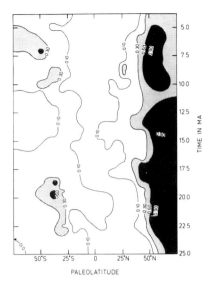
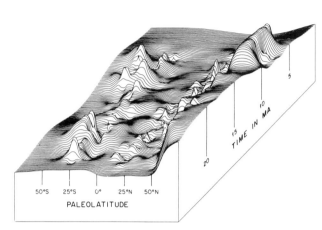

FIGURE 4

Contour plot and semi-perspective diagram of spatial and temporal dominance patterns of *Coccolithus pelagicus* (*s. ampl.*) assemblage (see caption of fig. 3 for explanation).

tinguishable from other, closely similar, taxa under the light microscope. The discoasters, sphenoliths and the reticulofenestrids fall under this category. In addition, many individual taxa have short time ranges, and plots of their temporal distributions would be meaningless for long-term biogeographic patterns.

For the above reasons it became necessary to group certain taxa; in this way *Discoaster* spp., *Sphenolithus* spp., and small *Reticulofenestra* spp. were grouped together. These broader groupings are also more likely than the various included species to remain stable in response to possible changes with time in their ecologic requirements (see Haq and Lohmann, 1976).

Another broad grouping is that of *Coccolithus pelagicus* (*s. ampl.*) and includes small to medium-sized oval placoliths (6–10 μm in length) with a distinctive morphology, characterized by a ring of crystals (collar) around the central opening visible from the proximal or the distal sides. In spite of the large variations in size, these placoliths are grouped together because their morphology is similar and because they are characteristic of high latitudes. Only the very large forms (up to 20 μm in length) are separated as *Coccolithus miopelagicus* from this broad category of placoliths.

The commonly occurring reticulofenestrids that can be distinguished under the light microscope from the broad category of *Reticulofenestra* spp. are the large

to medium-sized *R. pseudoumbilica* and the medium-sized form, *R. haqii. Reticulofenestra pseudournbilica*, however, has 3 distinct variants: 1) a larger form (length: 10–12 μm) with small central pore due to a thickened, central collar; 2) a large variant (average length: 9.5 μm) with a large central pore and thin collar; and 3) the typical *R. pseudoumbilica*, a medium-sized form (average length: 7 μm) with medium-sized central pore and thin central collar. These, however, may reflect purely preservational differences, since all variants have similar biogeographic distribution. *Reticulofenestra haqii* is a medium-sized form (average length: 4.5 μm) that seems to be closely related to *R. pseudoumbilica* biogeographically, but has an optical extinction pattern that is atypical of *R. pseudoumbilica*. *Reticulofenestra haqii* also has a different stratigraphic range.

Dictyococcites minutus includes small forms (3.5 μm average length), morphologically somewhat similar to the larger *D. hesslandii*, which is more common in Oligocene and Early Miocene sediments. *Dictyococcites minutus* has a greater stratigraphic range than *D. hesslandii*, and probably includes more than one form, but these are indistinguishable from each other under the light microscope due to their minute size. The apparent biogeography of the *D. minutus* assemblage (see fig. 6), which shows complex, uninterpretable patterns and a cosmopolitan

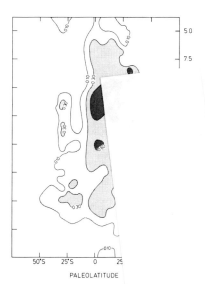

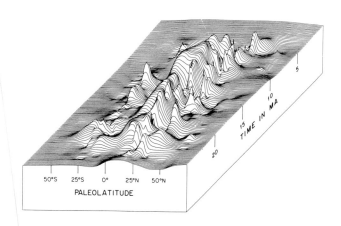

PALEOLATITUDE

FIGURE 5
Contour plot and semi-persp[...]nd temporal dominance patterns of *Discoaster-Sphenolithus* assemblage (see
caption of fig. 3 for explanatio[...]

occurrence, is probably d[...]   discriminate real taxa in this g[...]   complexity it would be necessar[...]   cally
the minute dictyococcids u[...]   lope,
which would be an extrem[...]   and
perhaps unrewarding, task.

MIOCENE NANNOPLANKTON BIOGE[...]

In this section the major Vari[...]   ges
identified by a Q-mode Facto[...]   ius
data are described. Their "idea[...]   si-
tions, which in the case of maj[...]   ki-
mate the actual sample comp[...]   id
Lohmann, 1976, p. 134), are giv[...]   al
distribution through the Mioce[...]   i)
are illustrated (figs. 3–7). For [...]   c
nannofossil biogeographic maps[...]   c
reconstructions of the Atlantic Oc[...]   [...]
(figs. 9–11) to show the actual g[...]
of various assemblages in selecte[...]

Varimax Factor Assemblages

The first 5 Varimax Factors explain[...]
variance in the census data retaine[...]
of these factors and the relationshi[...]
within each factor are shown in [...]
Appendix IV for Factor Loadings).

TABLE 2
Compositions (Factor Scores) of the first five Q-mode Varimax Factor Assemblages identified in 399 Miocene samples from the Atlantic Ocean.

TAXA	FACTORS				
	1	2	3	4	5
1. *Discoaster* spp.	-0.024	0.378	0.159	-0.212	2.916
2. *Sphenolithus* spp.	0.183	0.313	-0.260	-0.042	1.737
3. *C. floridanus*	-0.007	3.403	0.032	-0.165	-0.473
4. *C. pelagicus*	-0.045	0.198	0.142	3.352	0.138
5. *D. hesslandii*	0.021	0.165	-0.030	0.115	-0.075
6. *D. antarcticus*	0.003	0.011	0.338	0.159	-0.253
7. *D. minutus*	3.394	0.021	-0.034	-0.077	-0.144
8. *R. pseudoumbilica*	-0.126	0.002	3.297	-0.358	-0.069
9. *R. haqii*	0.560	-0.132	0.974	0.686	0.259
10. *Ret.* spp.	-0.003	0.149	0.017	0.026	0.011
11. *U. jafari*	0.337	-0.058	0.126	-0.057	0.273
12. *D. abisectus*	0.015	0.258	0.055	0.211	0.012
Variance (%)	35.89	20.19	15.15	8.77	9.04
Cumulative Variance (%)	35.89	56.09	71.24	80.01	89.05

The first Varimax Factor (VF) (no. 1, table 2) assemblage consists almost exclusively of *Dictyococcites minutus*; the second VF (no. 2, table 2) assemblage is dominated by *Cyclicargolithus floridanus*; in the third VF assemblage (no. 3, table 2), *Reticulofenestra pseudoumbilica* is overwhelmingly dominant, although *R. haqii* is associated in significant proportions; the fourth VF assemblage (no. 4, table 2) is dominated by *Coccolithus pelagicus* (*s. ampl.*) and the fifth VF assemblage (no. 5, table 2) consists of *Discoaster* spp.

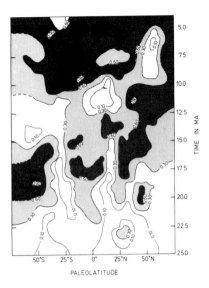

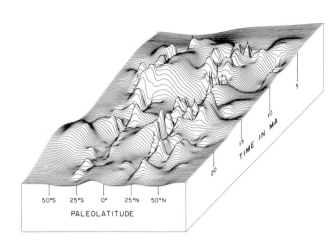

PALEOLATITUDE

FIGURE 6

Contour plot and semi-perspective diagram of spatial and temporal dominance patterns of *Dictyococcites minutus* assemblage (see caption of fig. 3 for explanation).

together with about half as much *Sphenolithus* spp. In the following sections these Varimax assemblages (and a minor, qualitatively defined, assemblage that is dominated by *Dictyococcites antarcticus* and occurs predominantly in the southern high latitudes) are described and illustrated in the order of their chronologic appearance and dominance in the Atlantic Ocean.

(1) The *Cyclicargolithus floridanus* assemblage (VF 2) consists of an overwhelming abundance of this taxon and explains some 20% of the variance in census data. The time and space distribution of this assemblage is shown in figure 3. The *C. floridanus* assemblage was a remnant of the early Cenozoic, and first achieved importance in the Early Oligocene (see Haq and Lohmann, 1976, p. 166, where it is listed as the *Cyclococcolithus neogammation* assemblage). Together with the *Dictyococcites hesslandii* assemblage, the *Cyclicargolithus floridanus* assemblage dominated and characterized the Oligocene Epoch (Haq et al., 1977). *Dictyococcites hesslandii*, however, lost prominence in the latest Oligocene-Early Miocene, whereas *C. floridanus* continued to be the important and generally dominant assemblage in the Atlantic Ocean during the Early Miocene.

The *C. floridanus* assemblage was clearly cosmopolitan in the earliest Miocene, occurring commonly in both the Northern and Southern Hemispheres (fig. 3). The dominance of this assemblage began to decrease within the Early Miocene at about 22.5 Ma,

remaining prominent in mid latitudes until about 18 Ma, and then decreased in importance until it almost completely disappeared at the Middle/Late Miocene boundary, around 16.5 Ma. Its sporadic low occurrence above the Early Miocene (fig. 3) may be entirely due to reworking.

(2) The *Coccolithus pelagicus* (s. ampl.) assemblage (VF 4) is dominated by the nominate taxon, but a smaller amount (about 18%) of *Reticulofenestra haqii* is ideally also associated with the assemblage. This factor explains slightly less than 9% of the total variance in the data matrix; but the assemblage is an important high latitude constituent, especially in the Northern Hemisphere. *Coccolithus pelagicus* (s. ampl.) first became an important nannofloral assemblage in the Eocene; however, in the Eocene and Oligocene it was relatively less dominant (see Haq and Lohmann, 1976), achieving prominence in the Miocene and a dominant status in the high latitudes.

The temporal and geographic distribution of the *C. pelagicus* assemblage is shown in figure 4. It is clear that this assemblage was dominant in the northern higher latitudes through most of the Miocene. In the Northern Hemisphere it was especially dominant between 22.5 and 20 Ma, around 14.5 Ma and between 8.5 and 5 Ma. For a short time, between 11 and 10 Ma, and after 5 Ma, it lost this dominance. In the Southern Hemisphere it occurred generally in low abundance, with the exception of relative prominence in mid lati-

214

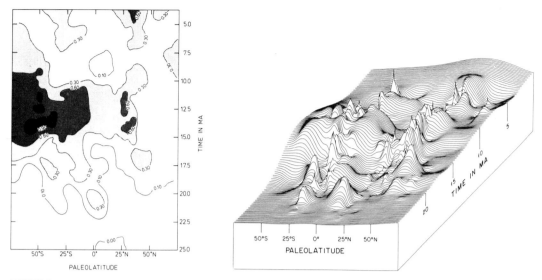

FIGURE 7
Contour plot and semi-perspective diagram of spatial and temporal dominance patterns of *Reticulofenestra pseudoumbilica-R. haqii* assemblage (see caption of fig. 3 for explanation).

tudes around 18 Ma and between 10 and 7.5 Ma (fig. 4).

The paleobiogeographic distribution of *Coccolithus pelagicus* assemblage shows similarities to the modern biogeographic distribution of this taxon. Today this species is found living only in the Northern Hemisphere, but the fossil form was found in post-glacial (< 12000 years B.P.) surface sediment of both Hemispheres (McIntyre et al., 1970); the present-day absence of *C. pelagicus* from the Southern Hemisphere may be due to post-glacial warming that was strong enough to interrupt the delicate and stenothermal life cycle of the species and cause a regional extinction (McIntyre et al., 1970).

The biogeographic patterns of the *C. pelagicus* assemblage (fig. 4) show that just such short-term, regional, restrictions may have occurred several times in the Miocene during times of peak warming. If very low traces of this species in the South Atlantic are discounted as due to reworking, then the more prominent occurrences in the Southern Hemisphere (around 18 Ma and between 10 and 7.5 Ma) correspond closely with the periods of dominant occurrences of this assemblage in the Northern Hemisphere, the assemblage becoming almost completely excluded from the Southern Hemisphere in the intervening time intervals.

From the Miocene example one is tempted to analogize and suggest that the present-day restriction of the *C.*

pelagicus to the Northern Hemisphere is temporary, and that the species is awaiting a return to the Southern Hemisphere during the next climatic cooling, or even during a short-term removal of the tropical thermal barrier.

(3) The *Discoaster-Sphenolithus* assemblage (VF 5) ideally consists of an almost 2 : 1 ratio between these 2 taxonomic categories. This assemblage explains some 9% of the total variance in the census data of the first 12 variables (see table 2) and was an important nannoplanktonic component of the Miocene low latitudes in the Atlantic Ocean.

Discoasters became numerically important in the Late Paleocene and remained prominent in the Eocene low to mid latitudes. A *Sphenolithus-Discoaster* assemblage was one of the more prominent assemblages in the low to mid latitudes of the Oligocene Atlantic Ocean (see Haq and Lohmann, 1976).

Text-figure 5 illustrates the temporal and paleolatitudinal distribution of the *Discoaster-Sphenolithus* assemblage in the Miocene Atlantic Ocean. The distribution pattern of this assemblage is antithetical to that of the *Coccolithus pelagicus* assemblage, which was mainly concentrated in the higher latitudes. Especially noteworthy is the dominance of low latitudes by the *Discoaster-Sphenolithus* assemblage between 12.5 and 10 Ma, at the same time that *C. pelagicus* shows its lowest occurrence in the northern high latitudes.

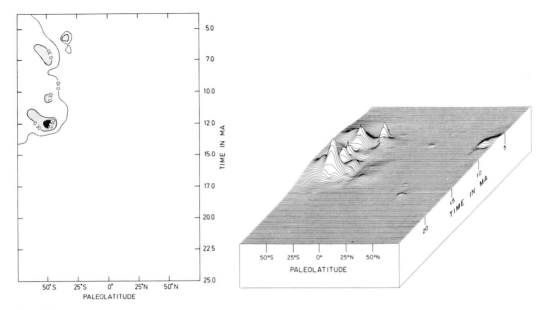

FIGURE 8

Contour plot and semi-perspective diagram of spatial and temporal dominance (based on relative abundance data) of *Dictyococcites antarcticus* (see caption of fig. 3 for explanation).

(4) The *Dictyococcites minutus* assemblage (VF 1) was the most dominant Miocene assemblage, responsible for the maximum share of total variance (over 35%) in the data (see table 2). It consists overwhelmingly of small dictyococcids that are treated as a single group here because they are indistinguishable under the light microscope (see Taxonomic Concepts).

The assemblage became numerically prominent in the later part of the Early Miocene ; however, the distribution patterns of the assemblage are irregular and confusing (fig. 6). In general the assemblage was more dominant in the Late Miocene than earlier, and showed a cosmopolitan distribution.

The complexity of the patterns indicate that further taxonomic splitting of this category with the help of the electron microscope may be necessary to unravel the individual patterns of forms lumped for this census (see Taxonomic Concepts). Alternatively, it is possible that in spite of the caution with which only well-preserved samples were selected for counting to minimize the preservational bias in the data, the patterns may still contain some of this bias. The complexity of *D. minutus* assemblage patterns may potentially contain a paleoenvironmental message which can, however, only be decoded after further detailed work.

(5) The *Reticulofenestra pseudoumbilica-R. haqii* as-

semblage (VF 3) consists of the nominate taxa in ratio of approximately 3 : 1, and is responsible for some 15% of the variance in the data (see table 2).

The distribution patterns of this assemblage (fig. 7) indicate that it became numerically important in the later part of the Early Miocene, after 18 Ma. It generally preferred the mid latitudes of the South Atlantic between 15 and 10 Ma during which time it dominated at DSDP Site 329 on the Falkland Plateau (fig. 7, also see fig. 12), and during the Late Miocene alternated in dominance with an endemic *Dictyococcites antarcticus* assemblage.

(6) *Dictyococcites antarcticus* has not been identified as a Varimax assemblage by the Q-mode factor analysis because it explains only a very small proportion of the total variance in the data. However, this species was very prominent in the southern high latitudes during Middle and Late Miocene. It was first described from around Antarctica (Haq, 1976).

Because of its prominence in the Late Miocene at DSDP Site 329 on the Falkland Plateau, the time and paleolatitudinal distribution of the relative abundance of *D. antarcticus* is reconstructed in figure 8. This taxon was clearly preferentially distributed around Antarctica between 12 and 5 Ma, showing 2 prominent peaks around 12 and 7 Ma (fig. 8). It alternated in

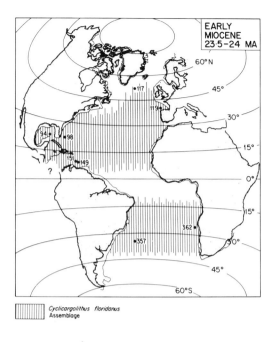
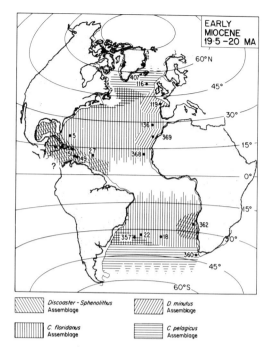

FIGURE 9

Synoptic paleobiogeographic maps of 2 Early Miocene time-slices (24–23.5 Ma and 20–19.5 Ma). Biogeographic patterns deduced from time-latitudinal distribution of assemblages. Paleogeographic reconstructions after Sclater et al. (1977).

---

dominance with the *Reticulofenestra pseudoumbilica-R. haqii* assemblage during the Late Miocene (see fig. 12). *Dictyococcites antarcticus* occurred in very small numbers in the northern high latitudes, except during the Middle Miocene (between 11 and 9 Ma) when it made the only incursion in these latitudes during a cold episode and when it occurred in significant numbers at DSDP Site 116 (Backman, personal communication; 1980).

In the southern high latitudes the appearance of *D. antarcticus* in the Middle Miocene (at about 15 Ma) coincides with the probable development of an extensive ice-cap on Antarctica (Shackleton and Kennett, 1975) and the oscillations of this assemblage may indicate variations in stability of the Antarctic ice-cap (see Discussion).

Synopsis of nannoplankton biogeographic evolution in the Miocene Atlantic Ocean

In the paleogeographic maps presented in this section, the Early and Late Miocene paleogeographic reconstructions of the Atlantic Basin by Sclater et al. (1977) were used as base-maps. In these reconstructions the paleopositions of the DSDP sites have been backtracked to account for plate movements.

These synoptic biogeographic maps show the dominant assemblages at the respective DSDP sites at a given time interval. Often 2 assemblages are equally prominent at a particular site, in which case the composite patterns of both dominant assemblages are drawn to represent the actual biogeographic composition. Maps are presented for 6 time-slices, with 2 each in the Early, Middle, and Late Miocene. Each time-slice has a resolution of about half a million years.

The Miocene nannoplankton biogeographic evolution proceeded from a relatively simple to progressively more complex distribution pattern. The patterns that developed in Late Oligocene time continued into the earliest Miocene. The first time-slice (fig. 9) shows a general lack of endemism; the cosmopolitan *Cyclicargolithus floridanus* assemblage dominated the biogeographic scene in most of that part of the Atlantic Ocean for which data are available. Discoasters and sphenoliths were generally common in the Gulf of Mexico sites at this time, but did not dominate. The distribution patterns are essentially similar to the latest Oligocene patterns delineated by Haq and Lohmann (1976).

In the late Early Miocene the biogeographic patterns began to change and a number of new assemblages

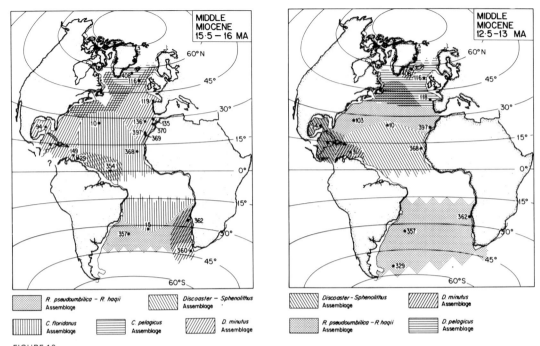

FIGURE 10
Synoptic paleobiogeographic maps of 2 Middle Miocene time-slices (16–15.5 Ma and 13–12.5 Ma). Biogeographic patterns deduced from time-latitudinal distribution of assemblages. Paleogeographic reconstructions after Sclater et al. (1977).

appeared during this interval. The second Early Miocene time-slice map (between 13 and 12.5 Ma; fig. 9) shows that the low and mid latitudes were still dominantly populated by the *Cyclicargolithus floridanus* assemblage, but · that the *Discoaster-Sphenolithus* assemblage had become equally prominent in the low latitudes of the Caribbean and Gulf of Mexico. The *Dictyococcites minutus* assemblage began to dominate the nannoflora in the eastern North Atlantic sites along the continental margins (Sites 119 and 369), and showed an equal prominence with *C. floridanus* assemblage along the eastern South Atlantic continental margin (Site 362). A high latitude *Coccolithus pelagicus* assemblage appeared in both Hemispheres and dominates the nannoflora at Sites 116 and 407 in the North and Site 360 in the South; it was prominent even at Site 357 on the Rio Grande Rise, where *C. floridanus* was still dominant.

Thus within the Early Miocene the biogeographic scene in the Atlantic changed from one characterized by a relative lack of provinciality amongst nannoplankton to one with at least 3 distinct latitudinal provinces, 2 characterizing the low and mid latitude marginal and

open-ocean areas and one characterizing the boreal and austral provinces of higher latitudes. The lack of endemism and dominance of cosmopolitan long-ranging taxa such as that in the earliest Miocene, is characteristic of periods of climatic deterioration and/or environmental stress (Haq et al., 1977; Fischer and Arthur, 1977), which seems to have been the case in Late Oligocene-earliest Miocene.

In the early Middle Miocene time-slice (16–15.5 Ma; fig. 10) the biogeographic patterns in the Northern Hemisphere remain essentially similar to the last mentioned time-slice, with the exception of the spread of the *Dictyococcites minutus* assemblage to the Gulf of Mexico. In the South Atlantic, however, 2 changes occur in the distribution patterns. The *Coccolithus pelagicus* assemblage lost importance in the southern high latitudes, becoming predominantly boreal and crossing over to the Southern Hemisphere for short durations during periods of climatic cooling. The second change in the Southern Hemisphere is the appearance and dominance of the new *Reticulofenestra pseudoumbilica-R. haqii* assemblage in mid latitudes. The *Cyclicargolithus floridanus* assemblage became

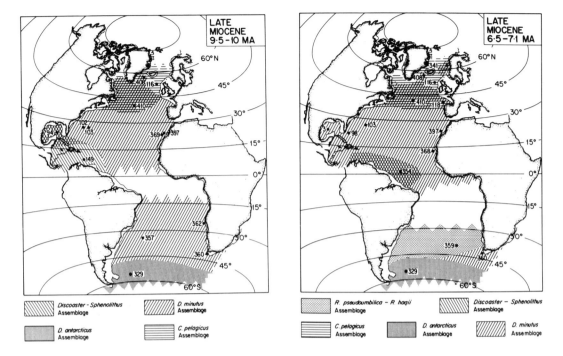

FIGURE 11
Synoptic paleobiogeographic maps of 2 Late Miocene time-slices (10–9.5 Ma and 7.1–6.5 Ma). Biogeographic patterns deduced from time-latitudinal distribution of assemblages. Paleogeographic reconstructions after Sclater et al. (1977).

even more restricted at this time and disappeared completely soon afterward.

In the later Middle Miocene (13–12.5 Ma, fig. 10) the *Reticulofenestra pseudoumbilica-R. haqii* assemblage spread throughout most of the South Atlantic and also moved into the northern mid latitudes. In the Gulf of Mexico this assemblage shared dominance with the *Discoaster-Sphenolithus* assemblage. The niche vacated by the disappearance of *Cyclicargolithus floridanus* was by this time occupied by the *Discoaster-Sphenolithus* assemblage. The *Coccolithus pelagicus* assemblage still maintained dominance in the northern high latitudes. The *Dictyococcites minutus* assemblage, however, was more restricted than in the last time-slice, sharing prominence with other assemblages at northern high latitude Site 408 and Caribbean Site 31. The prominence of the small dictyococcids in both high and low latitudes, but near-absence from mid latitudes suggests that *D. minutus* may represent at least 2 distinct ecotones (or taxa) that are morphologically indistinguishable under the light microscope.

In the early Late Miocene (10–9.5 Ma, fig. 11) the *D. minutus* assemblage once again spread through

most of the Atlantic Ocean, in both Hemispheres, with the exception of the higher latitudes and in the Gulf of Mexico and Caribbean where the *Discoaster-Sphenolithus* assemblage was still partially prominent. The *Coccolithus pelagicus* assemblage continued to be important in northern high latitude sites in the Norwegian-Greenland Sea, but with a mixture of the *D. minutus* assemblage at Sites 408 and 410. The new austral assemblage, *Dictyococcites antarcticus*, first dominated in the southern high latitudes at this time and remained dominant in the area around Antarctica in the Late Miocene. *Reticulofenestra pseudoumbilica-R. haqii* assemblage completely lost importance during this time-slice, although it was present in numerically small amounts through most of the mid latitudes.

The latest of the Miocene time-slices (7.1–6.5 Ma, fig. 11) showed a return to prominence of the *R. pseudoumbilica-R. haqii* assemblage in southern mid latitudes. Other patterns remain virtually unchanged from the last time-slice, with the exception of the loss of importance of the *Discoaster-Sphenolithus* assemblage in the Gulf of Mexico.

The biogeographic patterns described and illustrated

219

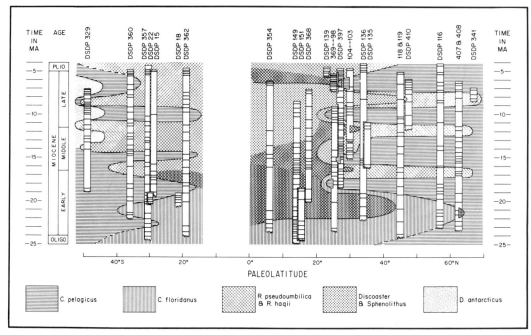

FIGURE 12

Summary of Miocene distribution patterns of 4 major and 1 minor assemblages in North and South Atlantic Ocean. To make migratory patterns more discernable, cosmopolitan and generally dominant *Dictyococcites minutus* assemblage has not been plotted. Migrations of assemblages characterizing low and mid latitudes (*Discoaster-Sphenolithus* and *Reticulofenestra pseudoumbilica-R. haqii*) into higher latitudes interpreted as response to climatic warming; likewise extensions of an assemblage preferring high latitudes (*Coccolithus pelagicus* assemblage) into mid and low latitudes interpreted as response to cooling trends in climate. DSDP sites and their paleolatitudinal positions identified by numbers at top of cores. Lines within core columns represent temporal levels of samples studied. Latitudinal shifts indicate generally cooler climates in earliest and latest Miocene and 4 warming and cooling cycles in North and South Atlantic (and additional minor warming episode in South Atlantic).

here clearly show that during the Miocene the Atlantic Ocean saw: a) disappearance of an older dominant assemblage and appearance of numerous new ones; b) repeated restrictions of a previously bi-Hemispheric assemblage to Northern Hemisphere, and the development of boreal and austral assemblages characterized by different taxa; and c) major fluctuations in the geographic distribution of the assemblages within the Atlantic Basin. Some of these fluctuations manifested themselves in distinct migration of assemblages through latitudes with time, and it is these migration patterns that have been used to reconstruct the paleoclimatic history of the Atlantic Ocean, discussed in the next section.

MIOCENE CLIMATIC HISTORY OF THE ATLANTIC OCEAN

The basic assumption in interpreting climatic conditions from biogeographic data is that the spatial migrations of assemblages are controlled mainly by major

fluctuations in climate. Numerous studies of modern nannoplankton biogeography have shown that these photosynthetic, planktonic calcareous algae are highly stenothermal (see e.g. McIntyre et al., 1970; Okada and Honjo, 1973), responding quickly to changes in thermal structure of the water masses. This analogy, when applied to the floral and faunal spatial shifts in the geologic past, has been used to reconstruct the paleoclimatic history of the oceans (see e.g. Imbrie and Kipp, 1971; Ingle, 1973; CLIMAP, 1976; Haq and Lohmann, 1976; McIntyre et al., 1976; and Haq et al., 1977).

The Miocene nannofloral migration patterns in the North and South Atlantic Ocean are shown in figure 12. In this figure the DSDP sites are arranged according to their paleolatitudinal position, irrespective of their position in the eastern or western part of the basin. In general this tends to simplify the patterns, e.g. in both Hemispheres the sites located in the west are likely to be warmer than sites at similar latitudes in the east due to the moderating influence of the western boundary currents. In order to make the patterns more easily discernable, the *Dictyococcites minutus* assemblage, which shows no clear distribution pattern, has not been plotted in figure 12. Four major and 1 minor assemblages have been plotted and all major assemblages show distinct migration patterns that can be used to interpret the Miocene climatic history of the Atlantic Ocean.

Figure 12 clearly shows that the maximum biogeographic variability in the Northern Hemisphere occurred in the mid latitudes, between about 25 and 35°N. The lower latitude sites show little nannofloral change, and the warm *Discoaster-Sphenolithus* assemblage seems to have remained dominant through most of the Miocene, with an admixture of the cosmopolitan *Cyclicargolithus floridanus* assemblage in the early Miocene and *Reticulofenestra pseudoumbilica-R. haqii* assemblage in the Middle Miocene. Sites that are < 35°N latitude show the most distinct shifts, alternating between the cold *Coccolithus pelagicus* assemblage and the mid-latitude *Reticulofenestra pseudoumbilica-R. haqii* assemblage, through most of the Miocene. In the Southern Hemisphere the patterns were apparently similar, although data are lacking in the low southern latitudes.

The climatic picture between 25 and 22 Ma is unclear. In general, the cosmopolitan *Cyclicargolithus floridanus* assemblage was distributed in low and mid latitudes, whereas *Coccolithus pelagicus* dominated the higher latitudes. The biogeographic patterns are similar to those in the latest Oligocene (see Haq and Lohmann, 1976), and the spread of the stress-adapted, cosmopolitan taxa indicates a general climatic deterioration. The first distinct nannoplankton migration is observed between 22 and 19 Ma, when the warm *Discoaster-Sphenolithus* assemblage invaded the northern mid latitudes at Sites 135 and 136. This interval is also a period when the *C. floridanus* assemblage shows its maximum and last extension into higher latitudes in the Northern Hemisphere and into mid latitudes in the Southern Hemisphere.

This migration of tropical nannofloral elements into mid latitudes was followed by a shift of the cold *Coccolithus pelagicus* assemblage into lower mid latitudes in both Hemispheres, with a peak between 18.5 and 18 Ma. In this interval *C. pelagicus* penetrated as far south as Sites 98, 139, and 369 in the North Atlantic and to Site 362 in the South Atlantic.

The cold pulse was followed by the first of a series of 3 invasions of the mid-latitude *Reticulofenestra pseudoumbilica-R. haqii* assemblage into higher latitudes.

This first incursion occurred between 17 and 16 Ma in the Northern Hemisphere, but was slightly offset in the South Atlantic where it peaks at 16 Ma, ending after 15.5 Ma. In both Hemispheres this event is followed by the second major invasion of mid latitudes by the *Coccolithus pelagicus* assemblage. The event peaks at 15 Ma in the North Atlantic and at 14.5 Ma in the South.

Between 14 and 11 Ma the migrationary patterns were somewhat different in the Northern and Southern Hemispheres. In the North the cold *C. pelagicus* assemblage held sway most of this time in the high latitudes, up to 12.5 Ma, whereas in the South the *R. pseudoumbilica-R. haqii* extended for a short time into the high latitudes at about 14 Ma, and *C. pelagicus* had almost completely disappeared. A colder event between 13.5 and 12 Ma may have occurred in southern high latitudes when the endemic *Dictyococcites antarcticus* flora first became prominent. In the North Atlantic during part of this time a major shift of the *R. pseudoumbilica-R. haqii* assemblage into high latitudes took place between 12.5 and 11 Ma; a similar shift occurred between 12 and 10.5 Ma in the southern high latitudes.

After 11 Ma a major shift of *C. pelagicus* into mid latitudes occurred in the Northern Hemisphere between 11 and 9 Ma, and in the south this assemblage reappeared, after an almost complete exclusion, between 14.5 and 11 Ma. The peak of this shift and the cold episode it reflects occurred at about 10 Ma.

This cold episode is followed by another invasion of higher latitudes by the *Reticulofenestra pseudoumbilica-R. haqii* assemblage in both Hemispheres. The maximum extension of this migration occurred between 9 and 7.5 Ma in the North Atlantic and between 9 and 8 Ma in the South, where this event was quickly followed by another invasion of higher mid latitudes by *C. pelagicus*. In the Northern Hemisphere the *C. pelagicus* invasion of mid latitudes did not begin until after 7.5 Ma. That the warm episode between 9 and 7.5 was a particularly strong one is indicated by the extension not only of the mid-latitude reticulofenestrid assemblage to Norwegian Sea Site 341, but also by the extension of the tropical *Discoaster-Sphenolithus* assemblage into the higher mid-latitude Sites 118 and 119 at this time.

The latest Miocene cold episode, evidenced by the invasion of *C. pelagicus* into mid latitudes, peaked at about 6 Ma, and lasted until the end of the Miocene (5 Ma).

From these patterns we conclude that there have been 4 distinct climatic cycles in the Miocene Atlantic Ocean. The duration of these cycles varied from 4 to 4.5 m.y.

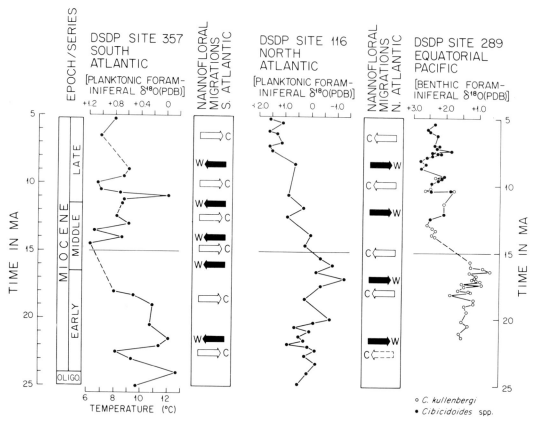

FIGURE 13

Comparison of nannofloral migrationary events (white arrows indicating cooler and black arrows warmer episodes) in South and North Atlantic, with $\delta^{18}O$ (isotopic paleotemperature) curves from various sources. Planktonic $\delta^{18}O$ curve from South Atlantic Site 357 according to Boersma and Shackleton (1977). Planktonic foraminiferal oxygen isotopic curve from Site 116 in North Atlantic after Rabussier-Lointier et al. (Vergnaud-Grazzini, written comm., 1980). Biochronology for Sites 116 and 357 modified by nannofossil biostratigraphic determinations. Benthic foraminiferal $\delta^{18}O$ curve from Site 289 from equatorial Pacific fide Woodruff (written comm., 1980). Line drawn at 15 Ma through isotopic curves to denote approximate time of development or enlargement of extensive ice-sheet on Antarctica and beginning of composite imprint of temperature and ice-volume effect on $\delta^{18}O$ signal.

DISCUSSION

In recent years the analyses of stable isotopic ratios of oxygen and carbon in marine calcareous microplankton and benthos have become powerful tools for paleo-environmental interpretations. An analysis of oxygen isotopes in well-preserved calcareous microfossils from a deep-sea core sample can yield information about such paleoceanographic variables as the surface water temperature through the analysis of shallow-dwelling planktonic foraminifera; bottom water temperature through analysis of benthic foraminifera; and the structure of the thermocline through analysis of deeper-dwelling planktonic species (Douglas and Savin, 1978; Matthews et al., 1980). In addition, interpretations of

former bottom water temperatures in deep ocean cores are also indicative of surface water temperatures that prevailed at higher latitudes where cool, denser, water sinks to form bottom water (Emiliani, 1954). Thus a comparison of planktonic and benthic isotopic temperatures also yields information about the past latitudinal thermal gradients (Savin, 1977). Carbon isotopic ratios have also recently been used as indicators of marine fertility and productivity and supply and storage of organic carbon to the oceanic realm (Kroopnick et al., 1977; Bender and Keigwin, 1979; Vincent et al., 1980). However, both oxygen and carbon isotopic composition of marine carbonates are subject to biological, chemical, and preservational biasing factors

(Savin, 1977), imparting numerous contraints to the interpretations based on these values (see summary in Arthur, 1979).

One of the factors that complicates the isotopic-temperature interpretations in the Miocene is the variation in isotopic composition due to the development of an extensive ice-cap on Antarctica in the Middle and Late Miocene (Shackleton and Kennett, 1975). The resolution of the $\delta^{18}O$ signal into components due to temperature on the one hand and ice accumulation on the other is difficult in the present state of the art (Savin, 1977). The timing of the extensive glaciation on Antarctica is still controversial and age estimates based on different data vary from 11.5 Ma (Keigwin, 1979) to 19.5 Ma (Stump et al., 1980). Most paleontologic and isotopic data, however, indicate an age of 15 to 14 Ma for this event (Kennett, 1977).

The available $\delta^{18}O$ curves from the North and South Atlantic generally show parallel trends for most of the Miocene (see fig. 13). Miocene isotopic curves from the Pacific Ocean also follow similar trends. These indicate the earliest Miocene was generally cool, with a climatic amelioration within the Early Miocene between 22 and 21 Ma. This event is clearly seen in the planktonic and benthic $\delta^{18}O$ data of Boersma and Shackleton (1977) from the South Atlantic DSDP Site 357 (see also fig. 13, col. 1), but it is not clear in the planktonic data of Rubussier-Lointier et al. (written communication, Vergnaud-Grazzini, 1980) from the North Atlsntic Site 116 (see fig. 13, col. 2). In the Pacific Ocean this event has been documented in both the planktonic and benthic $\delta^{18}O$ by Shackleton and Kennett (1975). This is followed by a decline in isotopic temperatures (increase in $\delta^{18}O$ values) at these sites, to be followed by a marked decrease in $\delta^{18}O$ values and indicating a climatic optimum between 17 and 15.5 Ma near the Early/Middle Miocene boundary. This warming event seems to have been global in its extent: all published curves covering this stratigraphic interval show a decrease of 0.5 to 1.0 ‰ in the $\delta^{18}O$ values, in most cases the lowest values on the respective curves for the entire Miocene. North Atlantic Sites 116 (see fig. 13), 398 and 400 (Vergnaud-Grazzini et al., 1978), and Pacific Sites 279 and 281 (Shackleton and Kennett, 1975) and 289 (Woodruff, personal communication, see fig. 13) record this sharp isotopic marker event. In South Atlantic Site 357 this event is represented by a depositional gap. This warming event is followed by the well-documented Middle Miocene rapid decline of isotopic temperatures that appear on all available isotopic curves, and which is especially distinct in the benthic data.

Most curves show another decrease in $\delta^{18}O$ (rise in temperature) that is relatively sharp (although not as marked as the 17–16 Ma decrease) within the Late Miocene between 9 and 8 Ma. This warming event is recorded in both the planktonic and benthic oxygen-isotopic curves from South Atlantic Site 357 (Boersma and Shackleton, 1977), planktonic curve from North Atlantic Sites 398 and 400 (Vergnaud-Grazzini et al., 1978), in the planktonic curve from Pacific Site 281 (Shackleton and Kennett, 1975), and the benthic curve from Site 289 of Woodruff et al. (personal communication, 1980; see fig. 13). In the higher latitude North Atlantic Site 116 the planktonic $\delta^{18}O$ signal does not show much variation for this interval (see fig. 13, col. 3). According to the reinterpreted biochronology at South Atlantic Site 357 (based on nannofossil biostratigraphy), Boersma and Shackleton's (1977) planktonic oxygen-isotopic curve shows another marked decrease in $\delta^{18}O$ value at about 11 Ma, prior to the increase at about 10 Ma and the subsequent decrease between 9 and 8 Ma (see fig. 13, col. 1).

A relatively sharp decline in isotopic temperatures in the latest Miocene (6.5 to 5 Ma) is witnessed by the trends in the oxygen isotopic data from most of the curves mentioned above and was apparently also global in scope.

Savin et al. (1975) documented the uncoupling of high and low latitudinal isotopic temperature curves in the early Middle Miocene. Prior to this time the low latitude surface and high latitude surface (as indicated by benthic data) isotopic temperature curves are sub-parallel. From the Middle Miocene they become divergent, leading to a steepening of equator-to-pole surface temperature gradients, enhancement of the tropical thermocline and intensification of abyssal circulation (Savin et al., 1975); the development is indicative of the establishment (or marked growth) of an extensive ice-cap on Antarctica (Savin, 1977). Keigwin (1979) has remarked on the significant variation in the $\delta^{18}O$ data in the late Miocene (11.5–5 Ma) of equatorial Pacific DSDP sites, which can be interpreted as a consequence of instability in the Antarctic ice-cap.

The North Atlantic has had a somewhat different isotopic history than the other oceans because both surface and deep waters in the North Atlantic resulted from mixtures of water of different origins, i.e., the cooler waters of the Northern basins in the deep, and the warm Tethyan influences in the surface waters (Vergnaud-Grazzini et al., 1978). The Tethyan influence finally ceased after the closure of the eastern Tethys at about 16 Ma, and at that time the Gulf Stream Gyre developed a form similar to that of the present. This event has caused the northward migration of warmer waters and significant moderation of higher latitude climates since the Middle Miocene.

223

In figure 13 the nannofloral migration events in the North and South Atlantic have been plotted against the planktonic foraminiferal oxygen-isotope curves of North Atlantic Site 116 and South Atlantic Site 357, respectively. The South Atlantic migrations show a remarkable agreement for the Early Miocene isotopic temperature trends from Site 357. In the Middle and Late Miocene the trends are also generally compatible, although there seems to be a slight time offset between some isotopic events and corresponding migrations. In the North Atlantic the only isotopic events in Site 116 cores that agree well with migration events are the 17 Ma warming and the coeval migration of temperate nannoflora into higher latitudes, and the 15 Ma reversal of these trends. The temporal offset between isotopic events and nannofloral migrations may be partially due to poor biochronologic resolution capability because of the high latitude location of Site 116.

In figure 13 the benthic $\delta^{18}O$ curve from the equatorial Pacific Site 289 (fide Woodruff and Savin, written communication, 1980) also has been plotted for comparison. This curve is the most detailed available for the Miocene and in general agrees with nannofloral migrationary events recorded in the South Atlantic.

The collective evidence for the event between 17 and 16 Ma seen in the isotopic data (the sharp and widespread drop in $\delta^{18}O$ values) and the nannofloral migration patterns in both the Northern and the Southern Hemispheres of the Atlantic, indicates that this was a period of marked climatic amelioration, apparently of global extent. The nannofloral migration patterns (see fig. 12) also show that the Late Miocene warming event centered at 8.5 Ma was equally, or perhaps more, significant than the warming event at 17 to 15.5 Ma, and also appears to have been of global extent from the isotopic data. This Late Miocene warming is indicated by the marked shift of the temperate assemblage into higher latitudes (up to the Norwegian Sea) and the tropical assemblage up to 45°N, at this time, although the oxygen isotopic data do not indicate a decrease in $\delta^{18}O$ values of similar amplitude.

The biogeographic patterns in figure 12 also imply a latitudinal shift of paleo-isotherms to a plane of symmetry some 10°N of the equator. This indicates that the Northern Atlantic was significantly warmer than the Southern Atlantic through most of the Miocene, as a consequence of the strong moderating influence of the Tethys (in the early Miocene) and the Gulf Stream Gyre (in the Middle and Late Miocene). By the same token, the migratory events in the Middle and Late Miocene North Atlantic reflect the variations in the intensity of the Gulf Stream, and corroborate the paleoceanographic interpretations of Cifelli (1976), who suggested northward incursions of tropical faunal

elements during this time as being due to intensification of the Gulf Stream flow. The thermal differences between the North and South Atlantic may also explain the apparent time lag between warm and cold migrationary events in both Hemispheres.

There is evidence of greater biogeographic variability in the Middle and Late Miocene of the South Atlantic, showing more shorter term variations than the North Atlantic (fig. 13, col. 2). This variability is most probably associated with variations in the volume of the Antarctic ice-cap, as also indicated by the Pacific isotopic data (Keigwin, 1979), and to which North Atlantic surface water conditions would show a relatively minor response. Another indicator of fluctuations in stability of the Antarctic ice-cap may prove to be the *Dictyococcites antarcticus* assemblage which first appears around 15 Ma, at about the same time as the development (or increase) of an extensive ice-cap on Antarctica, and which subsequently shows fluctuations similar to the oxygen-isotopic trends in the South Atlantic and Pacific Oceans.

SUMMARY AND CONCLUSIONS

1) There were distinct spatial and temporal changes in the distribution of major nannofossil assemblages during the Miocene. The biogeographic picture evolves from relatively simple distribution patterns in the Early Miocene to relatively more complex patterns in the Middle to Late Miocene. Most of this biogeographic variability developed in the mid latitudes (25–35°) ; in higher latitudes, there were repeated, distinct alternations of 2 major assemblages through most of the Miocene, while in the lower latitudes there was little change in the dominant assemblages.

2) The late Early Miocene saw the disappearance of Oligocene type assemblages and the appearance of more typical Miocene assemblages. This period also saw the development of greater latitudinal biogeographic provinciality, apparently in response to changes in the thermal structure of water masses and steepening of both latitudinal and vertical thermal gradients at this time.

3) The dominantly cold (higher latitude preferring) *Coccolithus pelagicus* assemblage showed periodic regional restrictions during which time it disappeared from the southern high latitudes. By analogy to the behavior of this assemblage in the Holocene, these disappearances are interpreted as having been caused by Middle and Late Miocene warming episodes and the intensification of the equatorial thermal barrier that hindered its subsequent spread in large numbers to the southern high latitudes.

224

4) The warm (low latitude preferring) *Discoaster-Sphenolithus* assemblage shows distribution patterns that were essentially the opposite of *Coccolithus pelagicus* patterns. The relative variations in the dominance of these 2 assemblages can be used as an effective paleoclimatic tool in the mid latitudes. However, the biogeographic patterns in the lower than 25° and higher than 35° latitudes in the Northern Hemisphere show that comparative abundance of the *C. pelagicus* and *Discoaster-Sphenolithus* assemblages cannot be used for climatic interpretations because of the relative inactivity of the *Discoaster-Sphenolithus* assemblage in these latitudes. In the higher latitudes the comparative abundance of the mid-latitude *Reticulofenestra pseudoumbilica-R. haqii* assemblage relative to the *C. pelagicus* assemblage can be used effectively as a paleoclimatic indicator.

5) The *Dictyococcites antarcticus* assemblage that first appeared in the southern high latitudes in the Middle Miocene and became dominant around Antarctica soon afterwards, shows fluctuations that may reflect variations in the volume of the Antarctic ice-cap. Three distinct periods of dominance by this assemblage in the Middle to Late Miocene also coincide approximately with isotopically determined coolings of the Southern Hemispheric Atlantic.

6) The patterns of migration of some of the major assemblages through latitudes have been interpreted as the result of major climatic fluctuations. These patterns reveal that starting in the earliest Miocene with relatively cool climates, there have been 4 climatic cycles of warming-cooling of approximately 4 to 4.5 m.y. duration in both Hemispheres of the Atlantic. In the North Atlantic the cycles show that the warming events occurred between 22 and 20 Ma, 17 and 16 Ma, 12.5 and 11.5 Ma and 9 and 7.5 Ma, alternating and ending (between 7.5 and 5 Ma) with cooling events. In the South Atlantic the major Miocene warming events occurred between 22 and 20 Ma, 17 and 15.5 Ma, 12 and 10.5 Ma and 9 and 8 Ma, with additional evidence of a minor warming between 14 and 13.5 Ma. In the Southern Hemisphere the Late Miocene also ends with a cooling episode between 8 and 5 Ma (fig. 12).

7) The migratory patterns summarized in figure 12 show a displacement of latitudinal symmetry in paleo-isotherms to about 10°N of the equator, indicating first the moderating influence of the Tethys (in Early Miocene) and subsequently that of the Gulf Stream (in Middle-Late Miocene). These patterns also suggest that the Southern Hemisphere was considerably cooler than the Northern Hemisphere during most of the Miocene.

8) Review of Miocene oxygen isotopic data from various sources reveals considerable agreement between paleoclimatic interpretations based on nannofloral migrationary events and major fluctuations in $\delta^{18}O$ values. In particular, the 17 to 16 Ma warming, the cooling trend at 15 Ma, the warming between 9 and 7 Ma, and the cooling that followed at 6.5 Ma stand out as the most significant, and probably global, climatic events of the Miocene. The peak cooling at 15 Ma and associated steepening of thermal gradients and biogeographic activity in the South Atlantic suggest the development of an extensive ice-cap on Antarctica at this time.

ACKNOWLEDGMENTS

I am indebted to many colleagues for their help during different phases of this project. Thanks are due to: C. Vergnaud-Grazzini and F. Woodruff, S. Savin and R. Douglas for their unpublished results on stable isotopic data from DSDP Sites 116 and 289, respectively; to J. Sclater for paleogeographic maps of the Miocene Atlantic Ocean on which figures. 9-11 are based; to B. Malmgren for help with factor analysis of census data, and T. Flodén for distributional plots and semi-perspective diagrams of assemblages. DSDP samples were obtained through the auspices of the National Science Foundation from the Deep Sea Drilling Project, East Coast Repository at Lamont-Doherty Geological Observatory. This research was supported by the National Science Foundation, Division of Submarine Geology and Geophysics, grant number OCE78–19769. This paper was reviewed by the following colleagues whose helpful comments improved the quality of the text: W. Berggren, B. Corliss, J. Van Couvering, A. McIntyre, P. Roth, S. Savin, H. Thierstein and T. Worsley.

This is Woods Hole Oceanographic Institution Contribution Number 4654.

[*Editor's Note:* Appendixes I-IV have been omitted at this point.]

REFERENCES

ADAMS, C. G., BENSON, R. H., KIDD, R. B., RYAN, W. B. F., and WRIGHT, R. C., 1977. The Messinian salinity crisis and evidence of Late Miocene eustatic changes in the World Ocean. Nature, 269: 383–386.

ARTHUR, M. A., 1979. Paleoceanographic events—Recognition, resolution and reconsideration. Rev. Geophys. Space Phys., 17: 1474–1494.

BACKMAN, J., 1978. Late Miocene-Early Pliocene nannofossil biochronology and biogeography in the Vera Basin SE Spain. Stockholm Contr. Geol., 32(2): 93–114.

BENDER, M. L., and KEIGWIN, L. D., JR., 1979. Speculations about the Upper Miocene change in abyssal Pacific dissolved bicarbonate δ^{13}C. Earth Planet. Sci. Letters, 45: 383–393.

BERGER, W. H., 1973. Deep-sea carbonate: evidence of a coccolith lysocline. Deep-Sea Res., 20: 917–921.

BERGGREN, W. A., and HOLLISTER, C. D., 1974. Paleogeography, paleobiogeography and the history of circulation in the Atlantic Ocean. In: Hay, W. W., Ed., Soc Econ. Pal. Mineral., Special Publ., 20: 126–186.

———, 1977. Plate tectonics and paleocirculation—commotion in the ocean. Tectonophysics, 38: 11–48.

BERGGREN, W. A., and VAN COUVERING, J. A., 1974. The Late Neogene. Palaeogeogr., Palaeoclimatol., Palaeoecol., 16(1/2): 1–216.

BOERSMA, A., and SHACKLETON, N., 1977. Tertiary oxygen and carbon isotope stratigraphies, Site 357 (mid latitude South Atlantic). In: Supko, P., Perch-Nielsen, K., et al., Initial Reports of the Deep Sea Drilling Project, volume 39: 911–924. Washington, D.C.: U.S. Government Printing Office.

BUKRY, D., 1975. Coccolith and silicoflagellate stratigraphy, northwestern Pacific Ocean, DSDP Leg 32. In Larson, R. L., Moberly, R., et al., Initial Reports of the Deep Sea Drilling Project, volume 32: 677–701. Washington, D.C.: U.S. Government Printing Office.

CIFELLI, R., 1976. Evolution of ocean climate and the record of planktonic foraminifera. Nature, 264: 431–432.

CLIMAP, 1976. The surface of the ice-age Earth. Science, 191: 1131–1137.

COOLEY, W. W., and LOHNES, D. R., 1971. Multivariate data analysis. New York: Wiley: 364 pp.

DOUGLAS, R. G., and SAVIN, S. M., 1978. Oxygen isotopic evidence for the depth stratification of Tertiary and Cretaceous planktic foraminifera. Marine Micropal., 3: 175–196.

EMILIANI, C., 1954. Temperatures of Pacific bottom waters and polar superficial waters during the Tertiary. Science, 119: 853–855.

FISCHER, A. G., and ARTHUR, M. L., 1977. Secular variations in the pelagic realm. In: Cook, H. E., and Enos, P., Eds. Deep water carbonate environments. Soc. Econ. Pal. Mineral., Special Publ., 25: 119–150.

FLODÉN, T., and HAQ, B. U., 1980. Computer generated three-dimensional plots in paleobiogeographic studies. Stockholm Contr. Geol., 34(3): 27–33.

HAQ, B. U., 1976. Coccoliths in cores from the Bellingshausen Abyssal Plain and Antarctic Continental Rise (DSDP Leg 35). In: Hollister, C. D., Craddock, C., et al., Initial Reports of the Deep Sea Drilling Project, volume 35: 557–567. Washington, D.C.: U.S. Government Printing Office.

HAQ, B.U., and BERGGREN, W. A., 1978. Late Neogene calcareous plankton biochronology of Rio Grande Rise (S. W. Atlantic Ocean). Jour. Pal., 52(6): 1167–1194.

HAQ, B. U., and LOHMANN, G. P., 1976. Early Cenozoic calcareous nannoplankton biogeography of the Atlantic Ocean. Marine Micropal., 1: 111–194.

HAQ, B. U., PREMOLI-SILVA, I., and LOHMANN, G. P., 1977. Calcareous plankton paleobiogeographic evidence for major climatic fluctuations in the Early Cenozoic Atlantic Ocean. Jour. Geophys. Res., 82: 3861–3876.

HAQ, B. U., WORSLEY, T. R., BURCKLE, L. H., DOUGLAS, R. G., KEIGWIN, L. D., OPDYKE, N. D., SAVIN, S. M., SOMMER, M. A., VINCENT, E., and WOODRUFF, F., in press. The Late Miocene marine carbon-isotopic shift and the synchroneity of some phytoplanktonic biostratigraphic datums. Geology.

HONJO, S., and OKADA, H., 1974. Community structure of coccolithophores in the photic layer of the mid-Pacific. Micropaleontology, 20(2): 209–230.

HSÜ, K. J., CITA, M. B., and RYAN, W. B. F., 1973. Origin of the Mediterranean evaporites. In: Ryan, W. B. F., and Hsü, K. J., et al., Initial Reports of the Deep Sea Drilling Project, volume 13: 1203–1231. Washington, D.C.: U.S. Government Printing Office.

HSÜ, K. J., MONTADERT, L., BERNOUILLI, D., CITA, M. B., ERICKSON, E., GARRISON, R. E., KIDD, R. B., MELIERES, F., MULLER, C., and WRIGHT, R., 1978. History of the Mediterranean salinity crisis. In: Hsü, K. J., Montadert, L., et al., Initial Reports of the Deep Sea Drilling Project, volume 42(I): 1053–1078. Washington, D.C.: U.S. Government Printing Office.

IMBRIE, J., and KIPP, N., 1971. A new micropaleontological method for quantitative paleoclimatology: Application to a Late Pleistocene Caribbean core. In: Turekian, K. K., Ed., Late Cenozoic glacial ages. New Haven, Connecticut: Yale Univ. Press: 71–181.

INGLE, J. C., JR., 1973. Summary comments on Neogene biostratigraphy, physical stratigraphy and paleoceanography in the marginal northeastern Pacific Ocean. In: Kulm, L. D., von Huene, R., et al., Initial Reports of the Deep Sea Drilling Project, volume 18: 949–960. Washington, D.C.: U.S. Government Printing Office.

JÖRESKOG, K. G., KLOVAN, J. E., and REYMENT, R. A., 1976. Geological factor analysis. Amsterdam, Oxford, New York: Elsevier Scientific Publishing Co.: 178 pp.

KEIGWIN, L. D., JR., 1979. Late Cenozoic stable isotope stratigraphy and paleoceanography of DSDP sites from the east equatorial and central Pacific Ocean. Earth Planet. Sci. Letters, 45: 361–382.

KENNETT, J. P., 1977. Cenozoic evolution of Antarctic glaciation, the circum-Antarctic Ocean, and their impact on global paleoceanography. Jour. Geophys. Res., 82: 3843–3860.

KLOVAN, J. E., and IMBRIE, J., 1971. An algorithm and FORTRAN-IV program for large scale Q-mode factor analysis and calculation of factor scores. Math. Geol., 3: 61–67.

KROOPNICK, P. M., MARGOLIS, S. V., and WONG, C. S., 1977. $\delta^{13}C$ variations in marine carbonate sediments as indicators of the CO_2 balance between the atmosphere and oceans. In: Anderson, N. R., and Malahoff, A., Eds., The fate of fossil fuel CO_2 in the ocean. Plenum Press: 295–322.

MATTHEWS, R. K., CURRY, W. B., LOHMANN, K. C., and SOMMER, M. A., 1980. Late Miocene paleoceanography of the Atlantic: Oxygen isotope data on planktonic and benthic foraminifera. Nature, 283: 555–557.

MCINTYRE, A., BÉ, A. W. H., and ROCHÉ, M. B., 1970. Modern Pacific coccolithophorida: A paleontological thermometer. New York Acad. Sci., Trans., Ser. 2, 32(6): 720–731, text-figs. 1–9.

MCINTYRE, A., KIPP, N. G., BÉ, A. W. H., CROWLEY, T., KELLOGG, T., GARDNER, J. V., PRELL, W., and RUDDIMAN, W. F., 1976. Glacial North Atlantic 18,000 years ago: A CLIMAP reconstruction. Geol. Soc. Amer., Mem., 145: 43–76.

OKADA, H., and HONJO, S., 1973. The distribution of oceanic coccolithophorids in the Pacific. Deep-Sea Res., 20(3): 355–374, text-figs. 1–13.

ROTH, P. H., and BERGER, W. H., 1975. Distribution and dissolution of coccoliths in the South and Central Pacific. Cushman Found. Foram. Res., Spec. Publ. 13: 87–113.

RYAN, W. B. F., CITA, M. B., DREYFUS RAWSON, M., BURCKLE, L. H., and SAITO, T., 1974. A paleomagnetic assignment of Neogene stage boundaries and the development of isochronous datum planes between the Mediterranean, the Pacific and Indian Oceans in order to investigate the response of the World Ocean to the Mediterranean "Salinity Crisis." Riv. Ital. Pal., 80: 631–688.

SAVIN, S., 1977. The history of the earth's surface temperature during the last 100 million years. In: Donath, F. A., Stehli, F. G., and Wetherill, G. W., Eds., Ann. Rev. Earth Planet. Sci., 5: 319–355.

SAVIN, S. M., DOUGLAS, R. G., and STEHLI, F. G., 1975. Tertiary marine paleotemperatures. Geol. Soc. Amer., Bull., 86: 1499–1510.

SCHNITKER, D., 1980. North Atlantic oceanography as possible causes of Antarctic glaciation and entroplication. Nature, 284: 615–616.

SCLATER, J. G., HELLINGER, S., and TAPSCOTT, C., 1977. The paleobathymetry of the Atlantic Ocean from the Jurassic to the present. Jour. Geol., 85: 509–522.

SHACKLETON, N. J., and KENNETT, J. P., 1975. Paleotemperature history of the Cenozoic and initiation of Antarctic glaciation: oxygen and carbon isotope analysis in DSDP Sites 277, 279 and 281. In: Kennett, J. P., Houtz, R. E., et al., Initial Reports of the Deep Sea Drilling Project, volume 29: 743–755. Washington, D.C.: U.S. Government Printing Office.

STUMP, E., SHERIDAN, M. F., and BORG, S. G., 1980. Early Miocene subglacial basalts, the East Antarctic Ice Sheet, and uplift of the transantarctic mountains. Science, 207: 757–759.

TALWANI, M., and UDINTSEV, G., 1976. Tectonic synthesis. In: Talwani, M., Udintsev, G., et al., Initial reports of the Deep Sea Drilling Project, volume 38: 1213–1240. Washington, D.C.: U.S. Government Printing Office.

VAN COUVERING, J. A., BERGGREN, W. A., DRAKE, R. E., AGUIRRE, E., and CURTIS, G. H., 1976. The terminal Miocene event. Marine Micropal., 1: 263–286.

VERGNAUD-GRAZZINI, C., PIERRE, C., and LETOLLE, R., 1978. Paleoenvironment of the N. E. Atlantic during Cenozoic: oxygen and carbon isotope analyses at DSDP Sites 398, 400A and 401. Oceanologica Acta, 1(3): 381–390.

VINCENT, E., KILLINGLEY, J. S., and BERGER, W. H., 1980. The magnetic epoch 6 carbon shift: A change in the ocean's $^{13}C/^{13}C$ ratio 6.2 million years ago. Marine Micropal., 5(2): 185–203.

Manuscript received June 17, 1980.

24

Reprinted from page 517 of *Soc. Econ. Paleontologists and Mineralogists Spec. Pub. 32,* 1980, pp. 517–546, by permission of the publisher

MIDDLE CRETACEOUS CALCAREOUS NANNOPLANKTON BIOGEOGRAPHY AND OCEANOGRAPHY OF THE ATLANTIC OCEAN

PETER H. ROTH AND JAY L. BOWDLER*
Department of Geology and Geophysics, University of Utah, Salt Lake City, Utah 84112

ABSTRACT

Quantitative studies of the preservation and biogeographic distribution of calcareous nannofossils in Mid-Cretaceous sediments from the Atlantic are combined with observations on the temporal and spatial distribution of major facies patterns to put constraints on speculations about the paleoceanographic evolution of the Atlantic about 110 to 90 m.y.BP. Organic-rich sediments in deep ocean and continental margin settings are indicative of organic carbon influx from terrestrial and marine sources and considerable carbonate dissolution largely caused by carbon dioxide production during catabolic breakdown of organic matter. Preservation of organic carbon in sediments is the result of rapid burial, especially along continental margins, and/or low concentrations of oxygen in warm and saline bottom waters derived largely from marginal evaporite basins. In the eastern basin of the Atlantic perhaps only the interstitial waters within the sediments were anoxic. There is an overall negative correlation between calcium carbonate and organic carbon content in organic-rich sediments. Coccolith preservation is generally poorest in the most organic carbon-rich shales. Semiquantitative preservation estimates are as sensitive as quantitative estimates using dissolution indices.

Paleogeographic patterns of coccolith distribution show weak latitudinal gradients and more pronounced neritic-oceanic gradients. *Boreal and austral assemblages* first observed in the later half of the Mid-Cretaceous are restricted to latitudes greater than 40°. Rapid spatial and temporal fluctuation in coccolith abundance, including monospecific blooms, are typical of higher latitudes and restricted basins (*e.g.,* Angola Basin). *Neritic assemblages* are found along the continental margin of the eastern Atlantic and over the Walvis-Rio Grande Ridge and are possibly indicative of increased advection of nutrient-rich waters. *Oceanic assemblages* characterize areas far removed from continents such as deep basin and Mid-Atlantic Ridge sites. At the beginning of the Upper Cretaceous, better deepwater connections, the rise in sea-level, and the development of stronger temperature gradients resulted in more vigorous mixing of the deep and surface waters and reduced detrital organic carbon input. Stagnation of deep basins ended, and surface-water temperature-gradients became stronger.

*Present address: Union Oil Company of California, Houston, Texas

25

Coccoliths as Paleosalinity Indicators—
Evidence from Black Sea[1]

DAVID BUKRY[2]

INTRODUCTION

Coccolithophyceae is an order[3] of golden-brown, unicellular, planktonic marine algae that live in greatest numbers from the sea surface to about 50 m depth. The cells are encased in generally spherical skeletons composed of many identical calcite scales, called "coccoliths," which range in size from 1 to 25 μ. These individual coccoliths are commonly preserved in the strata of marine areas and can be used to provide relative ages and ecologic information for most marine deposits from 180 m.y. old to the present.

Cores of Quaternary strata obtained by the R/V *Atlantis II* from the Black Sea (Fig. 1) contain diverse assemblages of coccoliths totaling more than 100 species. The coccoliths in the highest strata are chiefly of Holocene age, whereas reworked ones from the lowest levels sampled (12 m) are of Eocene and Cretaceous ages. This relation is typical of the 36 cores examined (Fig. 2); it results from a late Holocene change in which coccolith production became dominant over the deposition of reworked Eocene and Cretaceous coccoliths.

COCCOLITH STRATIGRAPHY

The distribution of coccoliths permits the recognition of three units within the 12 m penetrated. Unit 1, at the top, is mainly biogenic carbonate ooze composed predominantly of Holocene coccoliths and a few Eocene and Cretaceous taxa; unit 2 contains a mixture of

Holocene, Eocene, and Cretaceous taxa in roughly equal abundance; unit 3, at the base, is characterized by Eocene and Cretaceous coccoliths exclusively. This stratigraphy shows a change from a condition in which only old coccoliths eroded from rock outcrops surrounding the Black Sea were deposited to a condition in which Holocene coccolithophores were abundant in the surface water, producing a rain of coccoliths to the bottom. Rising world sea level following the melting of the last great Pleistocene ice sheets allowed an increase in marine circulation into the Black Sea.

Unit 1, of late Holocene age, is a biogenic sediment composed predominantly of a single coccolithophore species, *Emiliania huxleyi* (Lohmann). Other Holocene coccoliths present include *Acanthoica acanthos* Schiller, *Acanthoica* sp., *Braarudosphaera bigelowi* (Gran and Braarud), *Discolithina* sp. cf. *D. phaselosa* (Black and Barnes), *Scyphosphaera*? sp., *Syracosphaera mediterranea* Lohmann, and *Syracosphaera pirus* Halldal and Markali (Figs. 3-6).

The same Eocene and Cretaceous species that occur in the lower units are present in unit 1, although fewer individuals exist. The most typical and abundant of these species are the solution-resistant forms. Commonly occurring Eocene species of coccolithophores include: *Campylosphaera dela* (Bramlette and Sullivan), *Chiasmolithus grandis* (Bramlette and Riedel), *Chiasmolithus solitus* (Bramlette and Sullivan), *Coccolithus eopelagicus* (Bramlette and Riedel), *Coccolithus pseudogammation* Bouché, *Cyclococcolithina formosa* (Kamptner), *Cyclococcolithina neogammation* (Bramlette and Wilcoxon), *Dictyococcites bisectus* (Hay, Mohler, and Wade), *Dictyococcites scrippsae* Bukry and Percival, *Discoaster barbadiensis* Tan, *Discoaster lodoensis* Bramlette and Riedel, *Helicopontosphaera seminu-*

[1] Manuscript received, November 29, 1971. Publication authorized by the Director, U.S. Geological Survey.
[2] U.S. Geological Survey, La Jolla, California.

I thank Frank T. Manheim and George W. Moore, both of the U.S. Geological Survey, for helpful discussions during this project. Stanley A. Kling, Cities Service Oil Company, provided the scanning electron micrographs.

[3] The writer classifies algae by ICBN, not ICZN.

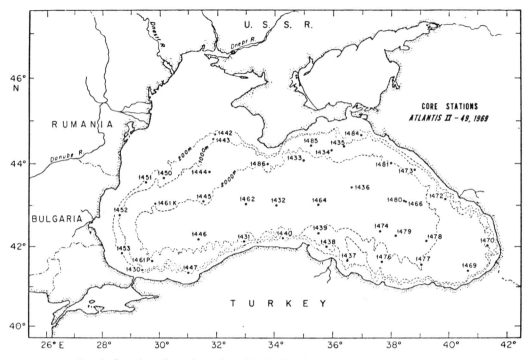

FIG. 1—Location of samples collected by R/V *Atlantis II*, cruise 49, in Black Sea.

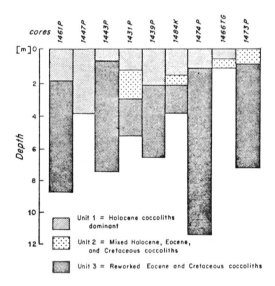

lum (Bramlette and Sullivan), *Isthmolithus recurvus* Deflandre, *Marthasterites tribrachiatus* (Bramlette and Riedel), *Reticulofenestra samodurovi* (Hay, Mohler, and Wade), *Reticulofenestra umbilica* (Levin), *Sphenolithus radians* Deflandre, and *Zygrhablithus bijugatus* (Deflandre).

The most widespread and abundant Cretaceous coccolithophore species of unit 1 include: *Arkhangelskiella cymbiformis* Vekshina, *Broinsonia parca* (Stradner), *Cretarhabdus crenulatus* Bramlette and Martini, *Cribrosphaera ehrenbergii* Arkhangelsky, *Eiffellithus*

FIG. 2—Occurrence of coccolith stratigraphic units in representative Black Sea cores, from west to east.

[*Editor's Note:* Figures 3 through 6 have been omitted.]

augustus Bukry, *Eiffellithus turriseiffeli* (Deflandre), *Gartnerago concavum* (Gartner), *Micula decussata* Vekshina, *Parhabdolithus embergeri* (Noël), *Prediscosphaera cretacea* (Arkhangelsky), *Watznaueria barnesae* (Black), and *Zygodiscus deflandrei* Bukry (Fig. 5).

The sediment of unit 1 is uniform light-gray clay and finely laminated clay with white and gray layers. In both sediment types, microscopic examination (Bukry *et al.*, 1970) revealed the sediment to be composed predominantly of the calcite coccoliths of *Emiliania huxleyi*. The regular laminations of the upper unit have been estimated to number 50–100 per centimeter; they probably represent annual accumulations (Degens *et al.*, 1970; Ross *et al.*, 1970). The fact that the white layers are composed almost exclusively of coccoliths of *E. huxleyi* suggests that they resulted from cyclic plankton blooms of this species.

The middle unit is recognized by the common occurrence of Eocene and Cretaceous coccoliths that, because of their larger size, are more prominent in the sediment than the Holocene species. All the Holocene coccoliths observed in unit 1 are present in unit 2 except for *Acanthoica* and *Discolithina*.

In unit 3 the only coccoliths present are reworked Eocene and Cretaceous fossil forms. The lack of any coccoliths younger than Eocene distinguishes this unit, which, on the basis of sedimentologic continuity with the radiometrically dated sediments of upper layers, probably represents late Pleistocene or early Holocene nonmarine or slightly brackish deposition.

Within all three units, the relative proportion of Eocene taxa generally increases toward the south across the basin, although the proportions of reworked Eocene and Cretaceous taxa differ in individual samples. This greater proportion of Eocene taxa results mainly from a great species diversification of the Coccolithophyceae during the Eocene. In core 1439, Eocene species constitute 79 percent of the total species; in core 1474, 68 percent. The few cores where Cretaceous taxa dominate are from the northern half of the Black Sea: core 1442, 62 percent Cretaceous, core 1484, 56 percent Cretaceous. Special conditions of local erosion may contribute to the Cretaceous dominance at station 1484, because it is located just offshore from the Novorossiysk coast, where Cretaceous calcareous flysch crops out in rapidly eroding seacliffs (Yesin and Savin, 1970).

All the fine-grained carbonate material (1–20 μ) of unit 1 is derived essentially from coccolithophores, whereas, in units 2 and 3, coccoliths represent only 30 percent or less of the fine-grained carbonate material. The preservation state of reworked coccoliths in all three units is similar—in general, fairly good; no significant secondary calcite overgrowths or solution thinning is present. Non-coccolith carbonate material in the lower units shows no clear evidence of being produced as a result of recrystallization or massive overgrowth of coccoliths.

PALEOSALINITY AND GENERAL PALEOECOLOGY

Eocene and Cretaceous coccoliths are present throughout the cored Quaternary Black Sea strata. Because these are eroded particles not contributed directly by organisms, they provide no direct Quaternary ecologic information themselves, but they do serve as a gauge to indicate that the predominance of Holocene pelagic sedimentation in the basin was achieved only about 3,000 years ago, at the beginning of deposition of unit 1.

The broad environmental regime of the Black Sea evidenced by the coccolith assemblages is such that marine salinities necessary for the minimum growth of the most tolerant Holocene species, *Emiliania huxleyi* (greater than 11 ‰), did not exist during deposition of unit 3, because the Black Sea was physically isolated from marine areas. Coccolith assemblages of unit 2 indicate the first marine conditions in the Black Sea sufficient to permit growth of the few coccolithophore species most tolerant of low salinity. During deposition of unit 1, optimal conditions for the growth of low-salinity species developed, resulting in regular blooms of *E. huxleyi*.

The presence of Holocene coccoliths representing such species as *Braarudosphaera bigelowi* and *Emiliania huxleyi* and the absence of certain other Holocene coccoliths from Black Sea sediment provide information on paleoecologic conditions of the Black Sea as contrasted with other bodies of water. Freshening of the Black Sea by major rivers and a restricted marine interchange at the Bosporus help to account for present-day low surface salinities of 17–18 ‰. Open oceans average about 35 ‰. In contrast, the Red Sea, an inland sea, receives little freshwater discharge and is located in a windy, arid evaporite setting; resulting high surface salinities are 37–41 ‰. Fossil Holocene and late Pleistocene coccolith assemblages of the Red Sea represent many wholly oceanic species that are absent from Black Sea sediment—for example, *Coccolithus pelagicus* (Wallich), *Cyclococcolithina leptopora* (Murray and Blackman), *Gephyrocapsa oceanica* Kamptner, *Helicopontosphaera kamptneri* Hay and Mohler, *Rhabdosphaera stylifera* Lohmann, and *Umbilicosphaera mirabilis* Lohmann (see McIntyre, 1969).

Braarudosphaera bigelowi significance—The absence of *Braarudosphaera bigelowi* in Red Sea assemblages and its presence in the Black Sea are particularly significant. Pentagonal coccoliths of this species in the Black Sea are widespread, although not abundant; they occur in 25 of the 36 cores. The assumption that *B. bigelowi* is deterred from developing in the Red Sea by high salinities is supported by evidence from living populations of other areas; the greatest concentrations of this species are

in coastal waters of low salinity, such as the Gulf of Maine and the Gulf of Panama. In the Gulf of Maine the highest concentration of *B. bigelowi* (680 cells per liter) was found in waters having salinities of 31.8–32.7 ‰ (Gran and Braarud, 1935). The greatest abundance of *B. bigelowi*—1,500 cells per liter— was from the Gulf of Panama in salinities of 26.8–32.5 ‰ (Smayada, 1966). Lower counts are recorded for open-ocean areas of normal salinity. Gaarder (1954) reported *B. bigelowi* at four North Atlantic stations; at three of the stations, only a single specimen was present. Hulburt (1962) recorded maximum concentrations of only 20 cells per liter at some western Atlantic stations with salinities of 36.8–37.2 ‰, and Hulburt and Rodman (1963) found *B. bigelowi* only at salinities below 34 ‰ between southern New England and Bermuda. Hulburt and Rodman omitted it from a longer phytoplankton list, including the coccolithophores *Rhabdosphaera stylifera* and *Umbilicosphaera mirabilis*, found in samples with salinities above 34.5 ‰. The concentration of *B. bigelowi* at Bermuda was only 34 cells per liter. In Pacific Ocean studies of equatorial to subantarctic coccolithophore populations, Hasle (1959, 1960) identified 34 species but reported no *B. bigelowi* in waters in which the salinity exceeded 34.7 ‰. Likewise, Ushakova's studies (1967, 1969) of Indian Ocean and Pacific Ocean sea-surface and bottom-sediment samples failed to reveal any *B. bigelowi*.

Although our data concerning the salinity tolerances of living coccolithophores are limited, the widespread occurrence of *B. bigelowi* in the Black Sea clearly indicates the ability of this species to thrive in unusually low salinities. The available information from other areas is in agreement, suggesting that this species is excluded from contemporary waters of high salinity.

Fossil forms of *B. bigelowi* are most abundant in Cenozoic coastal marine deposits. For example, the lower Cenozoic deposits of the Gulf Coast and the Paris basin (Bouché, 1962) contain abundant *Braarudosphaera* species. Conversely, midocean cores of lower Cenozoic sediment obtained by the Deep Sea Drilling Project typically contain no *Braarudosphaera* species. The only anomaly in the fossil record regarding a neritic dominance of *Braarudosphaera* is the report of a thin zone of braarudosphaerids in Oligocene strata from the South Atlantic (Saito and Percival, 1970). If an ability to thrive at lowered salinities was characteristic of Oligocene braarudosphaerids, as it is for modern Black Sea and Gulf of Panama forms, then a reduction of the salinity of South Atlantic near-surface water for a short time could have fostered a braarudosphaerid bloom. Although short-term blooms of low-salinity phytoplankton do occur in oceanic areas when other factors are optimal (Hulburt and Rodman, 1963), the widespread occurrence of this South Atlantic zone of braarudosphaerids, which has a thickness of 80 cm at the shallowest site and 3.5 cm and 0.5 cm at deeper sites, suggests a prolonged ecologic effect such as reduced surface salinities from increased regional rainfall.

Emiliania huxleyi significance—*Emiliania huxleyi*, which provided the greatest abundance of Black Sea coccoliths, is also the most abundant modern oceanic coccolithophore; it is the dominant species in tropic to subarctic waters. Large populations of more than 200,000 cells per liter have been reported from 62°S lat. in the subantarctic Pacific, where temperatures are 2–3°C (Hasle, 1960); and a population of 89,000 cells per liter was reported during the annual bloom in the Gulf of Panama, at 8°N lat., when the temperature was 23–26°C (Smayada, 1966). Berge (1962) reported an unusual bloom of *E. huxleyi* in the coastal water of Norway in June 1955 that reached a concentration of 115 million cells per liter at 11°C and 33 ‰ salinity. This summer bloom was considered to result from a combination of optimal conditions, including greatest annual sunlight. The annual *E. huxleyi* blooms are a well-documented part of the phytoplankton succession in coastal areas (Hasle and Smayada, 1960; Smayada, 1966).

Because of its adaptability to a wide variety of temperature and salinity conditions, *Emiliania huxleyi* has proved to be a convenient species to culture in the laboratory, and the extent of its environmental tolerance has been carefully measured. Mjaaland (1956) determined that growth proceeds in salinities from 16 to 45 ‰. Using a culture at 35 ‰ salinity, Watabe and Wilbur (1966) showed that growth occurred over their entire experimental range of temperatures, 7–27°C. A maximum temperature range of 2–29°C has been determined in nature (McIntyre and Bé, 1967), and *E. huxleyi* has been recorded from the Sea of Azov (Pitsyk, 1963), which has a mean salinity of 11.2 ‰ (Zenkevich, 1963, p. 471).

Several factors concerning the production of the calcite coccoliths of *E. huxleyi* have been determined. Braarud (1963) noted that ". . . the process of coccolith formation appears to take place within a few hours, and more than one layer of coccoliths may be produced [on the cell] during this period of time. Under natural conditions the outer layer may be detached rather rapidly. . . ." The form of the individual coccoliths is apparently controlled by the water temperature in which the population develops (Watabe and Wilbur, 1966; McIntyre and Bé, 1967). Cold-water forms have fewer elements (29) in the two coccolith rims, and the smaller rim is solid rather than spokelike. The warm-water form has a mean of 35 elements in each rim, and both rims are spokelike. Coccoliths of *E. huxleyi* from the Black Sea have from 32 to 38 rim elements, typically about 36, with spokelike arrangements in

Table 1. Comparison of Occurrences of Fossil Holocene Coccoliths in Marine Areas of Different Salinity

	Marine Area and Salinity			
	Sea of Azov (11 °/oo)	Black Sea (17–18 °/oo)	Atlantic Ocean (35 °/oo)	Red Sea (37–41 °/oo)
Emiliania huxleyi	X	X	X	X
Syracosphaera pirus		X	X	X
Braarudosphaera bigelowi		X	X	
Ceratolithus cristatus			X	
Gephyrocapsa oceanica			X	X
Helicopontosphaera kamptneri			X	X
Rhabdosphaera stylifera			X	X

both rims and narrow openings between the lower rim elements (Fig. 3D).

The geometry of *E. huxleyi* coccoliths indicates that the Black Sea population in bottom-sediment laminae is a product of annual summer blooms when the light intensity and temperature were optimal for maximum growth. The unique recording of these events in the bottom sediment is attributed to the exclusion of bottom-burrowing organisms owing to the nonoxygenated nature of Black Sea bottom waters (Degens *et al.*, 1970).

Members of Holocene oceanic coccolithophore populations that are missing from the Black Sea, such as *Cyclococcolithina leptopora*, provide further insight into the temperature and salinity history of the Black Sea. *C. leptopora* occurs in the Mediterranean Sea, Red Sea, and subantarctic to equatorial Pacific Ocean, and is therefore considered to be adapted to a wide temperature range. Thus, no temperature barrier should exist to the penetration of this species into the Black Sea. However, the low salinity of the Black Sea could be an effective barrier to such a characteristically oceanic coccolithophore. The report of Pitsyk (1963) noting *E. huxleyi* as the only coccolith form in the Sea of Azov, together with a comparison of the fossil Holocene assemblage of the Black Sea with assemblages from marine waters of progressively higher salinity, provides a general indicator for paleosalinity determination (Table 1).

CONCLUSIONS

Coccoliths in the upper layers of Black Sea sediment provide a graphic record of increased Black Sea salinity during the Holocene. The appearance of the dominant cosmopolitan species *Emiliania huxleyi* and its subsequent concentration in annual laminae provide a record of the uniformity of the sedimentary environment from about 3,000 years ago to the present. The specialized group of coccoliths able to grow in the low salinity of the Black Sea provides a guide for recognizing similar conditions of low salinity in the strata of other areas.

REFERENCES CITED

Berge, G., 1962, Discoloration of the sea due to *Coccolithus huxleyi* "bloom": Sarsia, v. 6, p. 27–40.

Bouché, P. M., 1962, Nannofossiles calcaires du Lutétien du bassin de Paris: Rev. Micropaléontologie, v. 5, no. 2, p. 75–103.

Braarud, T., 1963, Reproduction in the marine coccolithophorid *Coccolithus huxleyi* in culture: Napoli Zool. Sta. Pub., v. 33, p. 110–116.

Bukry, D., et al., 1970, Geological significance of coccoliths in fine-grained carbonate bands of postglacial Black Sea sediments: Nature, v. 226, no. 5241, p. 156–158.

Degens, E. T., S. W. Watson, and C. C. Remsen, 1970, From meter to centimeter to micron and finally to angstrom units: Oceanus, v. 15, no. 4, p. 11-15.

Gaarder, K. R., 1954, Coccolithineae, Silicoflagellatae, Pterospermataceae and other forms from the "Michael Sars" North Atlantic deep-sea expedition 1910: Sars North Atlantic Deep Sea Exped. Rept., v. 2, no. 4, p. 1–20.

Gran, H. H., and T. Braarud, 1935, A quantitative study of the phytoplankton in the Bay of Fundy and the Gulf of Maine (including observations on hydrography, chemistry, and turbidity): Biol. Board Canada Jour., v. 1, no. 5, p. 280–467.

Hasle, G. R., 1959, A quantitative study of phytoplankton from the equatorial Pacific: Deep-Sea Research, v. 6, p. 38–59.

——— 1960, Plankton coccolithophorids from the subantarctic and equatorial Pacific: Nytt Mag. Botanikk, v. 8, p. 77–88.

——— and T. J. Smayada, 1960, The annual phytoplankton cycle at Drøbak, Oslofjord: Nytt Mag. Botanikk, v. 8, p. 53–75.

Hulburt, E. M., 1962, Phytoplankton in the southwestern Sargasso Sea and north equatorial current, February 1961: Limnology and Oceanography, v. 7, no. 3, p. 307–315.

——— and J. Rodman, 1963, Distribution of phytoplankton species with respect to salinity between the coast of southern New England and Bermuda: Limnology and Oceanography, v. 8, no. 2, p. 263–269.

McIntyre, A., 1969, The Coccolithophorida in Red Sea sediments, *in* E. T. Degens and D. A. Ross, eds., Hot brines and recent heavy metal deposits in the Red Sea: New York, Springer-Verlag, p. 299–305.

——— and A. W. H. Bé, 1967, Modern Coccolithophoridae of the Atlantic Ocean—I. Placoliths and cyrtoliths: Deep-Sea Research, v. 14, p. 561–597.

Mjaaland, G., 1956, Some laboratory experiments on the coccolithophorid *Coccolithus huxleyi*: Oikos, v. 7, no. 2, p. 251–255.

Pitsyk, G. K., 1963, O kachestvennom sostave fytoplanktona Azovskogo morya (On the qualitative composition of phytoplankton in the Sea of Azov): Sevastopol' Biol. Sta. Trudy, v. 16, p. 71–89.

Ross, D. A., E. T. Degens, and J. MacIlvaine, 1970, The Black Sea: recent sedimentary history: Science, v. 170, no. 3954, p. 163–165.

Saito, T., and S. F. Percival, Jr., 1970, Paleontology: Deep Sea Drilling Project Initial Rept., v. 3, p. 444–445.

Smayada, T. J., 1966, A quantitative analysis of the phytoplankton of the Gulf of Panama—III. General ecological conditions, and the phytoplankton dynamics at 8°45'N, 79°23'W from November 1954 to May 1957: Inter-Am. Tropical Tuna Comm. Bull., v. 11, no. 5, p. 353–612.

Ushakova, M. G., 1967, Kokkolity vo vzvesi i v poverkhnostnom sloye osadkov Indiyskogo okeana (Coccoliths in suspension and their precipitation from the surface of the Indian Ocean), in Iskopayemyye vodorosli SSSR: Akad. Nauk SSSR Sibirskoye Otdeleniye Inst. Geologii i Geofiziki, p. 84–90.

———— 1969, Kokkolity vo vzvesi i v poverkhnostnom sloye osadkov Tikhogo i Indiyeskogo okeanov (Coccoliths in suspension and their precipitation from surfaces of the Pacific and Indian Oceans), in Osnovniye problemy mikropaleontologii i organogennogo osadkonakopleniya v okeankhi moryakh (Fundamental problems of micropaleontology and concentrating organisms in surface layers of ocean seas): Moscow, Izd. "Nauka," p. 96–104.

Watabe, N., and K. M. Wilbur, 1966, Effects of temperature on growth, calcification, and coccolith form in Coccolithus huxleyi (Coccolithinaea): Limnology and Oceanography, v. 11, no. 4, p. 567–575.

Yesin, N. V., and M. T. Savin, 1970, Abrasion of the flysch shore of the Black Sea: Okeanologiya (Am. Geophys. Union transl.), v. 10, no. 1, p. 96–100.

Zenkevich, L. A., 1963, Biologiya morei SSSR: Moscow, Izd. Akad. Nauk SSSR, 793, p.; Engl. transl. (Zenkevitch), 1963, Biology of the seas of the U.S.S.R.: New York, Interscience Pub., 955 p.

26

Reprinted from *Science* **189**:555–557 (1975)

OXYGEN AND CARBON ISOTOPES FROM CALCAREOUS NANNOFOSSILS AS PALEOCEANOGRAPHIC INDICATORS

Stanley V. Margolis, Peter M. Kroopnick, David E. Goodney,
Walter C. Dudley, and Maureen E. Mahoney

Calcareous nannofossils, principally Coccolithophoridae, a group of marine phytoplankton, are major contributors to pelagic sediments. They are widely used both in biostratigraphic age determinations and as paleoceanographic indicators (1) because of their rapid evolutionary changes, habitat, and apparently higher resistance to dissolution than planktonic foraminifera, as evidenced by their abundance in fossil carbonate sediments (2, 3). They have not, however, been the subject of any detailed stable isotope studies. We have determined $\delta^{18}O$ and $\delta^{13}C$ values (4) from calcareous nannofossils contained in three sediment cores collected in the Southern Ocean during Leg 29 of the Deep Sea Drilling Project (DSDP) and compared these data with similar published ones (5) for associated planktonic and benthic foraminifera contained in the same samples. The purpose of this comparison is to determine whether calcareous nannofossils preserve $\delta^{18}O$ and $\delta^{13}C$ values consistent with their euphotic habitat (3) or undergo exchange with bottom- or interstitial waters during or after burial.

The Leg 29 cores (5) are well suited for such a study as they represent a relatively continuous sequence spanning the last 55 million years (6) and contain a well-preserved calcareous fauna and flora deposited at depths between 1200 and 3300 m. During this time interval, profound changes in the configuration, circulation patterns, and temperature structure of the Southern Ocean occurred (7, 8), and extensive ice sheets developed on Antarctica (5, 7, 9).

Relatively pure but polyspecific calcareous nannofossil fractions consisting of isolated coccoliths, coccospheres, and discoasters were separated from the fraction of samples < 44 μm from the three cores by using short centrifuge techniques (10), and each sample was checked for purity of nannofossil content and state of preservation by scanning electron microscopy. Values of $\delta^{18}O$ and $\delta^{13}C$ for calcareous nannofossils were determined by using standard mass spectrometer techniques (11).

The $\delta^{18}O$ profile for calcareous nannofossils (Fig. 1) closely parallels the foraminiferal profiles for most of its length, including intervals of rapid change of apparent temperature in the late Cenozoic and at the Oligocene-Eocene boundary. Nannofossil $\delta^{18}O$ values are occasionally slightly higher than those of associated planktonic foraminifera, indicating either differential isotope (vital) fractionation during growth or postmortem reequilibration produced by dissolution or secondary encrustations within sediments (12). However, none of our nannofossil $\delta^{18}O$ values are as high as those of the colder-water benthic foraminifera, which would indicate complete isotopic exchange with bottom waters. The majority of nannofossil samples from site 277 exhibit $\delta^{18}O$ values that are equal to or slightly lower than those of the planktonic foraminifera, indicating that the nannofossils are apparently preserving surface water temperatures.

The $\delta^{13}C$ profile for calcareous nannofossils (Fig. 2) also shows a tendency to parallel the foraminiferal curves. The progressive increase in $\delta^{13}C$ values from the benthic and planktonic foraminifera to the nannofossils at sites 279A and 277 suggests that carbon isotopes reflect the $\delta^{13}C$ of the surrounding media and may be indicative of the water depth during growth (13). The $\delta^{13}C$ of the ΣCO_2 in present-day South Pacific waters varies from about +2.0 per mil at the surface to around +0.5 per mil in bottom waters (14, 15).

The effects of dissolution and secondary encrustation with resultant isotopic reequilibration of calcareous nannofossils should be related to either water depth or subsequent burial history. At the present time planktonic foraminifera show slight dissolution effects below 1000 m, with appreciable dissolution occurring below 3000 m in the Central Pacific. Etching, fragmentation, and dissolution features are evident on coccoliths found below 3000 m, and further deterioration increases rapidly below 4000 m. Overgrowths are most prevalent on coccoliths deposited between 3500 and 4800 **m** and apparently can be produced at or near the sediment-water interface (16). Among the cores studied here, only sediment samples from site 279A now lie below a water depth of 3000 m and should contain the most evidence for dissolution. The isotope data from site 279A (3341 m), when compared with sediments from sites 277 and 281 (1214 and 1591 m), do not support such a depth-dependent reequilibration model.

Reequilibration of the nannofossils during or after burial should produce higher $\delta^{18}O$ values because of colder bottom water temperatures and lower $\delta^{13}C$ values because of incorporation of ^{13}C-depleted organic carbon (14, 17). Several samples we analyzed do show possible indications of reequilibration with deeper water when compared to planktonic foraminifera. Detailed electron microscopic examinations

of nannofossils from site 277, core 28, and site 281, core 6, which have lower $\delta^{13}C$ and higher $\delta^{18}O$ values than planktonic foraminifera (Figs. 1 and 2), reveal that most of the nannofossil remains in these cores exhibit abundant evidence of dissolution and secondary calcite overgrowths and contain no whole coccospheres. However, samples where $\delta^{13}C$ values for nannofossils are higher than those for planktonic fora-

minifera, and $\delta^{18}O$ values are lower, such as from site 277, cores 4 and 9, contain numerous whole coccospheres and coccoliths which are well preserved. Thus, samples of nannofossils must be subjected to electron microscopic studies of preservation if meaningful paleoceanographic data are desired.

Examination of the Eocene portion of site 277 (Fig. 2) reveals a divergence in

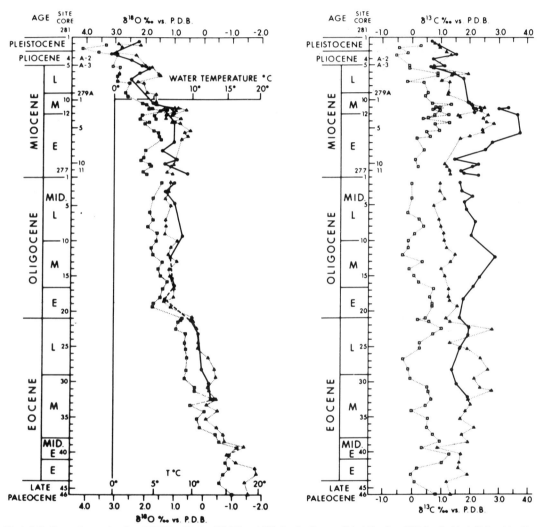

Fig. 1 (left). Oxygen isotope data from DSDP Leg 29, sites 277, 279, and 281, showing the per mil deviation from PDB for planktonic (▲) and benthic (☐) foraminifera (5) and calcareous nannofossils (●) (this study). Note the close correspondence between nannofossil and planktonic foraminifera curves. Stratigraphic plots (6) are not based on thickness or age differences between samples, but represent relative positions in each cored interval. Water temperature values (5) are valid from base of sequence to middle Miocene only. Above core 10 at site 281 water temperature estimations are complicated by seawater isotopic changes (5). Calcareous nannofossils are assumed to be equilibrating with surface waters during growth. Nannofossil values that are higher than planktonic foraminifera values represent possible isotope exchange with deeper waters.   Fig. 2 (right). Carbon isotope data from sites 277, 279, and 281 showing per mil deviation from PDB for planktonic (▲) and benthic (☐) foraminifera (5) and calcareous nannofossils (●) (this study). Nannofossil $\delta^{13}C$ trends parallel those for foraminifera, but are higher than planktonic foraminifera values in Oligocene to middle Miocene samples. Nannofossil $\delta^{13}C$ values significantly lower than those for associated planktonic foraminifera represent possible isotope exchange with deeper waters. Stratigraphic plots are not based on thickness or age differences between samples, but represent relative positions in each cored interval.

$\delta^{13}C$ values for benthic and planktonic foraminifera between cores 22 and 34. Individuals of the species *Globigerapsis index* were used to obtain the planktonic foraminiferal values from these samples, whereas in other samples mixed planktonic species have been used (5). It has been found that there are definite species-dependent departures from isotopic equilibrium in planktonic foraminifera (18), so that surface water $\delta^{13}C$ values cannot be derived from planktonic foraminiferal measurements even if the depth habitat of the species is known (and it is not for extinct species). Polyspecific samples of calcareous nannofossils may provide a more reliable indication of surface $\delta^{13}C$ changes than planktonic foraminifera (19). Measurement of changes in surface water carbon isotopic composition with time may yield useful information on changes in oceanic productivity.

The reliability of nannofossils as indicators of surface water paleotemperatures can be tested by growing pure cultures of coccolithophorids at varying temperatures (20). Preliminary oxygen isotope data for samples of these species indicate that during coccolith growth, oxygen isotopes are incorporated in equilibrium with the growth medium, when compared with the empirically determined paleotemperature curve (21). Similar isotopic results have been obtained for cultures of *Emiliania huxleyi* (22).

Our evidence indicates that $\delta^{18}O$ values obtained from well-preserved polyspecific samples of calcareous nannofossils can be used to estimate surface water paleotemperatures. The $\delta^{13}C$ values appear to reflect the depth of growth for benthic and planktonic foraminifera and calcareous nannofossils when compared with $\delta^{13}C$ profiles for today's oceans. These preliminary results add a new and potentially important method to the paleoclimatologist's arsenal for studying changes in the world's oceans. Many deep-water carbonate cores spanning critical intervals in the Mesozoic and Cenozoic lack planktonic foraminifera but often contain calcareous nannofossils, and these now can be used for detailed isotope paleotemperature studies.

References and Notes

1. E. Martini, in *Proceedings of the 2nd Plankton Conference, Rome 1970*, A. Farinacci, Ed. (Tecnoscienza, Rome, 1971), vol. 2, p. 739; A. McIntyre and R. McIntyre, in *The Micropaleontology of Oceans*, B. M. Funnell and W. R. Riedel, Eds. (Cambridge Univ. Press, Cambridge, England, 1971), p. 253; A. McIntyre, A. W. H. Bé, M. B. Roche, *Trans. N.Y. Acad. Sci* **32**, 720 (1970).
2. W. H. Berger, *Deep-Sea Res.* **20**, 917 (1973).
3. M. Black, *Endeavour (Engl. Ed.)* **24**, 131 (1965).
4. The δ values are per mil deviations from the Pee Dee belemnite (PDB) isotopic standard: $\delta = [(R/R_{std}) - 1] \times 1000$, where R is either $^{13}C/^{12}C$ or $^{18}O/^{16}O$ in the sample and R_{std} is the corresponding ratio in the standard [H. Craig, *Geochim. Cosmochim. Acta* **12**, 133 (1957)].
5. N. J. Shackleton and J. P. Kennett, in *Initial Reports of the Deep Sea Drilling Project* (Government Printing Office, Washington, D.C., 1975), vol. 29, pp. 743-755.
6. A. R. Edwards and K. Perch-Nielsen, in *ibid.*, pp. 469-539.
7. J. P. Kennett *et al.*, in *ibid.*, pp. 1155-1169.
8. ———, *Science* **186**, 144 (1974).
9. S. V. Margolis and J. P. Kennett, *ibid.* **170**, 1085 (1970).
10. A. McIntyre, A. W. H. Bé, R. Preikstas, *Prog. Oceanogr.* **4**, 3 (1967). Samples containing more than 10 percent nonnannofossil calcium carbonate after centrifugation were not isotopically analyzed, and the next closest samples where coccolith separations of high purity could be obtained were used. The most common calcareous biogenic contaminants were fragments of planktonic foraminifera. Nannofossil isotope analyses were not performed on samples from below core 33 from site 277 because of diagenetic changes.
11. N. J. Shackleton and N. D. Opdyke, *Quant. Res. (N.Y.)* **3**, 39 (1973). Reproducible isotopic results were achieved with each 25-mg sample by roasting in vacuo for 1 hour between 450° and 500°C to remove volatile organic components and char protecting organic films. The carbonate was then converted to CO_2 gas by reaction with 100 percent H_3PO_4 at 25°C. The gas was then analyzed with a Nuclide 3-60 RMS mass spectrometer, and the results are expressed as the per mil deviation from the PDB isotopic standard (4). The gas from several samples was analyzed twice by mass spectrometry, and for some samples two separate acid extractions were performed. Based on these experiments, the analytical reproducibility is at least ± 0.10 per mil.
12. R. G. Douglas and S. M. Savin, *Geol. Soc. Am. Publ. 6* (1974), p. 714.
13. T. Saito and J. Van Donk, *Micropaleontology* **20**, 152 (1974).
14. P. Kroopnick, *Deep-Sea Res.* **21**, 211 (1974).
15. H. Craig, *J. Geophys. Res.* **75**, 691 (1970).
16. P. H. Roth and W. H. Berger, *Cushman Found. Foraminiferal Res. Spec. Publ. 13* (1975), p. 87.
17. P. Kroopnick, *Earth Planet. Sci. Lett.* **22**, 397 (1974).
18. N. J. Shackleton *et al.*, *Nature (Lond.)* **242**, 177 (1973); R. G. Douglas and S. M. Savin, *J. Foraminiferal Res.*, in press; S. M. Savin and R. G. Douglas, *Geol. Soc. Am. Bull.* **84**, 2327 (1973).
19. N. J. Shackleton, personal communication.
20. Pure cultures of *Emiliania huxleyi* and *Cricospaera elongata* obtained from J. B. Jordan (Scripps Institution of Oceanography) and *Cricosphaera carterae* obtained from R. L. Guillard (Woods Hole Oceanographic Institution) have been successfully cultured at constant temperatures (± 0.5°C) ranging from 10° to 27°C using modified and buffered natural seawater, and light conditions recommended for optimum growth and plate production [R. L. Guillard and J. H. Ryther, *Can. J. Microbiol.* **8**, 299 (1962); K. M. Wilbur and N. Watanabe, *Ann. N.Y. Acad. Sci.* **109**, 82 (1963); W. F. Blankley, thesis, University of California, San Diego (1971)]. *Cricosphaera carterae* cultures have been harvested near the end of the logarithmic growth phase. After addition of distilled water and centrifugation to burst the cells, the concentrated carbonate plates were roasted at 500°C to remove residual organic matter before CO_2 extraction (11).
21. S. Epstein, R. Buchsbaum, H. A. Lowenstam, H. C. Urey, *Geol. Soc. Am. Bull.* **64**, 1315 (1953); H. Craig, in *Stable Isotopes in Oceanographic Studies and Paleotemperatures*, E. Tongiorgi, Ed. (Laboratori di Geologia Nucleare, Consiglio Nazionale della Ricerche, Pisa, 1965), p. 1.
22. A. McIntyre, personal communication.
23. The Deep Sea Drilling Project, a joint project of five oceanographic institutions in the United States, funded by the National Science Foundation, is responsible for collection of the Leg 29 cores analyzed here. We thank N. J. Shackleton, J. P. Kennett, W. H. Berger, and P. H. Roth for helpful suggestions. Supported by NSF grants GD-34270, GD-37514, and IDO71-04202.

18 April 1975

27

Reprinted from page 31 of *Marine Micropaleontology* **5**:31–42 (1980)

OXYGEN AND CARBON ISOTOPES OF RECENT CALCAREOUS NANNOFOSSILS AS PALEOCEANOGRAPHIC INDICATORS

DAVID E. GOODNEY[1], STANLEY V. MARGOLIS, WALTER C. DUDLEY[2], P. KROOPNICK and DOUGLAS F. WILLIAMS[3]

Department of Oceanography, Hawaii Institute of Geophysics, University of Hawaii, Honolulu, Hawaii 96822 (U.S.A.)

(Accepted August 16, 1979)

Abstract

Goodney, D.E., Margolis, S.V., Dudley, W.C., Kroopnick, P. and Williams, D.F., 1980. Oxygen and carbon isotopes of Recent calcareous nannofossils as paleoceanographic indicators. Mar. Micropaleontol., 5: 31—42.

$\delta^{18}O$ and $\delta^{13}C$ values for several species of planktonic foraminifera and calcareous nannofossils from Recent deep-sea sediments have been studied in order to evaluate their paleoceanographic and paleotemperature potential. Nannofossils from Indian Ocean core-tops reflect isotopic temperatures as warm as, or warmer than, the temperatures reported by Williams et al. (1977) for shallow-dwelling planktonic foraminifera from the same samples. In general, deep-sea sediment samples from the world's major oceans indicate that nannofossil $\delta^{18}O$ values are from 0.5 to 1 $^{o}/_{oo}$ heavier than shallow-dwelling planktonic foraminifera. Although nannofossil $\delta^{18}O$ values depart from thermodynamic equilibrium with oceanic surface water temperatures, the $\delta^{18}O$ temperature trend parallels that of surface-dwelling planktonic foraminifera.

Nannofossil $\delta^{13}C$ values also depart from equilibrium with surface water $\delta^{13}C$-ΣCO_2 values. A comparison of nannofossil $\delta^{13}C$ data with that from planktonic foraminifera suggests that the rate of primary productivity in different water masses may be influencing the $\delta^{13}C$ of carbonate-secreting phytoplankton and zooplankton.

[1] Present address: Department of Chemistry, Willamette University, Salem, Oreg. 97301 (U.S.A.).
[2] Present address: Natural Science Division, Hilo College, University of Hawaii, P.O. Box 1357, Hilo, Hawaii 96720 (U.S.A.).
[3] Department of Geology, University of South Carolina, Columbia, S.C. 29208 (U.S.A.).

239

28

Copyright ©1980 by Macmillan Journals Ltd
Reprinted from Nature 285:222-223 (1980)

Coccoliths in Pleistocene–Holocene nannofossil assemblages

**W. C. Dudley*, J. C. Duplessy†, P. L. Blackwelder‡,
L. E. Brand§ & R. R. L. Guillard§**

* University of Hawaii at Hilo, Hilo, Hawaii 96720
† Centre des Faibles Radioactivites CNRS-CEA, 91190 Gif-sur-Yvette France
‡ Nova University Ocean Science Center, Dania, Florida 33004
§ Woods Hole Oceanographic Institution, Woods Hole, Massachusetts 02543

Marine palaeotemperature studies are increasingly using oxygen isotope analyses of the minute calcium carbonate structures produced by a group of marine phytoplankton, the coccolithophores[1-6]. To provide a sound experimental basis for palaeotemperature calculations using the isotopic data from analyses of coccoliths, we have grown coccolithophores in laboratory batch culture in controlled environmental conditions, and determined the oxygen isotopic compositions of the coccoliths produced at known temperatures. The results reported here indicate that the oxygen isotopic composition of the coccoliths of all the species studied is strongly temperature dependent. A 'vital effect' was observed in all the species, with the isotopic values of different culture samples falling into two definite groups containing separate taxa. The difference in vital effect between different species suggests that calcification processes may vary among different taxa and indicates that a re-evaluation of coccolith oxygen isotope palaeoclimatic interpretations may be in order.

Coccolithophores possess certain unique advantages for oxygen isotope palaeoclimatic studies. As phytoplankters, they are restricted to the photic zone, with maximum population densities found between 50 and 100 m depth in most oceanic areas[7]. Consequently, coccolith secretion takes place only at near-surface water temperatures. On the other hand, planktonic foraminifera may produce their tests at the various temperatures found between the surface and depths as great as 1,500–2,000 m, although depth ranges for individual taxa may be more narrowly restricted[8]. In addition, coccolithophores show a wide geographic distribution, being found in all but polar seas[7]. Finally, coccoliths are more abundant than foraminiferal tests in many Tertiary and Cretaceous marine sediments[9].

We have maintained in laboratory culture the previously isolated *Emiliania huxleyi*, and successfully isolated and cultured for the first time *Gephyrocapsa oceanica*, *Cyclococcolithus leptoporus*, and *Thoracosphaera heimii*. These four species produce coccoliths frequently encountered in many Pleistocene and Holocene sediments.

The clonal algal cultures were grown in continuous batch culture and acclimated for 2 to 3 weeks at each temperature (12, 16, 20, 24, and 28 °C) before the coccoliths of the last batch culture were collected for analysis. The cultures were grown in 33‰ seawater enriched with $f/2$ levels of nutrients as described by Guillard[10] with silicic acid omitted. Light was provided by cool white fluorescent bulbs producing 0.023 langley min^{-1}. Coccoliths were concentrated by centrifugation in buffered distilled water, then oven dried and roasted under vacuum at 400 ± 25 °C to remove residual organic matter. The carbonate samples were reacted at 50 °C according to the method developed by Shackleton[11,12] and analysed by mass spectrometry. Samples of the water of the growth medium were analysed according to the method of Epstein and Mayeda[13] and corrections made according to Craig[14,15].

Analytical data obtained from over 70 samples showed two important trends. The $\delta^{18}O$ values of the coccoliths of *G. oceanica* and *E. huxleyi* are strongly temperature dependent (Fig. 1), but are ~1‰ positive (enriched in ^{18}O) relative to calcium carbonate precipitated at equilibrium. The $\delta^{18}O$ values obtained from coccoliths of *C. leptoporus* and from tests of *T. heimii* also show a strong temperature dependence but are ~2.5‰ depleted in ^{18}O relative to equilibrium (Fig. 1). The isotopic compositions of these latter species are similar to values reported for coccoliths of *Cricosphaera carterae*[5]. Cultures of *G. oceanica* were maintained under both 14:10 hour light/dark cycles and 24-h light regimes. The $\delta^{18}O$ data on this species (Fig. 1) indicate no significant difference in oxygen isotopic composition related to the period of illumination.

The experimental data show a temperature dependence in oxygen isotopic composition which in no case corresponds to equilibrium precipitation of calcium carbonate. This would indicate that in all the coccolithophore species studied, a not unexpected vital effect influences the fractionation of oxygen

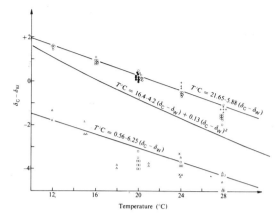

Fig. 1 The species plotted are *E. huxleyi*, 14 : 10 h light/dark cycle (+); *G. oceanica*, 14 : 10 h light/dark cycle (○), 24-h illumination (●); *C. leptoporus* 14 : 10 h light/dark cycle (□). *T. heimii*, 14 : 10 h light/dark cycle (X); and *C. carterae*[5], 12 : 12 h light/dark cycle (△). Each datum point represents an individual unialgal batch culture. The isotopic composition of the coccoliths are reported as $\delta_c - \delta_w$ versus temperature, where $\delta_c = \delta^{18}O$ of the coccoliths (‰ deviations from the PDB standard), and $\delta_w = \delta^{18}O$ of the water of the growth medium (‰ deviation from SMOW). The equations representing the least squares fit to the combined data of *E. huxleyi* and *G. oceanica* is $y = 3.68 - 0.17T$, ($r = 0.91$), and for *C. leptoporus*. *C. carterae*, and *T. heimii* is $y = 0.09 - 0.16T$, ($r = 0.87$).

isotopes during the secretion of the mineral. Isotopic analyses of both planktonic and benthonic foraminifera[16–19] have shown that their tests may be secreted out of isotopic equilibrium. In addition, the secretion of calcium carbonate in echinoderm tests and spines[20] and in parts of corals[21] has been found to reflect non-equilibrium conditions. Calcification in the coccolithophores is intracellular[22] in that coccoliths are produced inside each cell and extruded to the cell surface. Paasche[23] determined that bicarbonate rather than carbonate ions are used as a primary source of carbon in coccolith formation, further demonstrating that the deposition of calcium carbonate by coccolithophores is fundamentally different from any inorganic precipitation process. Therefore, comparison with products of extracellular calcification as described above can be made only with some reservation. The vital effect in the coccolithophores may reflect isotopic discrimination effects by cell membranes, as well as by the incorporation in the calcite of metabolic CO_2 or photosynthetic O_2. Much work obviously remains to be done to understand the process of calcification in these planktonic algae. Relatively little is known about pathways, although our data do imply that calcification processes in coccolithophores may vary among different taxa. This is further suggested by the close grouping of the isotopic data for *E. huxleyi* and *G. oceanica* (Fig. 1), as these two species are thought to lie in a phylomorphogenetic lineage[24].

Recently, several investigators[3,4] have published analyses of 'bulk' carbonate samples composed of mixtures of different coccolith species and small foraminiferal tests. The isotopic signal recorded in such samples is a complex mixture of the isotopic signals of the different species comprising the bulk sample. Apparently the isotopic signal cannot be interpreted without data on the vital effect of each species included.

The presence of a vital effect in the fractionation of oxygen isotopes in coccolithophores does not preclude the use of coccoliths for oxygen isotope palaeotemperature analysis, as long as the magnitude of the effect is known over the temperature range in question. Nanofossil assemblages composed of coccoliths of *E. huxleyi* and *G. oceanica*, both of which are small, may be separated by settling and filtering techniques, and analysed for the oxygen isotopic composition. The $\delta^{18}O$ values of such an assemblage reflect the temperature and the $^{18}O/^{16}O$ ratio of the water in which these coccolithophores lived, even though the mineral was not precipitated in equilibrium. Factors such as light intensity, nutrients, and salinity may also play a role in regulating fractionation. These variables and their effect on isotopic composition are being examined in our laboratories.

Our increased understanding of the vital effect in the fractionation of oxygen isotopes by coccolithophores should increase the value of coccolith oxygen isotopic data in palaeoclimatic studies. The $\delta^{18}O$ data from coccoliths used in conjunction with oxygen isotopes data from foraminifera should allow us to determine with greater reliability variations in surface water temperature during the Pleistocene.

This work was supported in part by an NSF post-doctoral research grant, the NSF-CNRS US-France Exchange Program, and NSF grants OCE 77-10876 and OCE 78-09643.

Note added in proof: The taxonomic position of the Thoracosphaerae, that is, whether they are coccolithophores or dinoflagellates, has been a matter for debate. Based on detailed examination of cell ultrastructure and particularly the presence of a typical dinoflagellate nucleus, we have concluded that *Thoracosphaera heimii* is a dinoflagellate and not a coccolithophore (L.E.B., P.L.B. and R.L.G., manuscript in preparation).

Received 2 January; accepted 14 March 1980.

1. Margolis, S. V., Kroopnick, P. M., Goodney, D. E., Dudley, W. C. & Mahoney, M. E. *Science* **189**, 555 (1975).
2. Anderson, T. F. & Cole, S. A. *J. Foram. Res.* **5**, 188 (1975).
3. Berger, W. H., Killingley, J. S. & Vincent, E. *Oceanol. Acta* **1**, 203 (1978).
4. Vergnaud-Grazzini, Pierre, C. & Letolle, R. *Oceanol. Acta* **1**, 381 (1978).
5. Dudley, W. C. & Goodney, D. E. *Deep-Sea Res.* **26A**, 495 (1979).
6. Savin, S. M. *A. Rev. Earth planet. Sci. Lett.* **5**, 319 (1977).
7. Okada, H. & Honjo, S. *Deep-Sea Res.* **20**, 355 (1973).
8. Bé, A. W. H. & van Donk, J. *Science* **173**, 167 (1971).
9. Berger, W. H. in *Chemical Oceanography* (eds Riley, J. P. & Chester, R.) (Academic, New York, 1976).
10. Guillard, R. R. L. in *Culture of Marine Invertebrate Animals* (eds Smith, W. L. & Chanley, M. H.) (Plenum, New York, 1975).
11. Shackleton, N. J. in *Stable Isotopes in Oceanographic Studies and Palaeotemperatures* (ed. Tongiorgi, E.) (Spoleto, Pisa, 1965).
12. Shackleton, N. J. in *Variation du climat au cours du Pleistocene* (ed. Labeyrie, J.) (CNRS, Paris, 1974).
13. Epstein, S. & Mayeda, T. *Geochim. cosmochim. Acta* **4**, 213 (1953).
14. Craig, H. *Geochim. cosmochim. Acta* **12**, 133 (1957).
15. Craig, H. in *Stable Isotopes in Oceanographic Studies and Palaeotemperatures* (ed. Tongiorgi, E.) (Spoleto, Pisa, 1965).
16. Duplessy, J. C., Lalou, C. & Vinot, A. C. *Science* **168**, 250 (1970).
17. Vinot-Bertouille, A-C. & Duplessy, J-C. *Earth planet Sci. Lett.* **18**, 247 (1973).
18. Shackleton, N. J., Wiseman, D. H. & Buckley, H. A. *Nature* **242**, 177 (1973).
19. Vergnaud-Grazzini, C. *Palaeogeogr. Palaeoclimat. Palaeoecol.* **20**, 263 (1976).
20. Weber, J. N. & Raup, D. M. *Geochim. cosmochim. Acta* **30**, 681 (1966).
21. Weber, J. N. & Woodhead, P. M. *J. Chem. Geol.* **6**, 93 (1970).
22. Watabe, N. & Wilbur, K. M. *Limnol. Oceanogr.* **11**, 567 (1966).
23. Paasche, E. *Physiol. Pl.* Suppl. III, 1 (1964).
24. McIntyre, A. *Deep-Sea Res.* **17**, 187 (1970).

Part IV

ORIGIN AND EVOLUTION
OF NANNOPLANKTON

Editor's Comments
on Papers 29 Through 33

29 DEFLANDRE
Présence de nannofossiles calcaires (coccolithes et Incertae
sedis) *dans le Siluro-dévonien d'Afrique du Nord*

30 BUKRY
Discoaster *Evolutionary Trends*

31 HAQ
Evolutionary Trends in the Cenozoic Coccolithophore Genus
Helicopontosphaera

32 GARTNER and BUKRY
Excerpt from *Morphology and Phylogeny of the*
Coccolithophycean Family Ceratolithaceae

33 HAQ
Transgressions, Climatic Change and the Diversity of Calcareous
Nannoplankton

Most nannoplankton workers agree that the earliest true coccoliths occur in the Lower Jurassic (Lower Liassic). However, there have been numerous reports recording the occurrence of nannofossil-like objects as far back as the Silurian (e. g., Paper 29). Gümbel (1870) has written the earliest report of coccolithid occurrence in Paleozoic sediments. Noël (1961) reported the presence of numerous coccolith forms from the Pennsylvanian of Oklahoma. These, however, are now considered to be Liassic contaminants. Deflandre (Paper 29) described numerous concentric forms from the Silurio-Devonian of North Africa which he considered to be related to coccoliths. Deflandre's objects are concentric, rather than radial like true coccoliths; however, they may represent forms ancestral to the Mesozoic coccoliths. Other occurrences of Paleozoic coccolithlike objects have been reported by Gartner and Gentile (1973), who described ovoid forms with lathlike imbricated structures from the Pennsylvanian of Missouri, and by Pirini-Radrizzani (1971), who recorded discolithids and scapholithidlike forms from the Permian of Turkey. There have also been reports of nannoconids in the middle and upper Triassic and schizospheres and zygodiscids in the upper Triassic of the Mediterranean region (Di Nocera and Scandone, 1977). The presence of the true coccolith *Crucirhabdus primulus* in the earliest Jurassic with its relatively complex construction, has led many nannopaleontologists to assume that coccoliths probably had an earlier origin to have been able to achieve such a degree of development by this time. However, there is no reason why this species could not have evolved rapidly in the early Jurassic from a noncoccolith bearing ancestor. After their first appearance on the scene, coccoliths evolved at a rapid pace (see Paper 33).

Two aspects of the evolution of calcareous nannofossils have been studied so far: phyletic evolutionary trends within morphologically related groups or within discrete lineages of nannoplankton, and rates of evolutionary change among nannoplankton as a whole.

The lineage studies attempt to identify phyletic changes in morphology with time and the ancestor-descendant relationships among species. Prins (1969) attempted to reconstruct the descendance relationships between Liassic coccoliths, and he identified the development of three major lineages from the first true coccolith, *C. primulus*. In his study of the *Discoaster* lineage, Bukry (Paper 30) found that in this group there was an evolutionary trend from more massive skeletons in the early Tertiary to less massive skeletons in the late Tertiary. This reduction in the amount of calcite was ascribed to the cooling of the oceans during the Tertiary. Prins (1971) speculated that the discoasters have evolved from the early Paleocene fasciculithids and heliolithids and he identified numerous cladistic offshoots in the discoasters, based on structural and temporal relationships.

Paper 31 by Haq and Paper 32 by Gartner and Bukry are examples of detailed lineage studies of Cenozoic nannofossil groups. Haq's study of the genus *Helicosphaera (= Helicopontosphaera)* identifies distinct structural modification trends in this coccolithid group. Gartner and Bukry identify morphological trends in the biostratigraphically useful group of ceratolithids that first appear in the late Miocene. Other lineage studies include: Gartner (1970) on the Cenozoic coccolithid genus *Chiasmolithus,* Roth et al. (1971) on the Tertiary sphenolithids, and Lauer (1974) on the upper Cretaceous arkhangelskiellids. Recently Romein (1979) has reconsidered numerous Paleogene lineages and presented cladograms of some of the Tertiary and some late Cretaceous lineages as well. Perch-Nielsen (1979) has also reviewed the status of the Cretaceous lineages and summarized various studies on important Mesozoic nannofossil groups.

The last paper in this section by Haq (Paper 33) documents the fluctuations in the diversity of nannoplankton from the Jurassic to the Recent. Such taxonomic frequency data serve as indirect measures of the total evolutionary process at work during a time interval, and reflect changes in the tempo of evolution, including rates of origination (diversification) and rates of extinction. Haq's data indicate that highest diversities (and evolutionary rates) occurred during times of extensive epicontinental seas. He suggests that widespread transgressions lead to equable climates with low seasonal fluctuations that are advantageous to the growth and diversification of marine plankton. Climate may thus be an important modulator of the evolution of nannoplankton.

REFERENCES

Di Nocera, S., and P. Scandone, 1977, Triassic Nannoplankton Limestones of Deep Basin Origin in the Central Mediterranean Region, *Palaeogeography, Palaeoclimatology, Palaeoecology* **21:**101–111.

Gartner, S., 1970, Phylogenetic Lineages in the Lower Tertiary Coccolith Genus *Chiasmolithus,* in *North American Paleontological Convention, Proceedings of the Part G, Ultra Microplankton,* pp. 930–957.

Gartner, S., and R. Gentile, 1973, Problematic Pennsylvanian Coccoliths from Missouri, *Micropaleontology* **18** (4):401–404.

Gümbel, C. W., 1870, Coccolithen (Bathybius) in allen Meerestiefen und in den Meeres-ablagerungen aller Zeiten, *Ausland* **43:**763–764.

Lauer, G., 1974, Evolutionary Trends in the Arkhangelskiellaceae (Calcareous Nannoplankton) of the Upper Cretaceous of Central Oman, S. E. Arabia, in Report on the Consultant Group on Calcareous Nannoplankton, Kiel, September 5–7, 1974, D. Noël and K. Perch-Nielsen, *Arch. Sci. Geneve* **28:**259–262.

Noël, D., 1961, Sur la présence de Coccolithophoridés dans les terrains primaires, *Acad. Sci. Comptes Rendus* **252:**3625–3627.

Perch-Nielsen, K., 1979, Calcareous Nannofossils from the Cretaceous between the North Sea and the Mediterranean, *Aspekte der Kreide Europas,* I. Wiedmann, ed., IUGS Series A., no. 6, E. Schweizerbart'sche Verlagsbuchhandlung, Stuttgart, pp. 223–272.

Pirini-Radrizzani, C., 1971, Coccoliths from Permian Deposits of Eastern Turkey, *Proceedings of the Second International Planktonic Conference,* Roma, 1970, vol. 2, A. Farinacci, ed., Edizioni Tecnoscienza, Rome, pp. 993–1001.

Prins, B., 1969, Evolution and Stratigraphy of Coccolithnids from the Lower and Middle Lias, *Proceedings of the First International Planktonic Conference,* Geneva, 1967, vol. 2, P. Bronimann and H. H. Renz, eds., E. J. Brill, Leiden, pp. 547–558.

Prins, B., 1971, Speculations on Relations, Evolution, and Stratigraphic Distribution of Discoasters, *Proceedings of the Second International Planktonic Conference, Roma, 1970,* vol. 2, A. Farinacci, ed., Edizioni Tecnoscienza, Rome, pp. 1017–1037.

Romein, A. J. T., 1979, Lineages in Early Paleogene Calcareous Nannoplankton, *Utrecht micropaleontological Bull. 22,* 231p.

Roth, P. H., H. E. Franz, and S. W. Wise, Jr., 1971, Morphological Study of Selected Members of the genus Sphenolithus Deflandre (Incertae Sedis, Tertiary), in *Proceedings of the II Planktonic Conference, Roma, 1970,* vol. 2, A. Farinacci, ed., Edizioni Tecnoscienza, Rome, pp. 1099–1117.

29

PRÉSENCE DE NANNOFOSSILES CALCAIRES (COCCOLITHES ET *INCERTAE SEDIS*) DANS LE SILURO-DÉVONIEN D'AFRIQUE DU NORD

Georges Deflandre

Jusqu'à présent, aucun document iconographique parfaitement objectif et indiscutable n'a été publié sur les coccolithes du Paléozoïque, ou, au sens plus large, sur les nannofossiles calcaires du Primaire : personne, à ma connaissance, n'a jamais donné de photographies de coccolithes *in situ* dans des lames minces, écartant ainsi toute discussion sur des possibilités de pollution ou de confusion. Les allusions que j'ai faites autrefois aux très anciens travaux de Gümbel (¹), qui avait affirmé l'existence de coccolithes dans diverses roches primaires, incitaient à la reprise et au contrôle de ses recherches par ses compatriotes, ce qui n'a pas encore été tenté. A cette époque (1952), les coccolithes indubitables les plus anciens étaient ceux que j'avais signalés dans le Lias. Par la suite D. Noël a publié une note, accompagnée de dessins, sur des coccolithes primaires (²) et présenté également plusieurs électromicrographies relatives à des formes du Pennsylvanien de l'Oklahoma (³). Cette documentation a été et est encore contestée par divers spécialistes, tant européens qu'américains. Ayant pu récemment examiner moi-même ce matériel, dans lequel l'état de conservation paraît varier considérablement, je pense qu'en l'occurrence, on ne saurait faire état ni de pollution, ni de confusion. Il faut simplement souhaiter, d'un part qu'une nouvelle recherche plus approfondie sur le terrain permette de retrouver et de situer définitivement le mince lit argileux découvert par le professeur Robert Laffitte, et d'autre part que le matériel existant soit réétudié selon des méthodes plus modernes, ce qui est dans les intentions de D. Noël.

Dans la même note, D. Noël mentionne, dans des lames minces d'un flysch du Viséen supérieur de Ben Zireg, en Algérie, l'existence de formes qui sont, dit-elle, un peu particulières, mais qu'elle n'a pas figurées. L'âge et la provenance de ces nannofossiles carbonifères sont bien établis et je reviendrai sur eux à la fin de cette note (⁴).

Il y a déjà plusieurs années que j'ai découvert dans des lames minces d'argilites du Siluro-dévonien du Sahara, de nombreux nannofossiles calcaires, en majorité des coccolithes (⁵). Leur âge s'échelonne du Silurien supérieur (Ludlowien) à la base du Dévonien.

L'état de conservation de ces coccolithes est extrêmement variable, encore qu'ils soient tous restés calcaires. La forme générale et la taille sont caractéristiques : corpuscules elliptiques ou arrondis, ou presque circulaires, de quelques microns. La structure de la calcite de la grande majorité d'entre eux est à peu près homogène et d'orientation unique : il reste habituellement peu ou pas de vestige de la structure héliolithique caractéristique (croix noire entre les nicols croisés).

Considérant l'importance de ce caractère quant au diagnostic d'un indubitable coccolithe, j'ai attendu d'avoir enfin, au terme de longues recherches, trouvé et repéré quelques exemplaires dans lesquels cette structure est parfaitement observable quoique quelque peu oblitérée. Un des meilleurs spécimens est figuré ici (*Pl.* I, *fig.* 1). L'examen entre nicols croisés avec la teinte sensible, où l'on distingue deux secteurs jaunes et deux secteurs bleus, est beaucoup plus démonstratif qu'une figure en noir : la couleur des secteurs permute en faisant tourner la teinte sensible de 90°.

Outre les coccolithes, la population de nannofossiles calcaires comporte quelques types *incertae sedis*, parmi lesquels de fréquents bâtonnets calcitiques, sur lesquels je reviendrai plus loin. Ces argilites à coccolithes contiennent aussi divers microfossiles, Chitinozoaires, Acritarches, Tintinnoïdiens, *Ampelitocystis*, *Salpingocryptum*, dont certains ont été déjà cités ([5]).

On sait que la systématique et la nomenclature des Coccolithophoridés fossiles [et actuels ([6])] subissent actuellement une crise fâcheuse, pratiquement engendrée par les progrès dus à la microscopie électronique. Je n'insiste pas ici sur le détail de ces problèmes auxquels a été consacrée une note récente ([7]) : pratiquement, il y a une sorte de divorce entre les systématiques fondées respectivement sur la microscopie photonique et sur la microscopie électronique, divorce en quelque sorte parallèle à celui, latent, qui divise la systématique des formes vivantes et celle des formes fossiles dans plusieurs domaines de la protistologie-paléoprotistologie (Coccolithophoridés, Dinoflagellés, Silicoflagellés, Radiolaires).

On ne saurait exclure *a priori* la possibilité d'étudier les coccolithes paléozoïques en microscopie électronique : D. Noël a d'ailleurs publié deux électromicrographies d'*Ellipsagelosphaera* spec. du Pennsylvanien. Mais seule la microscopie photonique permet l'étude des coccolithes des roches cohérentes et dures, d'où ils ne peuvent être extraits. Ainsi doit-on fatalement adopter une systématique et une nomenclature ne tenant pas compte des ultrastructures qui sont à la base du découpage générique présentement en usage, bien que discuté. Ceci n'est d'ailleurs pas particulier au Paléozoïque : dès 1942 ([8]) j'ai signalé l'abondance de coccolithes, observés pour la première fois *in situ* dans des silex mio-pliocènes d'Oranie, qui ne peuvent évidemment être étudiés qu'en microscopie photonique, comme d'ailleurs ceux signalés dans les phosphates par G. Lucas et D. Noël ([9]). A ce propos, je note en passant que la fossilisation des coccolithes, dans certains phosphates albiens, revêt des aspects très particuliers sur lesquels il y aura lieu de revenir.

Pour m'en tenir ici aux formes paléozoïques, leur introduction dans la systématique en usage et compte tenu de l'esprit qui y prévaut actuellement, incite à créer des genres nouveaux, ou plus exactement des noms de genres nouveaux. Leur discrimination peut s'appuyer sur de courtes diagnoses, leur validité étant garantie par l'existence et la conservation assurée des types. En fait, ce qui m'est actuellement connu se ramène à deux sortes de coccolithes, les uns en anneau, les autres à partie centrale pleine. Pour les premiers, l'existence d'assemblages indubitables correspondant à des vestiges de coccosphères, écarte l'attribution au genre *Cricolithus* Kamptner, où leur place était toute désignée. Il n'est, d'autre part, pas indiqué de les placer dans le genre des mers actuelles *Cricosphaera* Braarud, caractérisé précisément par ses coccolithes en anneaux *dont l'ultrastructure est bien établie*, laquelle reste inconnue pour le matériel paléozoïque. On comparera cependant — échelle mise à part — les figures données ici Planche II, figures 1 et 2, avec l'électro-micrographie Planche II, figures *d, e*, de Braarud et coll. ([10]). Par ailleurs, l'attribution au genre *Cricosphaera* impliquerait *ipso facto* que les représentants de ce genre *vivant* remontent au Siluro-dévonien, avec les conséquences que cela entraînerait du point de vue général et dont l'énoncé serait pour le moins prématuré. Il est donc logique de les classer dans un genre nouveau, *Eocricosphaera*.

Par contre, pour les autres formes, pleines, non attribuables honnêtement à aucun des genres décrits — plus ou moins bien — actuellement, je préfère, malgré les critiques prévues de certains systématiciens plus juristes que naturalistes, les classer « provisoirement » dans le paragenre *Coccolithites* Kamptner, précisément destiné à recevoir les coccolithes insuffisamment connus ou caractérisés.

Il n'est pas question de faire ici un inventaire exhaustif de tous les coccolithes rencontrés dans les quelques lames minces les plus riches, lames dont le nombre est encore très limité, par suite de leur origine. On se limitera donc à figurer et à décrire quelques-unes des formes les plus fréquentes ou les plus caractéristiques (tout en précisant qu'ont été vus bien d'autres types, en particulier plus petits et de contour allant du cercle à l'ellipse allongée). Ce sont :

Coccolithites antiquus n. sp. (*Pl.* I, *fig.* 1), *C. pontiferus* n. sp. (*Pl.* I, *fig.* 2), *C. fallax* (*Pl.* I, *fig.* 3 à 6), *C. impletus* (*Pl.* I, *fig.* 7 à 9), *Eocricosphaera* n. g. *perruchei* n. sp. ([11]) (*Pl.* II, *fig.* 1 à 4), *E. chennauxi* n. sp. (*Pl.* II, *fig.* 5).

Parmi les divers nannofossiles calcaires *incertae sedis*, je citerai une forme en anneau, sans rapport apparent avec les Coccolithophoridés, *Palmicricothus* n. g. *magnus* n. sp. et enfin une série de bâtonnets calcaires, qui peuvent trouver une place provisoire dans la famille des Microrhabdulidés Deflandre 1963 ([12]), ce qui d'ailleurs n'implique rien sur leur origine laquelle demeure énigmatique. Je n'en donnerai que deux exemples, *Archaeorhabdulites* n. g. *quadratus* n. sp. et *A. strangulatus* n. sp., en notant que leur fréquence relative motiverait une sérieuse étude de ces curieux types de microfossiles. Certains individus ont une structure ressemblant à celle des Nannoconidés mais il faut éviter de les confondre avec des petits fragments de coupes de loges de Tintinnides à paroi dédoublée.

Les lames minces du Viséen supérieur d'Algérie ([4]), auxquelles il a été fait allusion plus haut, contiennent des nannofossiles (*Pl.* IV) tout à fait différents des nôtres, dont la structure *sphérolithique* est admirablement visible. Très transparents et d'un indice de réfraction bien différent de celui de la calcite, certains ressemblent étonnamment à de petits grains d'amidon. Leur composition me reste inconnue et je ne les décris ici, sous le nom *Palamygdalithus* n. g. *noëlae* n. sp. que pour attirer l'attention sur eux. Je n'ai trouvé aucun intermédiaire entre ces *Palamygdalithus* et les nombreux grains et cristaux de calcite présents dans les mêmes lames.

Et ceci m'amène finalement à un problème qui se posait à moi depuis bien des années et qui me paraît maintenant résolu grâce à une simple sériation des aspects présentés par les éléments calcitiques étudiés ici.

Dès 1952 [([1]) p. 459] j'ai indiqué qu'un coccolithe (et j'ajoute ici, peut-être une coccosphère) était susceptible de se transformer *en un cristal de calcite*. Cela résultait de mes observations, très partiellement publiées, sur les nannofossiles calcaires des silex dits ménilites intercalés dans les diatomites mio-pliocènes d'Oranie ([13]).

A côté de coccolithes indubitables, dont des exemples viennent d'être donnés, les lames minces contiennent d'innombrables grains de calcite (*Pl.* III, *fig.* 1), de tailles et de formes variées, parmi lesquels les spécimens bien cristallisés ne sont pas rares. Or certains de ceux-ci, de contour anguleux, montrent au centre une zone différenciée, tout à fait analogue à ce qui se voit sur des coccolithes de taille semblable, très « corrodés » mais encore identifiables. Les figures 2, 3 et 4 de la Planche III, montrent ainsi successivement un coccolithe dont une partie des éléments marginaux a fait place à un cristal bien net, puis un autre coccolithe, dont le contour correspond en partie à un cristal en formation et enfin un cristal anguleux dans lequel la différenciation du centre témoigne seule de sa genèse à partir d'un coccolithe.

L'existence d'une apparente évolution parallèle des coccolithes dans les silex mio-pliocènes d'Oranie laisse supposer que de telles modifications de structure sont peut-être indépendantes de l'âge de la roche. Le phénomène en question peut-il être précoce ? Se poursuit-il au cours des temps ? D'autres recherches sont nécessaires dans ce domaine : présentement, je n'ai pas encore observé, dans des vases actuelles, de coccolithes évoluant vers un état cristallisé.

DIAGNOSES DES TAXONS NOUVEAUX.

Coccolithites antiquus n. sp. — Elliptique large ou subcirculaire, à bord formé d'éléments mal délimités, peu nombreux ; centre avec des vestiges de grains disposés concentriquement. Structure héliolithique conservée, plus ou moins nette suivant les spécimens. Longueur 17 μ, largeur 15 μ (*fig.* 1, *Pl.* I).

Coccolithites fallax n. sp. — Elliptique, en écuelle à fond bombé ; bord constitué d'une trentaine environ d'éléments jointifs ; fond plein, marqué de vermicules sur l'holotype. Longueur 17-19 μ, largeur 13-15,6 μ bord large de 2 à 2,6 μ (*fig.* 3 à 6, *Pl.* I).

EXPLICATION DES PLANCHES

Planche I

Fig. 1. — *Coccolithites antiquus* n. sp. Holotype Collection Micropaléontologie E. P. H. E., Muséum. Prép. DC 142. Repér. England Z' 11.

Fig. 2. — *Coccolithites pontiferus* n. sp. Holotype. Coll. DC 136, Rep. U 3 + U 4.

Fig. 3. — *Coccolithites fallax* n. sp. Paratype. Coll. DC 140, Rep. X 3.3.

Fig. 4 à 6. — *Coccolithites fallax* n. sp. Holotype. Coll. DC 140, Rep. Z' 3.2 ; 5, surface centrale ; 6, L. P.

Fig. 7 à 9. — *Coccolithites impletus* n. sp. 7 et 8, Holotype. Coll. DC 132. Rep. X 3.2 (*fig.* 8, L. P.) ; 9, paratype. Coll. DC 132. Rep. J 1. Siluro-dévonien, Sondage Meg 1, sauf fig. 2, Sond. Bel 1 Erg. occidental, Sahara.

Fig. 1 : G × 1 924 ; Fig. 3 à 6 : G × 1 540 ; Fig. 7 à 9 : G × 962.

Planche II

Fig. 1 à 4. — *Eocricosphaera* n. g. *perruchei* n. sp. 1 et 2, holotype. Coll. DC 136. Rep. Z 9.4 ; 3 et 4, paratypes. Coll. DC 136. Rep. Z 8.1 et Z 9.3.

Fig. 5. — *Eocricosphaera chennauxi* n. sp. Holotype Coll. DC 136. Rep. Z 8.4.

Fig. 6. — *Palmicricothus* n. g. *magnus* n. sp. Holotype. Coll. DC 140. Rep. Z' 3.1. Siluro-dévonien. Sondage Bel 1 sauf fig. 6, Sond. Meg. 1, Sahara.

Fig. 1, 3, 4, 6 : G × 962 ; Fig. 5 : G × 1 100 env.

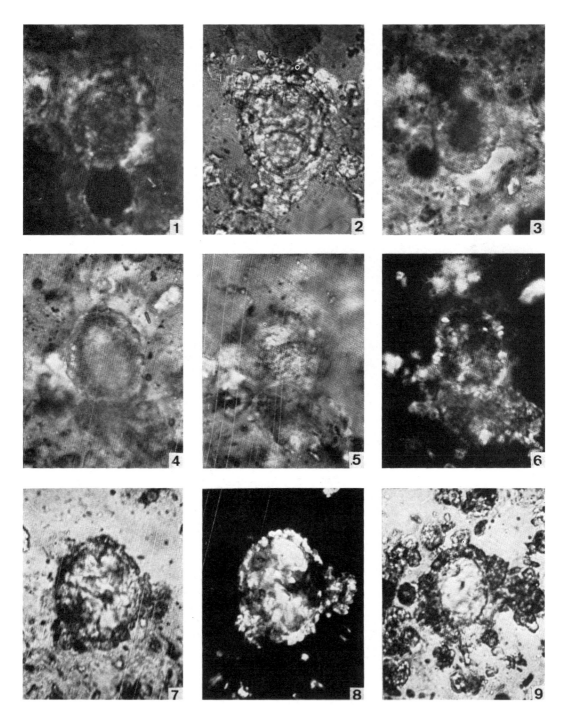

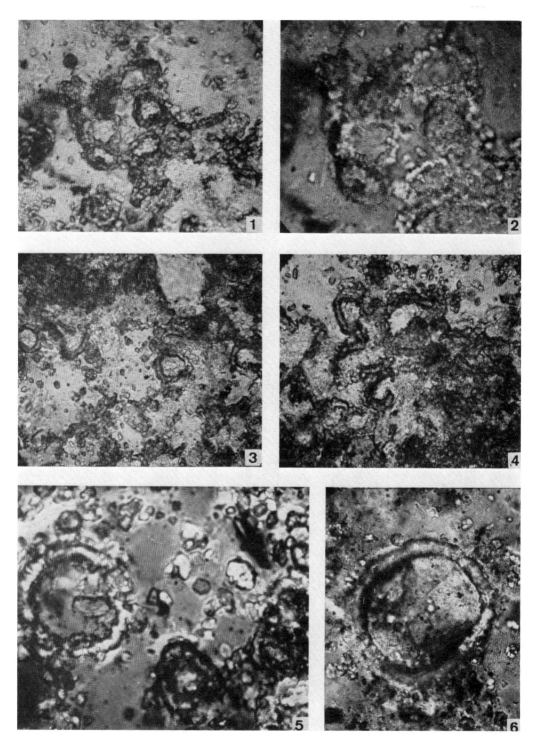

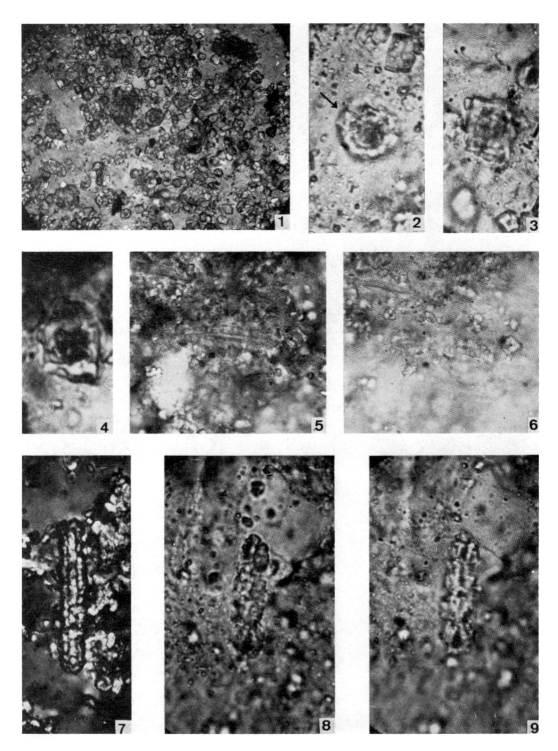

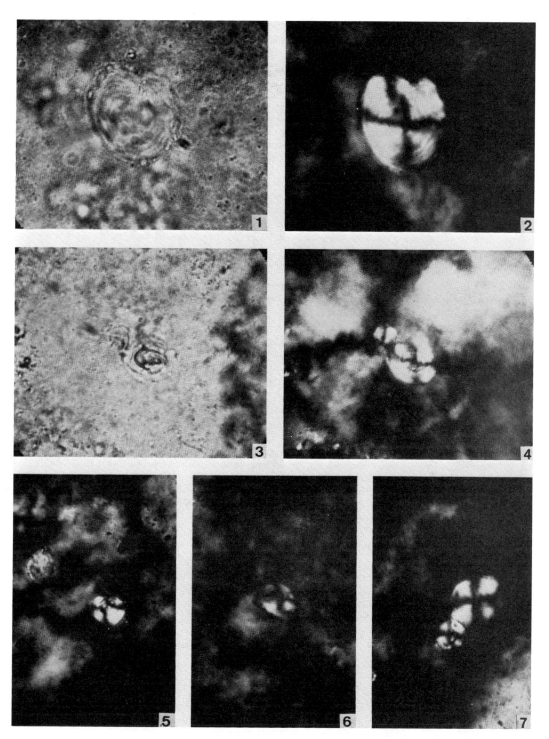

Planche III

Fig. 1. — Plage de cristaux de calcite avec un coccolithe au centre. Coll. DC 143. Rep. U 14.2.

Fig. 2. — Coccolithe avec un cristal de calcite (flèche) à la place d'éléments marginaux. Coll. DC 140. Rep. Z′ 3.1.

Fig. 3. — Coccolithe transformé en cristal de calcite, sauf sa partie centrale. Coll. DC 140. Rep. Z′ 3.1.

Fig. 4. — Cristal de calcite dont le centre évoque l'origine. Coll. DC 140. Rep. Z′ 3.1.

Fig. 5 à 7. — *Archaeorhabdulites* n. g. *quadratus* n. sp. 5 et 6, paratype en coupe médiane et en vue supérieure. Coll. DC 136. Rep. U 7.4 ; 7, holotype. Coll. DC 136. Rep. Z 8.3.

Fig. 8 et 9. — *Archaeorhabdulites strangulatus* n. sp. Holotype, Coll. 136. Rep. W 7.4. Siluro-dévonien. Sondage Meg 1, sauf 1 (Sond. DEA 1) et 5 à 8 (Sond. Bel 1). Sahara.

Fig. 1 : G × 386 ; Fig. 2, 3, 5, 6, 8, 9 : G × 1 540 ; Fig. 4 : G × 3 080 ; Fig. 7 : G × 962.

Planche IV

Fig. 1 à 7. — *Palamygdalithus* n. g. *noëlae* n. sp. — Fig. 1 et 2 : holotype. Coll. DC 129. Rep. P. 23.4. — Fig. 3 et 4 : paratype double. Coll. DC 129. Rep. U 23.3. — Fig. 5 et 6 : paratypes de petite taille. Coll. DC 130. Rep. K 15.3 et Coll. DC 129. Rep. V 22. — Fig. 7 : paratype double. Coll. DC 130. Rep. K 16.4. Fig. 2, 4 et 5 à 7 : L. P. Viséen supérieur. Flysch. Ben Zireg, Algérie (G × 1 540).

Coccolithites impletus n. sp. — Elliptique régulier, à bord relativement étroit ; centre régulièrement rempli par un mince bloc de calcite. Longueur 35 μ, largeur 30 μ (*fig.* 7 à 9, *Pl.* I).

Coccolithites pontiferus n. sp. — Elliptique, barré transversalement par un pont de calcite sur un fond plein. Longueur 22 μ, largeur 18 μ (*fig.* 2, *Pl.* I).

Eocricosphaera n. g. — Assemblage d'anneaux calcaires elliptiques ou circulaires, vestiges présumés d'une sorte de coccosphère. Anneaux de structure quasi homogène, non héliolithique (?) avec parfois traces d'éléments juxtaposés. Anneaux isolés fréquents. Générotype : *Eocricosphaera perruchei* n. sp.

Eocricosphaera perruchei n. sp. — Anneaux elliptiques souvent déformés, fréquemment réunis en groupes (coccosphères) atteignant 60 μ. Anneaux longs de 20 à 26 μ, larges de 14 à 18 μ (*fig.* 1 à 4, *Pl.* II).

E. chennauxi n. sp. — Anneaux circulaires ou subcirculaires, de 30 à 33 μ (*fig.* 5, *Pl.* II).

Palmicricothus n. g. — Anneaux calcaires d'orientation optique homogène quoique composés d'éléments juxtaposés. Générotype : *Palmicricothus magnus* n. sp.

P. magnus n. sp. — Diamètre de l'anneau égal à environ 10 à 12 fois son épaisseur. Forme générale largement elliptique ou presque circulaire, peut-être déformée secondairement. Dimensions 64-71 μ (*fig.* 6, *Pl.* II).

Archaeorhabdulites n. g. — Bâtonnets calcaires creux, formés de petits éléments transversaux, de tailles inégales ou presque égaux, de même orientation cristalline. Forme générale cylindrique ou affectée d'étranglements, parfois courbée. Générotype : *Archaeorhabdulites quadratus* n. sp.

A. quadratus n. sp. — Coupe transversale en croix à branches constituées d'éléments empilés en colonne. Holotype : longueur 46 μ, largeur 9 μ (*fig.* 5 à 7, *Pl.* I).

A. strangulatus n. sp. — Coupe transversale probablement non circulaire. Bâtonnet coupé par des étranglements en portions à contour parfois cordiforme. Longueur 24 μ, largeur 6,5 μ (*fig.* 8, 9, *Pl.* I).

Palamygdalithus n. g. — Nannofossile ellipsoïdal ou amygdaliforme irrégulier, constitué de couches concentriques comme un grain d'amidon. Structure sphérolithique très nette, à croix noire d'aspect variable (calcite ?). Générotype : *Palamygdalithus noëlae* n. sp.

P. noëlae n. sp. — Corpuscules isolés ou parfois en doublets de taille inégale. Couches concentriques souvent renforcées jusque vers le centre, y formant une sorte de noyau irrégulier. Longueur 7,5-22 μ, largeur 5,5-18 μ (*fig.* 1 à 7, *Pl.* IV).

(*) Séance du 1er juin 1970.

(1) G. DEFLANDRE, Classe des Coccolithophoridés, *in* : Traité de Zoologie de P.-P. Grassé, I (1), 1952, p. 439-470.

(2) D. NOËL, *Comptes rendus*, 252, 1961, p. 3625-3627, 11 fig.

(³) D. Noël, Sur les coccolithes du Jurassique... (Thèse), Ed. C. N. R. S., 1965, 209 p., 29 Pl., 74 fig.

(⁴) Don fait par Mˡˡᵉ Denise Noël à la collection du Laboratoire, de plusieurs de ses lames minces originales, enregistrées sous les numéros DC 129-130 et DC 147 à 149.

(⁵) C'est à Guy Chennaux que je dois ces matériaux (Sondages de Mehaiguene, Meg. 1, de Belketaief, Bel. 1, de Demrat el Acha, DEA 1, Grand Erg Occidental, Sahara algérien) qui ont déjà fait ici l'objet de plusieurs publications : *Comptes rendus*, 265, Série D, 1967, p. 1676 ; *Ibid.*, p. 1776 ; *Ibid.*, 268, Série D, 1969, p. 1482. Pour la stratigraphie, cf. L. Magloire, *Intern. Symp. Devon. System.* II, Calgary 1967, p. 473.

(⁶) La Classe des Coccolithophoridés, telle que je l'ai présentée en 1952 (¹), en marquant son caractère provisoire, est en voie de complète révision sur la base des progrès réalisés dans la cytologie des formes actuelles, planctoniques et benthiques.

(⁷) G. Deflandre et M. Deflandre-Rigaud, Commentaires sur la nomenclature et la systématique des nannofossiles calcaires II. III. IV. *Cah. Micropal.*, Sér. 2, N° 5 ; *Arch. Orig. Centre Docum. C. N. R. S.*, N° 470, 1970, 17 p.

(⁸) *Comptes rendus*, 214, 1942, p. 804-805.

(⁹) G. Lucas et D. Noël, *Comptes rendus*, 245, 1957, p. 94-96.

(¹⁰) T. Braarud, K. Gaarder, J. Markali et E. Nordli, *Nytt Mag. Bot.*, 1, (1952) 1953, p. 129-134, 4 fig., 2 Pl.

(¹¹) L'espèce est dédiée au Docteur Lucien Perruche, Ingénieur chimiste, collecteur dévoué à qui je dois de nombreux matériaux d'études recueillis en France et en Algérie.

(¹²) *Comptes rendus*, 256, 1963, p. 3484-3486, 25 fig.

(¹³) *Comptes rendus*, 214, 1942, p. 804-805.

(Laboratoire de Micropaléontologie de l'Ecole Pratique des Hautes Etudes, 8, rue de Buffon, 75-Paris, 5ᵉ.)

257

30

Reprinted from *Micropaleontology* **17**:43-52 (1971)

Discoaster evolutionary trends

David Bukry
United States Geological Survey
La Jolla, California

ABSTRACT

Examination of several well-known species, eight new species and one new subspecies of *Discoaster* indicates an evolutionary trend toward less massive skeletal elements in a genus that is assumed to be related to the calcareous nannoplankton family *Coccolithophyceae*. Reduction of the amount of calcite in the skeletons of *Discoaster* species during the Tertiary, which includes their entire time range from appearance to extinction, may reflect a cooling of oceanic waters.

INTRODUCTION

Discoasters are microscopic star-shaped calcite skeletal elements assumed to have been produced by an organism related to certain genera of the marine phytoplankton family Coccolithophyceae. The time range of Discoasters extends from their first known occurrence in the Upper Paleocene (approximately 60 m.y.b.p.) to their extinction at the top of the Pliocene (approximately 2 m.y.b.p.). During this time, which includes practically all of the Tertiary, there was a distinctive trend in morphology of the form genus *Discoaster*. Bramlette and Riedel, in a discussion of *Discoaster* morphology in a 1954 paper, first noted "that there is a tendency for the rays of some of these six-rayed forms to become progressively thinner from the early to late Tertiary, but the significance of this must remain a matter of speculation". In the years since, many more species of *Discoaster* have been described, and their stratigraphic ranges have been noted. Of special importance in this respect are the works of Bramlette and Sullivan (1961), Bramlette and Wilcoxon (1967), Hay and others (1967), Martini (1958, 1961), Martini and Bramlette (1963), Stradner (1959), and the monumental contribution of Stradner and Papp (1961). The extensive collections of Discoasters described in these works and the seven species of *Discoaster* described in this paper all point to a dominant trend in morphology throughout the Tertiary.

The earliest Discoasters from Paleocene deposits are compact, multi-rayed, heavily constructed forms with very short free length of the rays or very broad rays, producing a plan view in which the area encompassed by the greatest diameter of the fossil is largely a field of calcite. Mid-Tertiary Discoasters show fewer rays which are relatively broad, so that a substantial part of the area encompassed by the greatest diameter is filled by calcite rays. In the final stage of development, just before extinction in Late Pliocene time, Discoasters are very narrow-rayed and include, in addition to the usual six-rayed forms, three-, four-, or five-rayed forms of only two species. Only a minor part of the area encompassed by the greatest diameter is represented by the calcite skeleton of Upper Tertiary forms. This skeleton is a mere framework compared with the massive *Discoaster* "shields" of the lower Tertiary. The disappearance of Discoasters at the end of the Tertiary is confirmed by studies by many micropaleontologists of what now number thousands of samples of Quaternary oceanic sediments. These studies show that Discoasters are not present in the Quaternary, and that their presence can be used as a marker to distinguish Upper Tertiary from Lower Quaternary sediments.

It should be noted that there is only a slight reduction in the over-all diameter of *Discoaster* species during the Tertiary. It is the relative amount

TABLE 1
Age ranges of the species of *Discoaster* illustrated in plate. 1.

DISCOASTER SPECIES \ Age	PALEOCENE	EOCENE	OLIGOCENE	MIOCENE	PLIOCENE
D. MULTIRADIATUS	X				
D. BARBADIENSIS		XXX			
D. LODOENSIS		X			
D. SAIPANENSIS		XX			
D. STRADNERI		X			
D. DEFLANDREI		XX	XXX	XX	
D. NEOHAMATUS				X	
D. QUINQUERAMUS				X	
D. PENTARADIATUS				X	XX
D. BROUWERI				X	XXX
D. ASYMMETRICUS				X	XX

of calcite incorporated in each *Discoaster* that is substantially reduced. Sizes reported for abundant species of the Tertiary epochs are: Paleocene – *Discoaster multiradiatus*, 10–20 microns; Eocene – *Discoaster barbadiensis*, 6–18 microns, and *Discoaster lodoensis*, 15–25 microns; Oligocene – *Discoaster deflandrei*, 10–17 microns; Miocene and Pliocene – *Discoaster brouweri*, 13–21 microns, *Discoaster challengeri*, 4–18 microns, and *Discoaster pentaradiatus*, 9–23 microns.

There is little change in the absolute range, but accompanying the replacement of massively constructed Discoasters by more delicate forms is a general trend toward the reduction of the typical range in size of species of coccolith genera such as *Helicopontosphaera*, the Lower Tertiary forms being 10–15 microns in size and the Upper Tertiary forms 6–12 microns. The elimination of most of the very large coccoliths (10–25 microns in size), such as *Chiasmolithus grandis*, *Coccolithus bisectus* and *Reticulofenestra umbilica*, occurred in the Lower Tertiary. Miocene and Pliocene coccolith assemblages contain only moderately large coccoliths (6–12 microns in size), such as *Cyclococcolithus leptoporus*, *C. macintyrei*, *C. neogammation* and *Reticulofenestra pseudoumbilica*. The trend culminated in Quaternary assemblages, which

are dominated by tiny coccoliths (1–3 microns in size), such as *Emiliania huxleyi* and *Gephyrocapsa oceanica*. As noted previously, Discoasters are extinct by the Quaternary. The combined effect of these morphologic trends during the Tertiary and Quaternary is to produce skeletal elements of all calcareous nannoplankton with less calcite volume per unit.

Since these trends are worldwide, especially in oceanic sediments where coccoliths and Discoasters are abundant, broadly systematic changes in the physical environment during the Cenozoic Era may have influenced the skeletal trends. During this period major systematic changes in the environment included the general cooling of continental climates in the northern hemisphere (Dorf, 1955; Axelrod and Bailey, 1969) and the continued redistribution of land masses with respect to oceans as a consequence of sea-floor spreading (Fischer and Gealy, 1969; Heezen and others, 1969).

On the basis of comparative information on Pleistocene and Holocene coccolith ecology (Geitzenauer, 1969; Hasle, 1960; Hulburt, 1966; McIntyre, 1967; McIntyre and Bé, 1967), assemblages of species with limited diversity or with small size and simple shape are typical of cooler water. Geitzenauer, in his study of subantarctic Quaternary nannoplankton, reported that the two larger species, *Coccolithus pelagicus* and *Cyclococcolithus leptoporus*, formed a greater percentage of coccolith populations in samples where warm-water Radiolaria were most abundant. Application of these guides to the Tertiary trends of morphology in *Discoaster* and coccoliths points to a general cooling trend in the oceans throughout the period.

Thus the Tertiary cooling that has been interpreted from continental studies may also be interpreted from the record of the most abundant oceanic microfossils. In particular, Eocene assemblages in geographic range from Denmark (56° N.) to the South Atlantic (30° S.) and from the submarine Shatsky Plateau in the Pacific (32° N.) to New Zealand (45° S.) all contain large coccoliths, such as *R. umbilica*, and moderately massive Discoasters, such as *D. barbadiensis* and *D. saipanensis*, suggesting a warm surface environment in most oceanic areas. The worldwide nannoplankton zonation for the succeeding Tertiary is based in large measure on the worldwide trend toward reduced bulk in Discoasters and coccoliths and the resulting speciation. The zonal systems of Bramlette and Wilcoxon (1967), Bukry and Bramlette (1970), Gartner (1969), and Hay and others (1967), rely on *Discoaster* species, such as *D. multiradiatus*, *D. lodoensis*, *D. neohamatus* and *D. brouweri*, as guides to progressively younger assemblages (see illustrations on plate 1).

259

SYSTEMATIC PALEONTOLOGY
Genus DISCOASTER Tan, 1927

Discoaster araneus Bukry, **n. sp.**
Plate 2, figures 1–3

Description: This large *Discoaster* has a central knob and usually 7–9 tapering rays that have a variable free length from $\frac{1}{3}$ to $\frac{2}{3}$ of the total ray length. The length of the individual rays, even on the same specimen, is not uniform, and the angle between rays is likewise 'variable. The distinct taper of the rays, which terminate in points, and the large central knob are consistent characters in this variable species. Several paratypes are illustrated to show the range of forms assigned to this taxon.

Remarks: Its long tapering rays and large central knob, and the variability of the general proportions of the rays in a single specimen easily distinguish *Discoaster araneus* from other Lower Tertiary species, such as *D. lubinaensis, D. gemmeus, D. multiradiatus,* or *D. nobilis,* which show consistently regular ray arrangements and lack a massive central knob.

Distribution: This distinctive species has been observed only in Upper Paleocene samples. It is especially abundant in JOIDES core 4 at 91 meters on the submarine Blake Plateau.

Size: 12–21 microns.

Holotype: USNM 651535 (plate 2, figure 2).

Paratypes: USNM 651536–37.

Type locality: Blake Plateau, JOIDES core 4, 91 m.; 31° 02′ N., 77° 43′ W., Atlantic Ocean.

Discoaster berggrenii Bukry, **n. sp.**
Plate 2, figures 4–6

Discoaster quinqueramus GARTNER, 1969 (part), p. 598, pl. 1, fig. 7.
Discoaster quintatus BUKRY and BRAMLETTE, 1969 (part), p. 133, pl. 1, fig. 6.

Description: This species is a symmetric five-rayed asterolith with the free length of the tapering rays approximately equal to the diameter of the central area. The rays are radial and terminate simply. A prominent star-shaped knob on the concave side of the asterolith practically fills the central area and thus usually occupies a third of the diameter of the entire asterolith.

Remarks: On the basis of general morphology and distribution, *Discoaster berggrenii* appears to be the progenitor of *Discoaster quinqueramus. D. quinqueramus* is distinguished from this new species by long rays, with a much greater free length of rays in proportion to

central area diameter. The knob at the center of *D. berggrenii* is distinctly better developed.

Distribution: *D. berggrenii* first appears in Upper Miocene (Tortonian equivalent) marine sediments slightly below the first occurrence of *D. quinqueramus* and disappears from the record shortly afterward, still in the Upper Miocene. This new species is known to occur in sediments from the Gulf of Mexico, Atlantic Ocean and Pacific Ocean.

Size: 8–13 microns.

Holotype: USNM 651538 (plate 2, figure 5).

Paratypes: USNM 651539 and 651412.

Type locality: DSDP core 3–9–3, 75 cm.; 23° 01′ N., 92° 01′ W., Gulf of Mexico.

Discoaster braarudii Bukry, **n. sp.**
Plate 2, figure 10

Discoaster brouweri Tan. – HAY and others, 1967, p. 468, pl. 5, figs. 1–4.

Description: These small to medium-sized forms with slender untapering rays have little central-area development and are predominantly 6-rayed. The rays are radial and evenly spaced 60° apart, extending to tips that are rounded points or blunt. The rays all lie in a single plane.

Remarks: In emending *D. brouweri* in 1954, Bramlette and Riedel restricted that species to forms with rays that bend down like umbrella ribs so that the rays do not lie in a single plane. This emendation has been universally accepted, as it accurately characterizes an abundant Upper Tertiary species. However, there are simple 6-rayed Discoasters in Middle Miocene assemblages that do not have the characteristic umbrella-rib bending of the rays; instead, the rays lie in a single plane. This morphologic distinction between *Discoaster braarudii* and *D. brouweri* is stratigraphically useful.

The earliest forms of *D. brouweri,* as emended by Bramlette and Riedel, appear in Upper Miocene deposits assigned to the *Discoaster quinqueramus* Zone. Thus, the occurrence of *D. braarudii* in the absence of *D. brouweri* is a guide to Middle Miocene to lower Upper Miocene *Discoaster* assemblages.

Distribution: First appearing in the Middle Miocene, *D. braarudii* persists in assemblages through the Upper Pliocene and is identified from many marine cores penetrating sediments of these ages.

Size: 9–15 microns.

Holotype: USNM 651540 (plate 2, figure 10).

260

Type locality: DSDP core 54.0–6–1, 145 cm. ; 15° 37′ N., 140° 18′ E., Philippine Sea..

Discoaster calculosus Bukry, n. sp.
Plate 2, figures 7–9

Description: This compact 6-rayed species is characterized by the short free length of the broad bifurcate rays, the two lateral nodes near the distal end of each ray, the lack of any prominent central knob, the shallow, rounded interray areas, and the unique pebbly surface. The interray sutures are straignt.

Remarks: Although the general form of Discoaster calculosus is similar to that of the Discoaster deflandrei group, shallow rounded interray areas, small lateral nodes, and a pebbly surface distinguish it from the entire group (see Hay and others (1967) for illustrations of the species variety in this group). It is distinguished from Discoaster druggi by its short non-tapered rays.

Distribution: This unusual species is known only from two localities: 1) the Cipero Formation of Trinidad (Triquetrorhabdulus carinatus Zone), where it is abundant in some samples, and 2) the tropical western Pacific Ocean Caroline Ridge area in the late Oligocene (Sphenolithus ciperoensis Zone). As with other species of the D. deflandrei group, if the samples have been subjected to excess overgrowth by calcite, the specific morphology may be lost.

Size: 11–21 microns.

Holotype: USNM 651541 (plate 2, figure 8).

Paratypes: USNM 651542–43.

Type locality: Cipero Formation, Trinidad, Catapsydrax dissimilis Zone, TLL206264.

Discoaster moorei Bukry, n. sp.
Plate 2, figures 11–12 ; plate 3, figures 1–2

Description: This small species has five distinctly tapering rays that terminate in broad bifurcations. Typically, the five rays are asymmetrically arranged, one interray angle being distinctly smaller than the others. The rays are relatively short, with the result that the bifurcate terminations are emphasized. The central area is small and may have a small knob.

Remarks: The only other typically asymmetric species known, Discoaster asymmetricus Gartner, has rays that terminate simply. Discoaster pentaradiatus is a five-rayed species with bifurcating tips, has longer rays relative to the bifurcations, and is symmetric in the arrangement of rays.

Distribution: D. moorei is frequent, though not common, in samples thoughout the Middle Miocene from the western Pacific Ocean. It also occurs in the Middle Miocene Sphenolithus heteromorphus Zone portion of the Cipero Formation of Trinidad.

Size: 8–18 microns.

Holotype: USNM 651544 (plate 3, figure 2).

Paratypes: USNM 651545–47.

Type locality: DSDP core 54.0–2–4, 81 cm. ; 15° 37′ N., 140° 18′ E., Philippine Sea.

PLATE 1 ⟶
All figures ×2000 light micrographs

1–2, 7   Discoaster brouweri Tan
1, DSDP 12C–2R–1, 142 cm. (19° 42′ N., 26° 00′ W.).
2, 7, Core 64–A–9–5E, 150 cm. (23° 50′ N., 92° 25′ W.).

3–4   Discoaster asymmetricus Gartner
DSDP 12C–4–2, 73 cm. (19° 42′ N., 26° 00′ W.).

5   Discoaster pentaradiatus Tan
V16–21, 600 cm. (17° 16′ N., 48° 25′ W.).

6   Discoaster quinqueramus Gartner
J3, 3 m. (21° 30′ N., 77° 31′ W.).

8   Discoaster neohamatus Bukry and Bramlette
DSDP 25–4–1, 0–2 cm. (0° 31′ S., 39° 14′ W.).

9   Discoaster deflandrei Bramlette and Riedel
Cipero Formation, Trinidad.

10   Discoaster stradneri Noël
Cook Mountain Formation, Louisiana.

11   Discoaster saipanensis Bramlette and Riedel
DWBG–23B, core catcher (16° 42′ S., 145° 48′ W.).

12, 16   Discoaster lodoensis Bramlette and Riedel
Domengine Formation, California.

13–14   Discoaster multiradiatus Bramlette and Riedel
J4, 91 m. (31° 02′ N., 77° 43′ W.).

15   Discoaster barbadiensis Bramlette and Riedel
Navei Formation, Trinidad.

261

PLATE 1

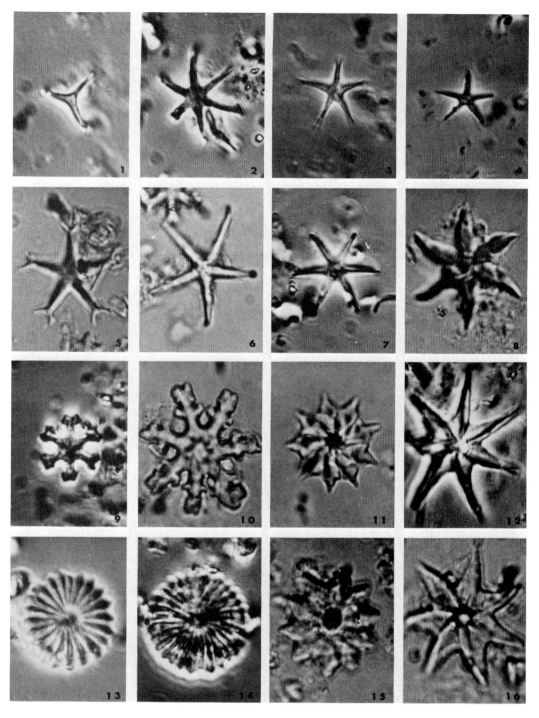

Discoaster signus Bukry, **n. sp.**
Plate 3, figures 3–4

Description: This six-rayed species is characterized by very slender rays that are untapering or only slightly tapering, and terminate in two even more slender and distinctly tapering bifurcating tips. The length of the bifurcation is equal to half or more of the unbifurcated ray length. Although no central area is developed, a prominent knob forms the hub for the six equally spaced rays.

Remarks: The long slender bifurcation at the end of the rays and the prominent central knob in association with the long slender rays combine to produce the diagnostic appearance of this species.

Discoaster signus is readily distinguished from the most similar species, *Discoaster exilis* Martini and Bramlette, by its longer and more acutely angled bifurcations, its consistetnly thinner rays, and lack of a central area development.

Distribution: D. signus is commonly found in Middle Miocene ash-rich sediments from the western Pacific Ocean and the Caribbean Sea, and in the Cipero Formation of Trinidad (*Sphenolithus heteromorphus* Zone).

Size: 16–23 microns.

Holotype: USNM 651548 (plate 3, figure 4).

Type locality: DSDP core 54.0–2–4, 81 cm.; 15° 37′ N., 140° 18′ E., Philippine Sea.

Discoaster variabilis decorus Bukry, **n. subsp.**
Plate 3, figures 5–6

Description: This subspecies is a large, symmetric, six-rayed (rarely five-rayed) form with narrowly bifurcate tips. The central area is small, the diameter being less than the free length of the individual rays.

Remarks: The new subspecies is distinguished from *Discoaster variabilis variabilis* by much longer free rays, and thus by a resulting higher ratio of ray length to central area measurements, and by narrower bifurcations. It is further characterized by a consistent larger size. Along with *Discoaster surculus*, which is distinguished by having trifurcate tips, it is the last of the large Discoasters.

Distribution: Most common in middle Pliocene samples from the western Pacific Ocean, it also occurs in the lower Pliocene of core V3–153 and the middle Pliocene of core DSDP–12C from the northern Atlantic Ocean. Its development may prove to be a useful indicator of lower to middle Pliocene sediment.

Size: 15–27 microns.

Holotype: USNM 651549 (plate 3, figure 5).

Paratype: USNM 651550.

Type locality: DSDP core 55.0–3, core catcher; 9° 18′ N., 142° 33′ E., Pacific Ocean.

PLATE 2 ⟶
All figures × 2000 light micrographs, unless noted otherwise

1–3   *Discoaster araneus* Bukry, n. sp.
  1, J4, 91 m., paratype, USNM 651536.
  2, J4, 91 m., holotype, USNM 651535.
  3, J4, 91 m., paratype, USNM 651537.

4–6   *Discoaster berggrenii* Bukry, n. sp.
  4, DSDP 3–9–3, 75 cm., paratype, USNM 651539.
  5, DSDP 3–9–3, 75 cm., holotype, USNM 651538.
  6, DSDP 3–9–3, 75 cm., paratype, USNM 651412.

7–9   *Discoaster calculosus* Bukry, n. sp.
  7, DSDP 57.1–3–1, 90 cm., ×2500, electron micrograph, paratype, USNM 651542.

  8, Cipero Formation, Trinidad, *C. dissimilis* Zone, TLL206264, holotype, USNM 651541.
  9, Cipero Formation, Trinidad, *C. dissimilis* Zone, TLL206264, paratype USNM 651543.

10   *Discoaster braarudii* Bukry, n. sp.
  DSDP 54.0–6–1, 145 cm., holotype, USNM 651540.

11–12   *Discoaster moorei* Bukry, n. sp.
  11, DSDP 60.0–6–1, 100 cm., paratype, USNM 651545.
  12, DSDP 54.0–2–3, 77 cm., paratype, USNM 651546.

263

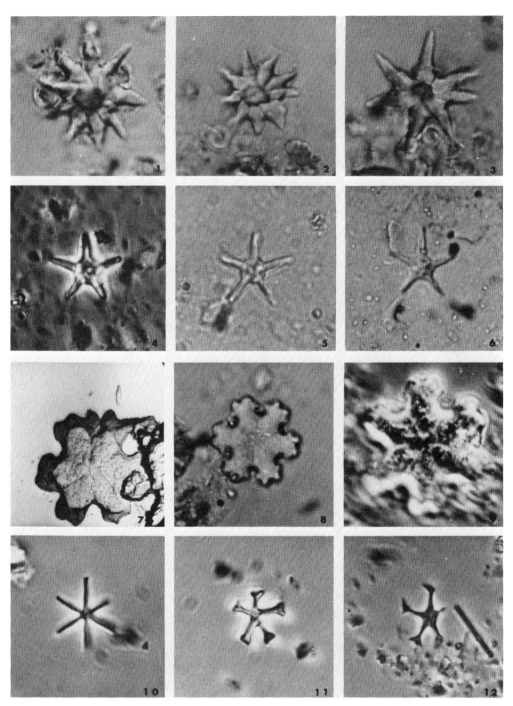

Genus CATINASTER Martini and Bramlette, 1963

Catinaster mexicanus Bukry, **n. sp.**
Plate 3, figures 7–9

Description: These small, thick, rosette-like nanno-fossils have an outer perimeter of six short bifurcate rays. The central area is large and contains a prominent central structure of six ribs extending radially to the interray areas at the perimeter.

Remarks: Catinaster mexicanus is distinguished from *Catinaster calyculus* Martini and Bramlette by its lack of long, curving, simple rays. It is distinguished from *Catinaster coalitus* Martini and Bramlette by its development of definite, short, bifurcate rays.

Distribution: C. mexicanus is known only from the Gulf of Mexico in the Upper Miocene portion of DSDP core 3.

Size: 5–8 microns.

Holotype: USNM 651551 (plate 3, figure 8).

Paratypes: USNM 651552–53.

Type locality: DSDP core 3–9–3, 75 cm.; 23° 01′ N., 92° 01′ W., Gulf of Mexico.

Catinaster? umbrellus Bukry, **n. sp.**
Plate 3, figures 10–13

Description: In plan view this species appears to be a regular rosette of about ten contiguous rays with broadly pointed or rounded terminations. In side view, the rays are seen to be uniformly bent or tilted down away from the center, like the ribs of a partly opened umbrella. There is no central knob or other ornamentation.

Remarks: While possessing the over-all basket shape of *Catinaster,* this form is less massive than and strati-graphically disjunct from the other species.

Distribution: Although common in some samples, it is never abundant. It occurs in Upper Oligocene samples from the Atlantic region in such areas as the Mid-Atlantic Ridge flank, Blake Plateau, and Trinidad.

Size: 8–12 microns.

Holotype: USNM 651554 (plate 3, figures 12–13).

Paratype: USNM 651555.

Type locality: Blake Plateau, JOIDES core 2d, 85 m.; 30° 20′ N., 80° 20′ W., Atlantic Ocean.

SAMPLE LOCALITIES OF NEW TAXA
Atlantic Ocean area
JOIDES core 2 (30° 20′ N., 80° 20′ W.)
JOIDES core 4 (31° 02′ N., 77° 43′ W.)
DSDP core 3 (23° 01′ N., 92° 01′ W.)
DSDP core 10 (32° 52′ N., 52° 13′ W.)
Cipero Formation, Trinidad, *C. dissimilis* Zone, TLL206264.

Pacific Ocean area
DSDP core 54.0 (15° 37′ N., 140° 18′ E.)
DSDP core 55.0 (9° 18′ N., 142° 33′ E.)
DSDP core 57.1 (8° 41′ N., 143° 32′ E.)
DSDP core 60.0 (13° 40′ N., 145° 42′ E.)

PLATE 3 ⟶
All figures × 2000, light micrographs

1–2   *Discoaster moorei* Bukry, n. sp.
1, DSDP 54.0–4–2, 92 cm., paratype, USNM 651547.
2, DSDP 54.0–2–4, 81 cm., holotype, USNM 651544.

3–4   *Discoaster signus* Bukry, n. sp.
3, DSDP 54.0–2–3, 77 cm., figured specimen.
4, DSDP 54.0–2–4, 81 cm., holotype, USNM 651548.

5–6   *Discoaster variabilis decorus* Bukry, n. sp.
5, DSDP 55–3–core catcher, holotype, USNM 651549.

6, DSDP 55–3–core catcher, paratype, USNM 651550.

7–9   *Catinaster mexicanus* Bukry, n. sp.
7, DSDP 3–9–3, 75 cm., paratype, USNM 651552.
8, DSDP 3–9–3, 75 cm., holotype, USNM 651551.
9, DSDP 3–9–3, 75 cm., paratype, USNM 651553.

10–13 *Catinaster? umbrellus* Bukry, n. sp.
10–11, DSDP 10–2–4, 73 cm., paratype, USNM 651555.
12–13, J2d, 85 m., holotype, USNM 651554.

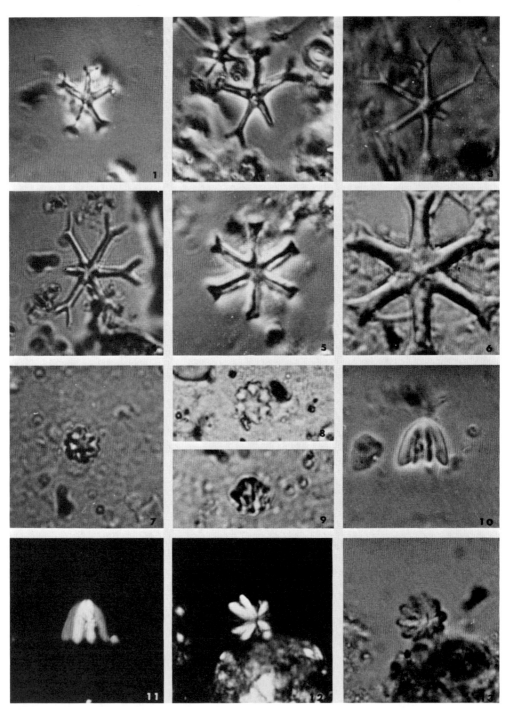

DSDP cores were recovered in a sequence of 9-meter barrels, which were then cut into 1.5 meter sections. Thus the core designation 57.1–3–1, 90 cm. begins with the core-hole number, followed by the core-barrel number, the section number, and finally the distance below the top of the section from which the sample was taken.

ACKNOWLEDGMENTS
I wish to thank William R. Evitt, Stanford University, and Paula Quinterno, U. S. Geological Survey, for constructive reviews of this paper. The samples used were made available by M. N. Bramlette, Scripps Institution of Oceanography, and by the National Science Foundation through the Deep-Sea Drilling Project. Publication has been authorized by the Director of the U. S. Geological Survey.

REFERENCES

AXELROD, D. I., and BAILEY, H. P.
1969   *Paleotemperature analysis of Tertiary floras.* Palaeogeogr, Palaeoclimatol. Palaeoecol., vol. 6, pp. 163–195, text-figs. 1–15, tables 1–4.

BRAMLETTE, M. N., and RIEDEL, W. R.
1954   *Stratigraphic value of Discoasters and some other microfossils related to Recent coccolithophores.* Jour. Pal., vol. 28, no. 4, pp. 385–403, pls. 38–39, text-figs. 1–3.

BRAMLETTE, M. N., and SULLIVAN, F. R.
1961   *Coccolithophorids and related nannoplankton of the early Tertiary in California.* Micropaleontology, vol. 7, no. 2, pp. 129–188, pls. 1–14, text-fig. 1, table 1.

BRAMLETTE, M. N., and WILCOXON, J. A.
1967   *Middle Tertiary calcareous nannoplankton of the Cipero section, Trinidad, W.I.* Tulane Stud. Geol., vol. 5, no. 3, pp. 93–131, pls. 1–10, tables 1–2.

BUKRY, D., and BRAMLETTE, M. N.
1969   *Some new and stratigraphically useful calcareous nannofossils of the Cenozoic.* Tulane Stud. Geol. Pal., vol. 7, no. 3–4, pp. 131–142, pls. 1–3.

1970   *Coccolith age determinations, Leg 3, Deep-Sea Drilling Project.* Deep-Sea Drilling Project, Initial Repts., vol. 3, pp. 589–611, text-fig. 1, tables 1–3.

DORF, E.
1955   *Plants and the geologic time scale.* Geol. Soc. Amer., Spec. Paper, no. 62, pp. 575–592, text-figs. 1–3, table 1.

FISCHER, A. G., and GEALY, E. L.
1969   *Summary and comparison of lithology and sedimentary sequence in northwest Atlantic and northwest Pacific.* Geol. Soc. Amer., Abstract in Programs for 1969, pt. 7 p. 65.

GARTNER, S., JR.
1969   *Correlation of Neogene planktonic foraminifer and calcareous nannofossil zones.* Gulf Coast Assoc. Geol. Soc., Trans., vol. 19, pp. 585–599, pls. 1–2, text-figs. 1–7.

GEITZENAUER, K. R.
1969   *Coccoliths as late Quaternary palaeoclimatic indicators in the subantarctic Pacific Ocean.* Nature, vol. 223, no. 5202, pp. 170–172, text-figs. 1–2, table 1.

HASLE, G. R.
1960   *Plankton coccolithophorids from the subantarctic and equatorial Pacific.* Nytt Mag. Bot., vol. 8, pp. 77–88, pls. 1–3, text-figs. 1–2, tables 1–5.

HAY, W. W., and MOHLER, H. P.
1967   *Calcareous nannoplankton from early Tertiary rocks at Pont Labau, France, and Paleocene-early Eocene correlations.* Jour. Pal., vol. 41, no. 6, pp. 1505–1541, pls. 196–206, text-figs. 1–5.

HAY, W. W., MOHLER, H. P., ROTH, P. H., SCHMIDT, R. R., and BOUDREAUX, J. E.
1967   *Calcareous nannoplankton zonation of the Cenozoic of the Gulf Coast and Caribbean-Antillean area and transoceanic correlation.* Gulf Coast Assoc. Geol. Soc., Trans., vol. 17, pp. 428–480, pls. 1–13, text-figs. 1–13.

HEEZEN, B. C., FISCHER, A. G., BOYCE, R. E., BUKRY, D., DOUGLAS, R. G., GARRISON, R. E., KLING, S. A., KRASHENINNIKOV, V., LISITZIN, A. P., and PIMM, A. C.
1969   *Preliminary scientific results of cruise VI D/V "Glomar Challenger", West Pacific.* Geol. Soc. Amer., Abstract in Programs for 1969, pt. 7, pp. 96–97.

HULBURT, E. M.
1966   *The distribution of phytoplankton, and its relationship to hydrography, between southern New England and Venezuela.* Jour. Mar. Res., vol. 24, no. 1, pp. 67–81, text-figs. 1–5, tables 1–2.

MARTINI, E.
1958   *Discoasteriden und verwandte Formen im NW-deutschen Eozän (Coccolithophorida).* Senckenb. Lethaea, vol. 39, no. 5/6, pp. 353–389, pls. 1–6.

1961   *Nannoplankton aus dem Tertiär und der obersten Kreide von SW-Frankreich.* Senckenb. Lethaea, vol. 42, no. 1/2, pp. 1–41, pls. 1–5, text-figs. 1–3.

MARTINI, E., and BRAMLETTE, M. N.
1963   *Calcareous nannoplankton from the experimental Mohole drilling.* Jour. Pal., vol. 37, no. 4, pp. 845–856, pls. 102–105, text-figs. 1–2.

MCINTYRE, A.
1967   *Coccoliths as paleoclimatic indicators of Pleistocene glaciation.* Science, vol. 158, no. 3806, pp. 1314–1317, text-figs. 1–3, tables 1–2.

MCINTYRE, A., and BÉ, A. W. H.
1967   *Modern Coccolithophoridae of the Atlantic Ocean I. Placoliths and Cyrtoliths.* Deep-Sea Res., vol. 14, no. 5, pp. 561–597, pls. 1–12, text-figs. 1–17, tables 1–10.

STRADNER, H.
1959   *First report on the Discoasters of the Tertiary of Austria and their stratigraphic use.* World Petrol. Congr., 5th, New York, Proc., sec. 1, pp. 1081–1095, text-figs. 1–30.

STRADNER, H., and PAPP, A.
1961   *Tertiäre Discoasteriden aus Österreich und deren stratigraphische Bedeutung mit Hinweisen auf Mexiko, Rumänien und Italien.* Austria, Geol. Bundesanst., Jahrb., spec. vol. 7, pp. 1–159, pls. 1–42, text-figs. 1–24, tables 1–4.

TAN, S. H.
1927   *Discóasteridae incertae sedis.* K. Akad. Wetensch. Amsterdam, Proc., Sec. Sci., vol. 30, no. 3, pp. 411–419, text-figs. 1–14.

31

Reprinted from *Micropaleontology* **19**:32–52 (1973)

Evolutionary trends in the Cenozoic coccolithophore genus *Helicopontosphaera*

Bilal Ul Haq
Woods Hole Oceanographic Institution
Woods Hole, Massachusetts

ABSTRACT

Light and electron microscopic examinations of known species of *Helicopontosphaera* from topotype and other material indicate three distinct trends in the evolution of coccoliths of this genus: 1) changes in the outline of the coccolith from ovoid to broad-elliptical and subrectangular in the Lower Tertiary, then to long-elliptical in the Upper Tertiary; 2) changes in the shape and size of the terminal flange of the distal shield from smooth and merged in the Eocene to sharply terminated in the Oligocene and concurrently to broad, extended and rounded in the Oligocene and subsequently in the Miocene, with a return to small, compressed flanges in two of the three modern species; and 3) changes in the construction of the central area from generally more complex types in the Eocene to less complex ones in later periods.

INTRODUCTION

The Helicopontosphaeraceae belong to the marine calcareous algal order *Coccolithophorales*. The calcareous plates or coccoliths of the genus *Helicopontosphaera* consist of two unequal shields, the larger of which shows a helicoid construction. The coccoliths of the species of this genus first appear in the geological record during the Middle Paleocene and seem to have diversified relatively rapidly during the Eocene but less rapidly in the Oligocene and Miocene, a few species surviving to the present day. The coccoliths of more than 20 species of *Helicopontosphaera* have been recorded and described. Their relatively well-delineated vertical ranges make them important marker fossils for marine biostratigraphy. Although not many data are available on the ecological adaptations of these species, the work of McIntyre and Bé (1967) shows that the extant species *Helicopontosphaera kamptneri* (ex *Helicosphaera carteri*) is adapted to the tropical and subtropical waters of the Atlantic, with a range of tolerance between 16° and 26° C. Also, the data presented by Bukry *et al.* (1971) and by Haq and Lipps (1971) show that *Helicopontosphaera* is rare in the open ocean sediments of the Tertiary, with some species, e. g., *H. reticulata*, *H. seminulum* and *H. lophota,* showing a distinct adaptation to near-shore and shelf areas (Bukry *et al.*, 1971, pp. 1262–1265).

Most species of *Helicopontosphaera* are easily recognized under phase contrast and cross-polarization in the light microscope. However, the study of ultrastructure necessitates the use of electron microscopy. Many authors have presented good electron micrographs of various fossil species of *Helicopontosphaera*, e. g., Haq (1966, 1971c), Hay *et al.* (1967), Boudreaux and Hay (1969), Clocchiatti (1969), Müller (1970), Bukry (1971), Gartner (1971) and Perch-Nielsen (1971).

In this biostratigraphically important genus of calcareous nannoplankton, distinct trends in evolution can be followed throughout the Tertiary. The main purpose of the present study is to offer a picture of the phylogenetic evolutionary trends in this genus. For the purpose of avoiding the existing confusion in the taxonomy and to ascertain the original concept of each species, topotype materials were examined in most cases. The study was carried out both with the light microscope (LM) and with the transmission electron microscope (TEM) or the scanning electron microscope (SEM) to establish correlative criteria.

EVOLUTIONARY TRENDS

The origin of *Helicopontosphaera* can be only a subject of speculation, as no clear-cut morphological relationships exist between the coccoliths of other genera and this helicoid form. The most likely

TABLE 1

Vertical ranges of known *Helicopontosphaera* species plotted against standard Tertiary calcareous nannoplankton zones of Martini (1971).

		Ranges of Helicopontosphaera Species	Helicopontosphaera sp	H. lophota	H. seminulum	H. papillata	H. dinesenii	H. heezenii	H. reticulata	H. compacta	H. bramlettei	H. wilcoxonii	H. euphratis	H. intermedia	H. perch-nielsenae	H. recta	H. obliqua	H. ampliaperta	H. kamptneri	H. rhomba	H. granulata	H. sellii	H. wallichi	H. hyalina

The table plots vertical ranges (shown as vertical bars) against the Calcareous Nannoplankton Zones:

Epoch	Sub	Zone	Zone species
			Calcareous Nannoplankton Zones
			HOLOCENE
PLEIST.		NN21	E. huxleyi
PLEIST.		NN20	G. oceanica
PLEIST.		NN19	P. lacunosa
PLIOCENE	UPPER	NN18	D. brouweri
PLIOCENE	UPPER	NN17	D. pentaradiatus
PLIOCENE	UPPER	NN16	D. surculus
PLIOCENE	LOWER	NN15	R. pseudoumbilica
PLIOCENE	LOWER	NN14	D. asymmetricus
PLIOCENE	LOWER	NN13	C. rugosus
MIOCENE	UPPER	NN12	C. tricorniculatus
MIOCENE	UPPER	NN11	D. quinqueramus
MIOCENE	UPPER	NN10	D. calcaris
MIOCENE	MIDDLE	NN9	D. hamatus
MIOCENE	MIDDLE	NN8	C. coalitus
MIOCENE	MIDDLE	NN7	D. kugleri
MIOCENE	MIDDLE	NN6	D. exilis
MIOCENE	LOWER	NN5	S. heteromorphus
MIOCENE	LOWER	NN4	H. ampliaperta
MIOCENE	LOWER	NN3	S. belemnos
MIOCENE	LOWER	NN2	D. druggii
MIOCENE	LOWER	NN1	T. carinatus
OLIGOCENE	UP	NP25	S. ciperoensis
OLIGOCENE	MI	NP24	S. distentus
OLIGOCENE	MI	NP23	S. predistentus
OLIGOCENE	LO	NP22	H. reticulata
OLIGOCENE	LO	NP21	E. subdisticha
EOCENE	UPPER	NP20	S. peudoradians
EOCENE	UPPER	NP19	I. recurvus
EOCENE	UPPER	NP18	C. oamaruensis
EOCENE	UPPER	NP17	D. saipanesis
EOCENE	MIDDLE	NP16	D. tani nodifer
EOCENE	MIDDLE	NP15	C. alatus
EOCENE	MIDDLE	NP14	D. sublodoensis
EOCENE	MIDDLE	NP13	D. lodoensis
EOCENE	LOWER	NP12	M. tribrachiatus
EOCENE	LOWER	NP11	D. binodosus
EOCENE	LOWER	NP10	M. contortus
P	UP	NP9	D. multiradiatus

269

ancestor of the coccolith of this genus would be an elliptical placolith. The earliest known occurrence of *Helicopontosphaera* has been reported by Haq (1971*a*, p. 24, pl. 3, fig. 2) from the Paleocene *Fasciculithus tympaniformis* (NP5) Zone of Iran. This species, referred to as *Helicopontosphaera* sp. in the present paper, shows some affinity to *H. lophota* (Bramlette and Sullivan). The exact structure of this species is not clear, as the material from which it is described is poorly preserved. It is a subrectangular form with a large central opening spanned by an oblique bridge. Although no intermediate forms between *Helicopontosphaera* sp. and the Eocene forms *H. lophota* and *H. seminulum* have been observed, the last two species may have originated from the first species. *H. lophota* in particular has an oblique bridge closely resembling that of *Helicopontosphaera* sp. From the Lower Eocene upwards this lineage is relatively easy to follow, and definite trends of changes in the shape of the shields, in the central area, and in the shape and size of the terminal flange of the distal shield can be observed (text-figure 2).

The ranges of all *Helicopontosphaera* species recorded to date are presented in table 1. The coccoliths of Lower and Middle Eocene species are all relatively large in size and egg-shaped in outline. The central areas are also relatively large, and the terminal flanges are rounded and merge into the initial part of the distal shield. The *H. dinesenii, H. heezenii* and *H. compacta* group is probably derived from *H. lophota. H. heezenii* and *H. compacta* may have originated directly from *H. lophota* or indirectly through *H. dinesenii.*

Although the ultrastructure of the coccoliths of *H. papillata* is different from that of any other species of *Helicopontosphaera*, it still maintains the essentially ovoid outline and the merged terminal flange of the other species of the Lower and Middle Eocene, and may have developed independently from *H. lophota.*

H. wilcoxonii, which makes its first appearance in the Upper Eocene, has broadly elliptical coccoliths with a broad, sharply terminated and serrate flange. The central area bears close resemblance to that of *H. seminulum*, from which it may have evolved independently.

In the Upper Eocene a new group of species appears (*Helicopontosphaera reticulata, H. bramlettei* and *H. euphratis*), which possesses coccoliths with intermediate characteristics between the large ovoid coccoliths of the Lower and Middle Eocene and the long-elliptical ones of the Oligocene and Miocene. The coccoliths of *H. reticulata* and *H. bramlettei* are roughly rhomboid in outline, and their terminal flanges

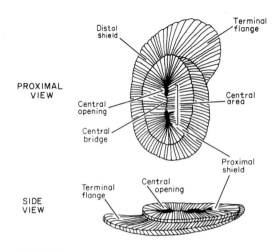

TEXT-FIGURE 1

Line drawing of the type species of *Helicopontosphaera, H. kamptneri* Hay and Mohler, and the application of the descriptive terms used in text.

are truncate and do not merge into the distal shield. In *H. reticulata* the central area of the coccolith remains reticulate, and in *H. bramlettei* an oblique bridge separates two kidney-shaped central openings. The most likely ancestor of these forms is *H. dinesenii.* In *H. euphratis* variations between oval outline and long-elliptical outline of the coccolith can be observed. The terminal flange is broad and extends outside the periphery of the distal shield. The central area is closed by a group of small crystals in this species.

H. intermedia has long-elliptical coccoliths with a prominent outward-extending terminal flange that make their first appearance in the lowermost Oligocene (?uppermost Eocene), developing either from the coccoliths of *H. bramlettei* or from those of *H. euphratis.* They have an oblique bridge separating two semicircular elliptical openings.

In the Middle Oligocene another distinct group of species with coccoliths having a subrectangular outline and a sharply terminated flange evolved, probably from *H. euphratis.* In this group the features of the central area vary from two oblique furrows on a closed central area as in *H. perch-nielseniae*, to two oblique slitlike openings as in *H. obliqua*, or to two large semicircular openings separated by a nearly straight to slightly oblique bridge as in *H. recta.*

H. ampliaperta evolved in the Lower Miocene from *H. intermedia.* Its coccolith is a subrectangular form with a compressed terminal flange and a large central

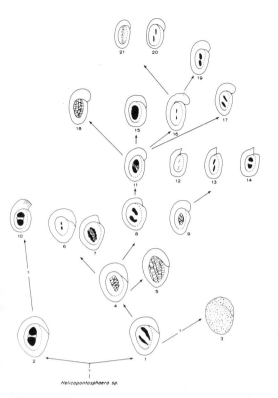

TEXT-FIGURE 2

Suggested phylogenetic lineages in the genus *Helicoponto-sphaera*. 1, *H. lophota*; 2, *H. seminulum*; 3, *H. papillata*; 4, *H. dinesenii*; 5, *H. heezenii*; 6, *H. compacta*; 7, *H. reticulata*; 8, *H. bramlettei*; 9, *H. euphratis*; 10, *H. wilcoxonii*; 11, *H. intermedia*; 12, *H. perch-nielseniae*; 13, *H. obliqua*; 14, *H. recta*; 15, *H. ampliaperta*; 16, *H. kamptneri*; 17, *H. rhomba*; 18, *H. granulata*; 19, *H. sellii*; 20, *H. wallichi*; and 21, *H. hyalina*.

opening. In the Lower Miocene *H. kamptneri*, another long-elliptical form with a broad, extended terminal flange and a nearly closed central area with two small, narrow openings along the long axis, developed from the same stem. *H. intermedia* is probably also the ancestor of the two Middle Miocene species possessing coccoliths with long-elliptical outlines and broad, outward-extended flanges, *H. rhomba* and *H. granulata*. The former has coccoliths with two oblique, slitlike openings separated by a narrow bridge, and in the latter the central area is almost completely closed by a group of small, irregular crystals giving the impression of a granulated surface in the light microscope.

H. kamptneri may have given rise to the Upper Miocene form *H. sellii* and the Recent species *H. wallichi* and *H. hyalina*. *H. sellii* has relatively small coccoliths with a compressed and merged flange and two large, elliptical central openings placed at a slight angle to the long axis and separated by a narrow bridge which is nearly parallel to the short axis. *H. wallichi* is a similar species, but its coccoliths possess narrower and more oblique openings. In the other extant species, *H. hyalina,* the central area of the coccolith is completely closed and traversed by a long, narrow groove along the long axis.

In short, three distinct trends of evolution can be followed in the coccoliths of *Helicopontosphaera*: 1) changes in the outline from typically ovoid in the Lower and Middle Eocene, to rhomboid or broad-elliptical in the Upper Eocene, to subrectangular in the Oligocene, and to long-elliptical in the Miocene; 2) changes in the shape and size of the terminal flange of the distal shield from a compressed and merged flange in the Lower and Middle Eocene, to a truncate one in the Upper Eocene, to a sharply terminated flange in the Oligocene and a large, outward-extended one in the Oligocene and Miocene, and in the two Recent species to a flange small and rounded at the extremity; 3) changes in the construction of the central area from a generally more complex construction (e. g., in *H. dinesenii*, *H. heezenii* and *H. reticulata*) in the Eocene to less complex ones (either closed or with two openings) in the Oligocene-Miocene to Recent.

A general trend towards reduction in the average coccolith size from the Eocene upwards can also be noticed in this genus. However, exceptions to this trend occur in the Miocene, when species such as *H. kamptneri*, *H. granulata* and *H. rhomba* show relatively increased average lengths.

The Late Eocene, Oligocene and Early Miocene are times of relatively high diversity for *Helicopontosphaera*. The highest number of species recorded in a discrete time span is in the Middle Oligocene. However, these forms evolved more rapidly in the Late Eocene and Early Miocene, when relatively more new species made their appearance, and less rapidly in the Early and Middle Eocene, as well as in the Oligocene and the remainder of the Miocene.

SYSTEMATIC PALEONTOLOGY
Order COCCOLITHOPHORALES Schiller, 1926

Family **Helicopontosphaeraceae** Haq, **n. fam.**

Description: Coccolithophores bearing coccoliths

with two asymmetrical shields, the larger showing a helicoid construction.

Type genus: Helicopontosphaera Hay and Mohler.

Remarks: Helicopontosphaera has been variously placed in the families Coccolithaceae Kamptner and Pontosphaeraceae Lemmermann. However, its coccoliths are distinct from those of the other genera of these families in their helicoid construction of the distal shields and their thickening of one extremity when viewed from the side (text-figure 1). The present new family is therefore created for this discrete genus of coccolithophores.

Genus **Helicopontosphaera** Hay and Mohler, 1967

Remarks: The genus *Helicopontosphaera* was created by Hay and Mohler to replace the genus *Helicosphaera* Kamptner, because the type species of the latter, *Coccosphaera carteri,* as illustrated by its author (Wallich, 1877, pl. 17, figs. 3–4, 6–7, 7a, 17) is clearly not a coccolith with a helicoid distal shield. Clocchiatti (1969) defended the usage of the names *Helicosphaera* and *H. carteri* on the premises that Kamptner's clear definition of the genus in 1954 constituted a revision, that the continued use in literature establishes these names, and that thus new names are not warranted. However, the erroneous designation of *Coccosphaera carteri* Wallich (which is clearly not distally helicoid but a placolith with two symmetrical shields) as the type species of the genus by Kamptner invalidated the name *Helicosphaera* for taxa that are not placoliths, and its repeated use by other authors does not validate it as a correct name.

Type species: Helicopontosphaera kamptneri Hay and Mohler. Proximal and side views of the type species and the relevant descriptive terms are shown in text-figure 1.

Helicopontosphaera ampliaperta (Bramlette and Wilcoxon, 1967) Bukry, 1970
Plate 6, figures 4–5; plate 7, figures 3–4

Helicosphaera ampliaperta BRAMLETTE and WILCOXON 1967, p. 105, pl. 6, figs. 1–4.
Helicopontosphaera ampliaperta Bramlette and Wilcoxon).— BUKRY, 1970, p. 377.

Description: The coccoliths have an oval form with a large central opening and a reduced terminal flange of the distal shield. Under the LM this species can be confused with other species in which the central areas are missing (plate 7, figures 3–4). Under the SEM (plate 6, figures 4, 6), it shows an ultrastructure similar to that of other species of this genus. The length varies from 7 to 13 μ.

Occurrence: This species has a relatively short range, extending from the *Discoaster druggii* (NN2) Zone to the *Helicopontosphaera ampliaperta* (NN4) Zone of the Lower Miocene. It has been recorded from Lower Miocene sediments of the Cipero section in Trinidad, from Indonesia, from the type Langhian of northern Italy, from the Relizian Stage of California, and from the Lower Miocene of the Caribbean (Hole DSDP 29B).

Helicopontosphaera bramlettei Müller, 1970
Plate 3, figures 3–4

Helicosphaera aff. *H. seminulum* Bramlette and Sullivan. — BRAMLETTE and WILCOXON, 1967, p. 106, pl. 5, figs. 11–12. *Helicosphaera bramlettei* MULLER, 1970, p. 114, pl. 5, figs. 4–6.

Description: The coccoliths have an oval form with a shortened terminal flange of the distal shield and two openings in the central area separated by an oblique bridge. In the LM the two openings appear characteristically kidney-shaped. In her original description of the species, Müller presented an electron micrograph of a form (pl. 5, fig. 4 of the reference cited in the synonymy) which shows intermediate characteristics between this species and *H. intermedia* (Martini). Length 8 to 11 μ.

Occurrence: This species ranges from the *Chiasmolithus oamaruensis* (NP18) Zone of the Upper Eocene to the *Sphenolithus ciperoensis* (NP25) Zone of the Upper Oligocene. It has been recorded from the Oligocene of the Cipero section in Trinidad, the Middle Oligocene of Germany, and the Upper Oligocene of Syria.

Helicopontosphaera compacta (Bramlette and Wilcoxon, 1967) Hay, 1970
Plate 2, figure 6; plate 7, figures 1–2

Helicosphaera compacta BRAMLETTE and WILCOXON, 1967, p. 105, pl. 6, figs. 5–8.
Helicopontosphaera compacta (Bramlette and Wilcoxon). — HAY, 1970, p. 458.

Description: The coccoliths are egg-shaped in outline, with the terminal flange of the distal shield merging into the initial part of the shield. The shields are relatively thick and closely appressed. Two small, narrow openings are situated near the center of the proximal shield. The distal shield shows weak birefringence in contrast to the proximal shield under the LM (plate 7, figures 1–2). Under the SEM (plate 2, figure 6), these forms show that the terminal flange is slightly protruding to give the distal shield an asymmetrical ovoid outline. The shields are composed of relatively large numbers of thin, slightly overlapping rays. For numerous revealing electron micrographs of both the proximal and distal views of this species, see Haq, 1971c. Length 9 to 13 μ.

Occurrence: This species ranges from the *Discoaster saipanensis* (NP17) Zone of the Upper Eocene to the *Sphenolithus distentus* (NP24) Zone of the Oligocene. It has been reported from the Cipero section in Trinidad, the Upper Eocene of Mississippi, the Lower and Middle Oligocene of Syria, and the equivalent strata of DSDP cores from the Atlantic and Pacific Oceans.

Helicopontosphaera dinesenii Perch-Nielsen, 1971
Plate 1, figure 6; plate 2, figures 4–5; plate 3, figures 11–12

Helicopontosphaera dinesenii PERCH-NIELSEN, 1971, pp. 42–43, pl. 35, figs. 3–4; pl. 36, figs. 3, 6, 9, 11; pl. 61, figs. 6–7.

Description: The coccoliths are ovoid with relatively thick shields not closely appressed and a terminal flange which completely merges into the initial part of the distal shield. The central area is occupied by an obliquely placed, subrectangular region, pierced with numerous small pores, which may extend along the long axis of the proximal shield out of this region. In the LM, coccoliths of this species can be confused with those of *H. lophota* (cf. plate 3, figures 9–12). However, under the SEM (plate 2, figures 4–5), it shows distinguishing features. The shields are composed of numerous thin rays. The distal view shows the distal shield to be composed of flat, overlapping crystals surrounding a central area pierced with numerous pores. This species shows close relationship to *H. lophota* (cf. plate 1, figures 3, 6), from which it may have originated. Length 5 to 15 μ.

Occurrence: Reported from the Middle Eocene *Chiphragmalithus alatus* (NP15) Zone and the *Discoaster tani nodifer* (NP16) Zone of Denmark.

Helicopontosphaera euphratis (Haq, 1966) Martini, 1969
Plate 4, figure 4; plate 5, figures 11–12

Helicosphaera euphratis HAQ, 1966, p. 33, pl. 2, figs. 1, 3.
Helicosphaera parallela BRAMLETTE and WILCOSON, 1967, p. 106, pl. 5, figs. 9–10.
Helicopontosphaera euphratis (Haq). – MARTINI, 1969, p. 136.

Description: The outline of the coccoliths varies from oval to long-elliptical. The terminal flange of the distal shield is prominent and rounded at the end but rarely extends beyond the outline of the distal shield. The central area is characteristically closed by a group of crystals, roughly oval in outline. Under the LM, the coccoliths of this species show a close resemblance to those of *H. intermedia*. However, the central bridge in the latter is quite distinct (cf. plate 5, figures, 9–12). Length 7 to 13 μ.

Occurrence: Ranges from the *Chiasmolithus oamaruensis* (NP18) Zone of the Upper Eocene to the *Heli-copontosphaera ampliaperta* (NN4) Zone of the Lower Miocene. Reported from the Upper Eocene to Lower Miocene sediments of many tropical, subtropical and temperate regions of the world.

Helicopontosphaera granulata Bukry and Percival, 1971
Plate 6, figures 1–2; plate 7, figures 7–8

Helicopontosphaera granulata BUKRY and PERCIVAL, 1971, p. 132, pl. 5, figs. 1–2.

Description: The coccoliths are long-elliptical with a large, outward-extending terminal flange and a central area closed by small equidimensional crystals that give the impression of a granular surface under the LM. This species resembles *H. kamptneri* in outline, but differs in the ultrastructure of the central area. Under the SEM (plate 6, figures 1–2), it shows the proximal shield not closely appressed to the distal shield and a central area of roughly equidimensional crystals. The flange is large and extends beyond the peripheral limits of the distal shield. The distal view (plate 6, figure 2) shows an open central area occupied by small irregular crystals. Length 9 to 12 μ.

Occurrence: Ranges from the *Sphenolithus heteromorphus* (NN5) Zone of the Middle Miocene to the Upper Miocene (?*Ceratolithus tricorniculatus* (NN12) Zone). Reported from DSDP cores from the Atlantic and Pacific.

Helicopontosphaera heezenii Bukry, 1971
Plate 1, figure 1; plate 3, figures 5–6

Helicopontosphaera heezenii BUKRY, 1971, pp. 308–310, pl. 5, figs. 1–5.

Description: A species having large coccoliths with an ovoid outline, a closed central area and a terminal flange that merges into the initial part of the distal shield. Under the LM the central area gives the impression of having a central bar along the long axis, but, under the SEM (plate 1, figure 5), the true nature of the central area is revealed. It is found to be composed of rows of small crystals arranged along the long axis, the crystals in each row being roughly equidimensional. Length 12 to 18 μ.

Occurrence: Reported from the upper Middle Eocene of the DSDP cores of the northwest Pacific and coeval strata in the northwest Atlantic. Range: ?*Chiphragmalithus alatus* (NP15) Zone and *Discoaster tani nodifer* (NP16) Zone.

Helicopontosphaera hyalina (Gaarder, 1970) Haq, **n. comb.**

Helicosphaera hyalina GAARDER, 1970, pp. 113–114, text-figs. 1–3.

Description: Long-elliptical coccoliths with a wide, closed, central area and a small, rounded flange. The central area is composed of very numerous, thin rays joining along a faint central groove. For both LM and EM illustrations of this species, see the reference cited above. Length 5.2 to 7.6 μ.

Occurrence: Recent, Gulf of Mexico.

Helicopontosphaera intermedia (Martini, 1965) Hay and Mohler, 1967
Plate 4, figure 3; plate 5, figures 9–10

Helicosphaera intermedia MARTINI, 1965, p. 404, pl. 35, figs. 1–2.
Helicopontosphaera intermedia (Martini). – HAY and MOHLER, in Hay and others, 1967, p. 448.

Description: Long-elliptical coccoliths with closely appressed shields, a medium to large terminal flange of the distal shield, and a central area with an oblique bridge separating two semicircular openings. In the LM, under cross-polarized light, the oblique bridge becomes prominent. The coccoliths of this species are similar to those of *H. euphratis,* in which the central area is completely closed by a group of crystals arranged obliquely in the central area and looking like an oblique bar. However, under phase-contrast light, the closed nature of the central area of *H. euphratis* is more obvious. Length 9 to 14 μ.

Occurrence: A long-range species occurring from the *Sphenolithus pseudoradians* (NP20) Zone of the uppermost Eocene to the *Discoaster kugleri* (NN7) Zone of the Middle Miocene. Reported by various authors from widely different parts of the world.

Helicopontosphaera kamptneri Hay and Mohler, 1967
Plate 7, figures 11–12

Coccolithus carteri (Wallich). – KAMPTNER, 1941, pp. 93, 111, pl. 13, fig. 136 (not *Coccosphaera carteri* WALLICH, 1877, p. 348, pl. 17, figs. 3–4, 6–7, 17).
Helicosphaera carteri (Wallich). – KAMPTNER, 1954, p. 21, text-figs. 17–19.
Helicopontosphaera kamptneri HAY and MOHLER, in HAY and others, 1967, p. 448, pls. 10–11, fig. 5.

Description: Long-elliptical coccoliths with a prominent medium-sized to large terminal flange extending beyond the periphery of the distal shield, and a closed central area with two small slitlike pores placed along the long axis. The shields are thin and composed of very numerous thin rays. For electron micrographs of this species, see Hay and others (1967).

Occurrence: The longest-range species of this genus, appearing first in the Lower Miocene *Discoaster druggii* (NN2) Zone and persisting to the present. Reported from Miocene and younger sediments of widely different parts of the world.

Helicopontosphaera aff. **H. kamptneri** Hay and Mohler, 1967
Plate 6, figure 5

Remarks: This small oval coccolith is composed of relatively few, slightly overlapping rays. The terminal flange is small and rounded, and does not extend beyond the periphery of the distal shield. The central area is closed and possesses a central groove and

PLATE 1

All figures are scanning electron micrographs. Bar scale on each figure represents 1 μ.

1–3 *Helicopontosphaera lophota* (Bramlette and Sullivan) Bukry et al.
Topotype, Canoas OC-4. 1, 3, proximal views; 2, distal view.

4 *Helicopontosphaera seminulum* (Bramlette and Sullivan) Bukry
Topotype, Lodo 67. Proximal view.

5 *Helicopontosphaera heezenii* Bukry
Topotype, DSDP 44.0-4-6, 145–150 cms. Proximal view.

6 *Helicopontosphaera dinesenii* Perch-Nielsen
Topotype, Skansebakken 160. Proximal view.

PLATE 1

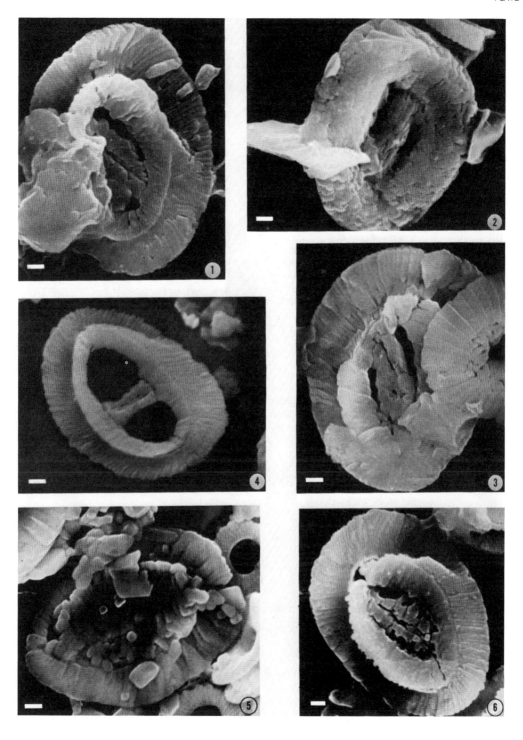

impressions of one or two small pores along this groove. This coccolith shows some affinity to those of *H. kamptneri.* Length 7 to 8 μ.

Occurrence: Recorded in rare numbers from the Lower Miocene *H. ampliaperta* (NN4) Zone of the Cipero section, Trinidad.

Helicopontosphaera lophota (Bramlette and Sullivan, 1961) Bukry *et al.*, 1971
Plate 1, figures 1–3; plate 3, figures 9–10

Helicosphaera seminulum lophota BRAMLETTE and SULLIVAN, 1961, p. 144, pl. 4, figs. 3–4.
Helicopontosphaera lophota (Bramlette and Sullivan). – BUKRY *et al.*, 1971, p. 1300.

Description: Large coccoliths, ovoid in outline, having a central area with a thick, oblique bridge separating two long, narrow, oblique openings. The terminal flange merges into the initial part of the distal shield. The electron micrographs show the shields to be composed of numerous thin rays, slightly overlapping near the flange, and an oblique bridge of two rows of crystals separated by a medial groove. The distal view (plate 1, figure 2) shows flat crystals covering the distal shield and long, narrow ones covering the central bridge. In the LM, this species looks similar to *H. seminulum,* but the latter has a larger central opening, and the bridge is more nearly parallel to the short axis. Length 10 to 15 μ.

Occurrence: Ranges from the *Marthasterites tribra-chiatus* (NP12) Zone to the *Isthmolithus recurvus* (NP19) Zone of the Eocene. Reported from Lower to Upper Eocene strata of various regions of the world. Probably restricted to near-coastal regions.

Helicopontosphaera obliqua (Bramlette and Wilcoxon, 1967), Haq, **n. comb.**
Plate 4, figure 6; plate 5, figures 7–8

Helicosphaera obliqua BRAMLETTE and WILCOXON, 1967, p. 106, pl. 5, figs. 13–14.

Description: Small, long-elliptical to roughly rectan-gular coccoliths with a truncate and sharply pointed, rather than a rounded terminal flange. Central area spanned by a narrow oblique bridge separating two long, narrow, slitlike openings. This species is similar in outline and flange shape to *H. recta* and *H. perch-nielseniae.* In *H. perch-nielseniae* the central area is also similar, but the two slitlike openings are reduced to faint grooves. Length 8–10 μ.

Occurrence: Ranges from the *Sphenolithus distentus* (NP24) Zone of the Oligocene to the *S. heteromorphus* (NN5) Zone of the Lower Miocene. Reported occur-ring in rare numbers in the Cipero section of Trinidad and coeval strata of the DSDP cores from the Pacific.

Helicopontosphaera papillata Bukry and Bramlette, 1969
Plate 2, figures 2–3; plate 7, figure 9

Helicopontosphaera papillata BUKRY and BRAMLETTE, 1969, pp. 133–134, pl. 2, figs. 1–2.

Description: Ovoid coccolith with a closed central area, the surfaces of the shields and the central area covered with fine, papillate ornamentation. Under the SEM (plate 2, figures 2–3), these papillae are visible as small equidimensional pores. This species is dis-tinguishable from all other species of the genus *Helicopontosphaera* by its possession of papillae. Length 11 to 15 μ.

PLATE 2

All figures are scanning electron micrographs. Bar scale on each figure represents μ.

1   *Helicopontosphaera reticulata* (Bramlette and Wilcoxon) Bukry *et al.*
Topotype, T7N, R10W. Proximal view.

2–3   *Helicopontosphaera papillata* Bukry and Bramlette
Topotype, Formación Aragón, Mexico. Distal views.

4–5   *Helicopontosphaera dinesenii* Perch-Nielsen
Topotype, Skansebakken 160. 4, proximal view; 5, distal view.

6   *Helicopontosphaera compacta* (Bramlette and Wilcoxon) Hay
Syria VB40, 94.00 m. Proximal view.

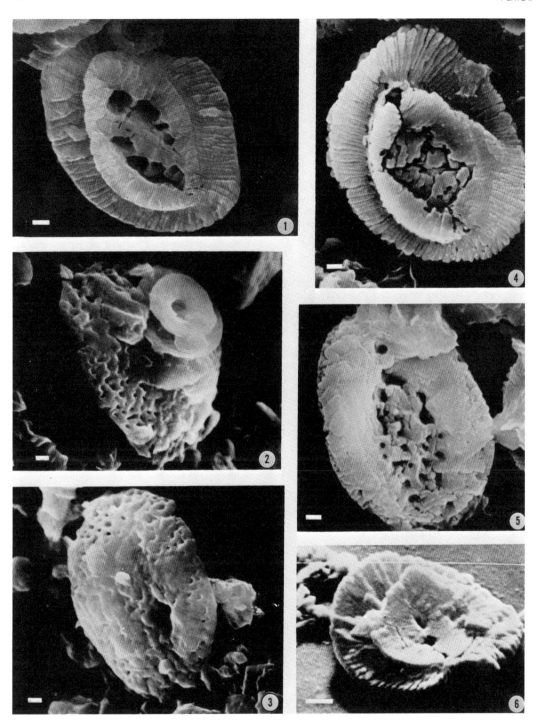

Occurrence: Reported as rare in the lower Middle Eocene type Aragón Formation of Mexico. Range: *?Discoaster lodoensis* Zone and *D. sublodoensis* Zone.

Helicopontosphaera perch-nielseniae Haq, 1971
Plate 4, figure 5; plate 5, figures 5–6

Helicopontosphaera perch-nielsenasae HAQ, 1971, p. 116, pl. 10, figs. 5–7.

Description: A subrectangular coccolith with a sharply pointed terminal flange and a closed central area showing two oblique grooves. The coccoliths of this species closely resemble those of *H. obliqua,* but the latter possess two narrow, slitlike openings separated by a narrow, oblique bridge. The extinction pattern of coccoliths of this species under cross-polarized light also differs from that of *H. obliqua. H. recta* has a similar outline and terminal flange, but possesses larger openings, and the central bridge is more nearly parallel to the short axis. Length 7.5 to 11.5 μ.

Remarks: This species was originally published in honor of Dr. Katharina Perch-Nielsen under the name *H. perch-nielsenasae* (Haq, 1971c). It is here corrected to *H. perch-nielseniae* to conform with the proper Latin orthographic ending of the specific epithet.

Occurrence: Reported from the Middle Oligocene of Syria. Range: From a probable *Sphenolithus predistentus* (NP23) Zone to the *S. distentus* (NP24) Zone.

Helicopontosphaera recta (Haq, 1966) MARTINI, 1969.
Plate 4, figure 6; plate 5, figures 3–4

Helicosphaera seminulum recta, HAQ, 1966, p. 34, pl. 2, fig. 6; pl. 3, fig. 4.
Helicosphaera truncata BRAMLETTE and WILCOXON, 1967, pp. 106–107, pl. 6, figs. 13–14.
Helicopontosphaera recta (Haq). – MARTINI, 1969, p. 136.

Description: Subrectangular coccolith with a truncate, sharply pointed, terminal flange. The central area has a bridge which is nearly parallel to the short axis of the coccolith and separates two subcircular openings, which vary in size from relatively large to very small in a single population. Differentiated from *H. obliqua* and *H. perch-nielseniae,* to which it is similar in outline and shape of flange, by a nearly straight central bridge and larger, more oval openings. Length 8 to 15 μ.

Occurrence: Restricted to the *Sphenolithus distentus* (NP24) and *S. ciperoensis* (NP25) Zones of the Oligocene. Reported from coeval strata of various parts of the world.

Helicopontosphaera reticulata (Bramlette and Wilcoxon, 1967) Bukry *et al.,* 1971
Plate 2, figure 1; plate 3, figures 1–2

Helicosphaera reticulata BRAMLETTE and WILCOXON, 1967, p. 106, pl. 6, fig. 15.
Helicopontosphaera reticulata (Bramlette and Wilcoxon). – BUKRY *et al.,* 1971, p. 1300.

PLATE 3

All figures are light micrographs x 4500.

1–2 *Helicopontosphaera reticulata* (Bramlette and Wilcoxon) Bukry *et al.*
Topotype, T7N, R10W. Proximal view. 1, phase-contrast; 2, X-nicols.

3–4 *Helicopontosphaera bramlettei* Müller
Syria VB40, 9.15 m. Proximal view. 3, phase-contrast; 4, X-nicols.

5–6 *Helicopontosphaera heezenii* Bukry
Topotype, DSDP 44.0-4-6, 145–150 cms. Proximal view. 5, phase-contrast; 6, X-nicols.

7–8 *Helicopontosphaera seminulum* (Bramlette and Sullivan) Bukry
Topotype, Lodo 67. Proximal view. 7, phase-contrast; 8, X-nicols.

9–10 *Helicopontosphaera lophota* (Bramlette and Sullivan) Bukry *et al.*
Topotype, Canoas OC-4. Proximal view. 9, phase-contrast; 10, X-nicols.

11–12 *Helicopontosphaera dinesenii* Perch-Nielsen
Topotype, Skansebakken 160. Proximal view. 11, phase-contrast; 12, X-nicols.

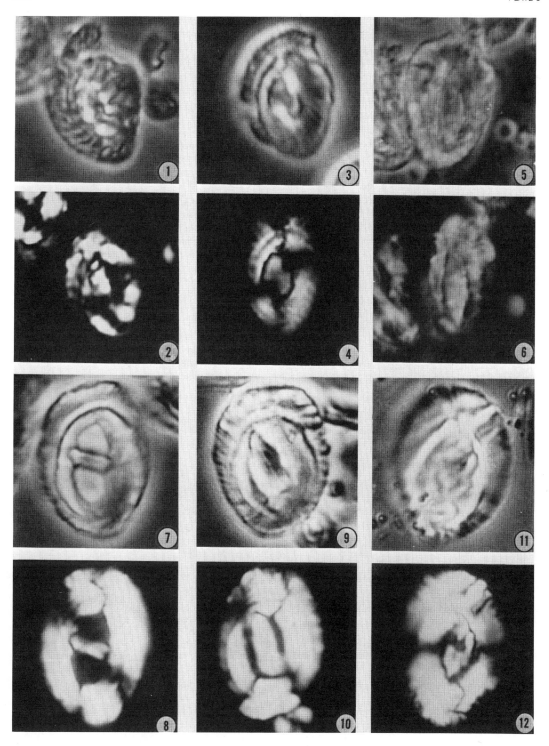

Helicopontosphaera salebrosa PERCH-NIELSEN, 1971, pp. 43–44, pl. 34, fig. 5; pl. 36, figs. 5, 10; pl. 37, fig. 5.

Description: The coccolith of this species is sub-rhomboid in outline with a truncate terminal flange of the distal shield. The central area possesses an oblique central bridge surrounded by two rows of pores. The central area of the coccolith of this species is somewhat similar to the central area of the cocco-lith of *H. dinesenii* (cf. plate 2, figures 1–2), but the latter is more completely covered with pores. Also, the outline of the latter is ovoid and not subrhomboid as in the present species. Length 10 to 13 μ.

Occurrence: Ranges from the *Discoaster tani nodifer* (NP16) Zone of the Middle Eocene to the *Helicoponto-sphaera reticulata* (NP22) Zone of the Lower Oligo-ene. Reported restricted to near-shore basin deposits.

Helicopontosphaera rhomba Bukry, 1971
Plate 4, figure 2; plate 7, figure 10

Helicopontosphaera rhomba BUKRY, 1971, p. 320, pl. 5, figs. 6–9.

Description: Long-elliptical coccolith, with a rounded, medium-sized, terminal flange and a central area possessing an oblique bridge, separating two long-elliptical openings. Under cross-polarized light in the LM, the central bridge is bright when the form is aligned with a polarization direction. Under the SEM (plate 4, figure 2), the coccolith shows similarity to that of *H. lophota* in the construction of the oblique

bridge and openings, but the outline and shape of the flange are different in the coccolith of this species. Length 11 to 18 μ.

Occurrence: Reported from the Middle Miocene of DSDP cores from the Philippine Sea. Range: *Spheno-lithus heteromorphus* (NN5) Zone to a probable *Discoaster exilis* (NN6) Zone.

Helicopontosphaera sellii Bukry and Bramlette, 1969
Plate 6, figure 3; plate 7, figures 5–6

Helicopontosphaera sellii BUKRY and BRAMLETTE, 1969, p. 134, pl. 2, figs. 3–7.

Description: Small, elliptical coccolith, with a reduced and rounded terminal flange and two elliptical to oval central openings separated by a bridge which is either parallel to the short axis or slightly oblique. The bridge is in optical continuity with the surrounding structure of the proximal plate. The SEM figure presented here (plate 6, figure 3) is a distal view of the coccolith of this species. Bukry (1971*b*, pl. 2, fig. 3) has presented an excellent electron micrograph of the proximal view of the coccolith of this species. Length 6 to 12 μ.

Occurrence: Ranges from the Upper Miocene *Dis-coaster calcaris* (NN10) Zone to the Pliocene *D. brouweri* (NN18) Zone. Reported from Italy, the Atlan-tic Ocean, the Gulf of Mexico and the tropical Pacific.

PLATE 4

Figures 1, 3–5 are transmission electron micrographs; figures 2, 6 are scanning electron micrographs. Bar scale on each figure represents 1 μ.

1   *Helicopontosphaera recta* (Haq) Martini
    Topotype, Syria VB40, 9.15 m. Proximal view.

2   *Helicopontosphaera rhomba* Bukry
    Topotype, DSDP 54-0-2.4, 81–82 cms., Philippine Sea. Proximal view.

3   *Helicopontosphaera intermedia* (Martini) Hay and Mohler
    Syria VB40, 9.15 m. Proximal view.

4   *Helicopontosphaera euphratis* (Haq) Martini
    Topotype, Syria VB40, 9.15 m. Proximal view.

5   *Helicopontosphaera perch-nielseniae* Haq
    Topotype, Syria VB40, 11.50 m. Proximal view.

6   *Helicopontosphaera obliqua* (Bramlette and Wil-coxon), n. comb.
    Topotype, TTOC 178888. Distal view.

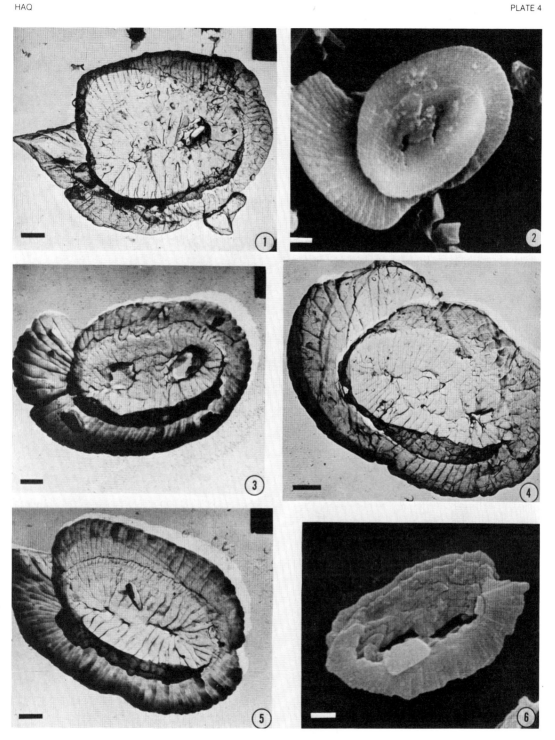

Helicopontosphaera seminulum (Bramlette and Sullivan, 1961) Bukry, 1970
Plate 1, figure 4; plate 3, figures 7–8

Helicosphaera seminulum seminulum BRAMLETTE and SULLIVAN, 1961, p. 144, pl. 4, figs. 1–2.
Helicopontosphaera seminulum seminulum (Bramlette and Sullivan). – BUKRY, 1970, p. 377.

Description: A coccolith ovoid in outline, with a large central opening spanned by a bridge that is either parallel to the short axis or slightly oblique, and a terminal flange which merges into the initial part of the distal shield. In the SEM, it can be seen (plate 1, figure 4) that the central bridge is not in continuity with the structure of the surrounding shield, but forms a distinctly independent structure of long, thin laths. This species is very similar to the younger *H. wilcoxonii*, but the latter is a broadly elliptical form and has a larger and sharply pointed terminal flange. Length 10 to 14 μ.

Occurrence: Ranges from the *Marthasterites tribrachiatus* (NP12) Zone of the Lower Eocene to the Middle Eocene ?*Discoaster tani nodifer* (NP16) Zone. Reported occurring widely in the Lower and Middle Eocene strata of various parts of the world. Not described from open-ocean sediments, probably a nearshore species.

Helicopontosphaera wallichi (Lohmann, 1902) Boudreaux and Hay, 1969

Coccosphaera wallichi LOHMANN, 1902, p. 138, pl. 5, figs. 58–60.
Coccolithus wallichi (Lohmann). – SCHILLER, 1930, in Rabenhorst, pp. 247–248, fig. 124c.

Helicopontosphaera wallichi (Lohmann). – BOUDREAUX and HAY, 1969, pp. 272–273, pl. 6, fig. 9.

Description: The coccolith of this species is similar in construction to that of *H. kamptneri,* but the flange in this form is reduced, and the two slitlike central openings are distinctly oblique in orientation with reference to the long axis. Boudreaux and Hay (1969) have presented a good electron micrograph of the proximal view of this form.

Occurrence: Recent. Reported from the Adriatic and Caribbean Seas.

Helicopontosphaera wilcoxonii Gartner, 1971
Plate 5, figures 1, 4

Helicopontosphaera wilcoxonii GARTNER, 1971, p. 110, pl. 2, figs. 1–4.

Description: A broadly elliptical coccolith with a large central opening spanned by a bridge that is either parallel to the short axis or slightly oblique. The bridge is not in structural continuity with the surrounding shield but constitutes a separate feature composed of thin, parallel laths. The terminal flange of the distal shield is a distinctive character in this species. It expends rapidly, ends sharply, and shows a serrate margin at the uppermost periphery. The older *H. seminulum* has a very similar coccolith, especially in the construction of the central area, but it has a rounded flange which merges into the distal shield. For excellent electron micrographs of this species, see Gartner (1971). Length 10 to 12 μ.

PLATE 5

All figures are light micrographs x 4500.

1–2   *Helicopontosphaera wilcoxonii* Gartner Topotype, JOIDES core J-6B, 233′, 1″. Proximal view. 1, phase-contrast; 2, X-nicols.

3–4   *Helicopontosphaera perch-nielseniae* Haq [recta ?] Topotype, Syria VB40, 11.50 m. Distal view. 5, phase-contrast; 6, X-nicols.

5–6   *Helicopontosphaera perch-nielseniae* Haq Topotype, Syria VB40, 11.50 m. Distal view. 5, phase-contrast; 6, X-nicols.

7–8   *Helicopontosphaera obliqua* (Bramlette and Wilcoxon), n. comb. Topotype, TTOC 178888. Distal view. 7, phase-contrast; 8, X-nicols.

9–10   *Helicopontosphaera intermedia* (Martini) Hay and Mohler Syria VB40, 9.15 m. Proximal view. 9, phase-contrast; 10, X-nicols.

11–12   *Helicopontosphaera euphratis* (Haq) Martini Topotype, Syria VB40, 9.15 m. Proximal view. 11, phase-contrast; 12, X-nicols.

ERRATUM

Plate 5, the Figure legend for Plates 3–4 should read:

3—4   *Helicopontosphaera recta* (Haq) Martini topotype, Syria VB40, 9.5 m. Distal view. 3, phase contrast; 4, X-nicols.

PLATE 5

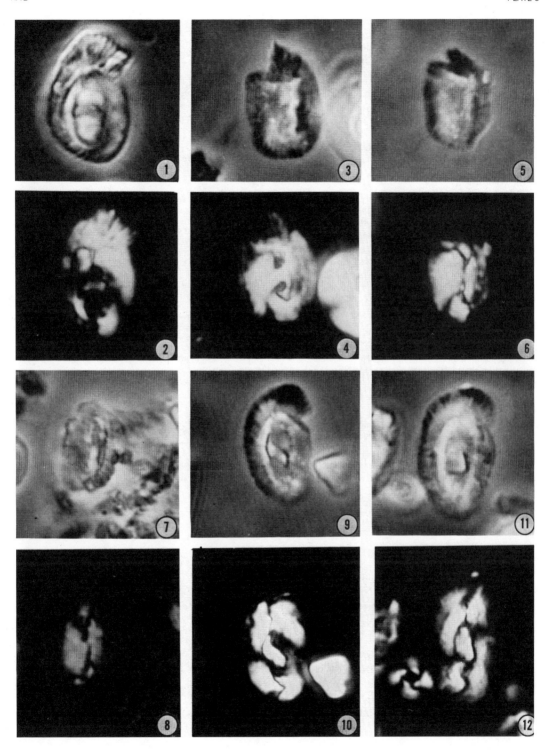

Remarks: This species is distinct from *H. bramlettei* (= *H.* aff. *H. seminulum* of Bramlette and Wilcoxon 1967). The latter has a subrhomboid outline and shows two kidney-shaped central openings when viewed under the LM.

Occurrence: Ranges from the *Chiasmolithus oamaruensis* (NP18) Zone of the Upper Eocene probably up to the *Sphenolithus distentus* (NP24) Zone of the Oligocene. Reported from JOIDES deep sea cores from the Blake Plateau. Also observed by the present author in the NP23 and NP24 Zones of Syria.

ACKNOWLEDGMENTS

The author is grateful to M. N. Bramlette, D. Bukry, S. Gartner, E. Martini, K. Perch-Nielsen and F. R. Sullivan for providing the topotype materials studied here. W. A. Berggren and C. D. Hollister kindly read the manuscript and suggested valuable improvements. This research was supported by National Science Foundation Grant GA 30723. This paper is Woods Hole Oceanographic Institution Contribution number 2865.

REFERENCES

BOUDREAUX, J. E., and HAY, W. W.
1969 *Calcareous nannoplankton and biostratigraphy of the late Pliocene-Pleistocene-Recent sediments in the Submarex cores.* Rev. Española Micropal., vol. 1, no. 3, pp. 249–292, pls. 1–10, text-fig. A.

BRAMLETTE, M. N., and SULLIVAN, F. R.
1961 *Coccolithophorids and related nannoplankton of the early Tertiary in California.* Micropaleontology, vol. 7, no. 2, pp. 129–188, pls. 1–14, text-fig. 1, table 1.

BRAMLETTE, M. N., and WILCOXON, J. A.
1967 *Middle Tertiary calcareous nannoplankton of the Cipero section, Trinidad, W. I.* Tulane Univ., Tulane Stud. Geol., vol. 5, no. 3, pp. 93–131, pls. 1–10, tables 1–2.

BUKRY, D.
1970 *Coccolith age determinations, Leg 4, Deep Sea Drilling Project.* In: Bader, R. G., et al., *Initial reports of the Deep Sea Drilling Project, Volume IV.* Washington: U. S. Government Printing Office, pp. 375–381, text-fig. 1.

1971a *Cenozoic calcareous nannofossils from the Pacific Ocean.* San Diego Soc. Nat. Hist., Trans., vol. 16, no. 14, pp. 303–327, pls. 1–7, tables 1–3.

1971b *Coccolith stratigraphy, Leg 6, Deep Sea Drilling Project.* In: Heezen, B. C., et al., *Initial reports of the Deep Sea Drilling Project, Volume VI.* Washington: U. S. Government Printing Office, pp. 965–1004, pls. 1–8.

BUKRY, D., and BRAMLETTE, M. N.
1969 *Some new and stratigraphically useful calcareous nannofossils of the Cenozoic.* Tulane Univ., Tulane Stud. Geol. Pal., vol. 7, no. 3, pp. 131–142, pls. 1–3.

BUKRY, D., DOUGLAS, R. G., KLING, S. A., and KRASHENINNIKOV, V. A.
1971 *Planktonic microfossil biostratigraphy of the northwestern Pacific Ocean.* In: Heezen, B. C., et al., *Initial reports of the Deep Sea Drilling Project, Volume VI.* Washington: U. S. Government Printing Office, pp. 1253–1300, pls. 1–3.

BUKRY, D., and PERCIVAL, S. F., JR.
1971 *New Tertiary calcareous nannofossils.* Tulane Univ., Tulane Stud. Geol. Pal., vol. 8, no. 3, pp. 123–146, pls. 1–7.

CLOCCHIATTI, M.
1969 *Contribution à l'étude de Helicosphaera carteri (Wallich) Kamptner (Coccolithophoridae).* Rev. Micropal., vol. 12, no. 2, pp. 75–83, pls. 1–3.

PLATE 6

All figures are scanning electron micrographs. Bar scale on each figure represents 1 μ.

1–2 *Helicopontosphaera granulata* Bukry and Percival
Topotype, DSDP 15-3-2, 78–79 cms. Atlantic Ocean. 1, proximal view; 2, distal view.

3 *Helicopontosphaera sellii* Bukry and Bramlette
Topotype, le Castella 81, Calabria. Distal view.

4–6 *Helicopontosphaera ampliaperta* (Bramlette and Wilcoxon) Bukry
Topotype, TTOC 178888. 4, proximal view, 6, distal view.

5 *Helicopontosphaera* aff. *H. kamptneri* Hay and Mohler
Topotype, TTOC 178888. Proximal view.

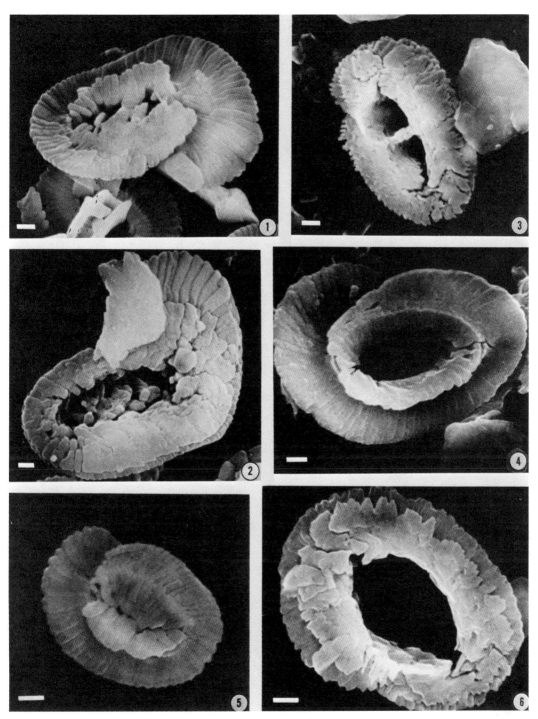

GAARDER, K.
1970   *Three new taxa of Coccolithineae.* Nytt Mag. Bot., vol. 17, no. 2, pp. 113–126, text-figs. 1–9.

GARTNER, S., JR.
1971   *Calcareous nannofossils from the JOIDES Blake Plateau cores and revision of Paleogene nannofossil zonation.* Tulane Univ., Tulane Stud. Geol. Pal., vol. 8, no. 3, pp. 23–37, pls. 1–6, text-fig. 1.

HAQ, B. U.
1966   *Electron microscope studies on some Upper Eocene calcareous nannoplankton from Syria.* Acta Univ. Stockholmiensis, Stockholm Contr. Geol., vol. 15, no. 3, pp. 23–37, pls. 1–6, text-fig. 1.

1971a   *Paleogene calcareous nannoflora, Part I: The Paleocene of west-central Persia and the Upper Paleocene-Eocene of West Pakistan.* Acta Univ. Stockholmiensis, Stockholm Contr. Geol., vol. 25, no. 1, pp. 1–56, pls. 1–14.

1971b   *Paleogene calcareous nannoflora, Part II: Oligocene of western Germany, Ibid.*, vol. 25, no. 2, pp. 57–97, pls. 1–18, text-fig. 1.

1971c   *Paleogene calcareous nannoflora, Part III: Oligocene of Syria. Ibid.*, vol. 25, no. 3, pp. 99–127, pls. 1–25.

HAQ, B. U., and LIPPS, J. H.
1971   *Calcareous nannoplankton and silicoflagellates, sites 69 to 75.* In: Tracey, J. I., Jr., et al., *Initial reports of the Deep Sea Drilling Project, Volume VIII.* Washington: U. S. Government Printing Office, pp. 68, 71, 143–146, 155–156, 298, 307–309, 464, 476–477, 541, 545, 551–552, 626–627, 631, 680–681, 683.

HAY, W. W.
1970   *Calcareous nannofossils from cores recovered on Leg 4.* In: Bader, R. G., et al., *Initial reports of the Deep Sea Drilling Project, Volume IV.* Washington: U. S. Government Printing Office, pp. 455–501.

HAY, W. W., MOHLER, H. P., ROTH, P. H., SCHMIDT, R. R., and BOUDREAUX, J. R.
1967   *Calcareous nannoplankton zonation of the Cenozoic of the Gulf Coast and Caribbean-Antillean area and transoceanic correlation.* Gulf Coast Assoc. Geol. Socs., Trans., vp,. 17, pp. 428–459, pls. 1–3.

KAMPTNER, E.
1941   *Die Coccolithineen der Südwestküste von Istrien.* Naturhist. Mus. Wien, Ann., vol. 51, pp. 54–149, pls. 1–15.

1954   *Untersuchungen über den Feinbau der Coccolithen.* Archiv Protistenk., vol. 100, no. 1, pp. 1–90, text-figs. 1–50.

PLATE 7

All figures are light micrograph x 4500.

1–2   *Helicopontosphaera compacta* (Bramlette and Wilcoxon) Hay
Topotype, TTOC 19.785. Proximal view. 1, phase-contrast; 2, X-nicols.

3–4   *Helicopontosphaera ampliaperta* (Bramlette and Wilcoxon) Bukry
Topotype, TTOC 178888. Proximal view. 3, phase-contrast; 4, X-nicols.

5–6   *Helicopontosphaera sellii* Bukry and Bramlette
Topotype, le Castella 81, Calabria. Proximal view. 5, phase-contrast; 6, X-nicols.

7–8   *Helicopontosphaera granulata* Bukry and Percival
Topotype, DSDP 15-3-2, 78–79 cms. Atlantic Ocean. 7, phase-contrast; 8, X-nicols.

9   *Helicopontosphaera papillata* Bukry and Bramlette
Topotype, Formación Aragón, Mexico. Proximal view. Phase-contrast.

10   *Helicopontosphaera rhomba* Burky
Topotype, DSDP 54-0-2.4, 81–82 cms., Philippine Sea. Proximal view. X-nicols.

11–12   *Helicopontosphaera kamptneri* Hay and Mohler
DSDP 72A-6A-6, 64–65 cms. Pacific Ocean. Proximal view. 11, phase-contrast; 12, X-nicols.

286

PLATE 7

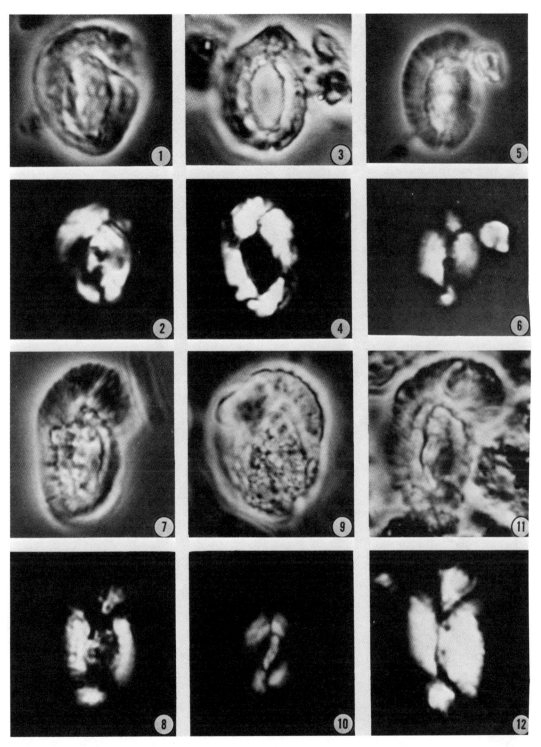

LOHMANN, H.
1902   *Die Coccolithophoridae, eine Monographie der Cocco-
       lithen bildenen Flagellaten.* Archiv Protistenk., vol. 1,
       pp. 89–165, pls. 4–6.

MARTINI, E.
1965   *Mid-Tertiary calcareous nannoplankton from Pacific
       deep-sea cores.* In: Whittard, W. F., and Bradshaw, R. B.,
       Eds., *Submarine geology and geophysics.* London:
       Butterworth, pp. 393–411, pls. 33–37, text-fig. 153.

1969   *Nannoplankton aus dem Latdorf (locus typicus) und
       weltweite Parallelisierungen im oberen Eozän und
       unteren Oligozän.* Senckenbergischen Naturforsch. Ges.,
       Senckenbergiana Lethaea, vol. 50, no. 2/3, pp. 117–159,
       pls. 1–4, text-figs. 1–4.

1971   *Standard Tertiary and Quaternary calcareous nanno-
       plankton zonation.* Internat. Conf. Planktonic Microfossils,
       2nd. Rome, Proc., pp. 739–777, pls. 1–4.

MCINTYRE, A., and BÉ, A. W. H.
1967   *Modern Coccolithophoridae of the Atlantic Ocean. —
       I. Placoliths and cyrtoliths.* Deep-Sea Res., vol. 14, no. 5,
       pp. 561–597, pls. 1–12, text-figs. 1–17.

MÜLLER, CARLA
1970   *Nannoplankton-Zonen der Unteren Meeresmolasse
       Bayerns.* Geol. Bavarica, vol. 63, pp. 107–118, pls. 1–9,
       table 1.

PERCH-NIELSEN, KATHARINA
1971   *Elektronenmikroskopische Untersuchungen an Cocco-
       lithen und verwandten Formen aus dem Eozän von
       Dänemark.* K. Danske Vidensk. Selsk., Biol. Skr., vol. 18,
       no. 3, pp. 1–76, pls. 1–61, text-figs. 1–2, table 1.

SCHILLER, J.
1925   *Die planktonische Vegetationen der Adriatischen
       Meeres. A: Die Coccolithophoriden-Vegetation in den
       Jahren 1911-14.* Archiv Protistenk., vol. 51, no. 1, pp.
       1–130, pls. 1–9, text-figs. a–y, tables 1–11.

1926   *Über Fortpflanzung, geissellose Gattungen und die
       Nomenklatur der Coccolithophoraceen nebst Mitteilung
       über Copulation bei Dinobryon. Ibid.,* vol. 53, no. 2, pp.
       326–342, text-figs. A–H.

1930   *Coccolithineae.* In: Rabenhorst, L., Ed., *Kryptogamen-
       Flora von Deutschland, Österreich und der Schweiz.*
       Leipzig: Akademische Verlagsgesellschaft, vol. 10,
       pt. 2, pp. 89–267, text-figs. 1–137.

WALLICH, G. C.
1877   *Observations on the coccosphere.* Ann. Mag. Nat. Hist.,
       ser. 4, vol. 19, pp. 342–348, pl. 17.

Manuscript received May 2, 1972.

32

Reprinted from pages 451–454, 455, and 460 of *U.S. Geol. Survey Jour. Research*
3:451–465 (1975)

MORPHOLOGY AND PHYLOGENY OF THE COCCOLITHOPHYCEAN FAMILY CERATOLITHACEAE

By STEFAN GARTNER [1] and DAVID BUKRY,

Miami, Fla., La Jolla, Calif.

Abstract.—The family Ceratolithaceae includes a group of horseshoe-shaped calcareous nannofossils and contains ten species which are assignable to two genera : *Amaurolithus* n. gen. and *Ceratolithus*. Species of *Amaurolithus* are characterized by showing faint or no birefringence in cross-polarized light when viewed in preferred orientation. Included in *Amaurolithus* are *A. amplificus* (Bukry and Percival), *A. bizzarus* (Bukry), *A. delicatus* n. sp., *A. primus* (Bukry and Percival), and *A. tricorniculatus* (Gartner). Species of *Ceratolithus* are characterized by their strong birefringence in cross-polarized light when viewed in preferred orientation. *Ceratolithus* includes *C. acutus* Gartner and Bukry, *C. armatus* Müller, *C. cristatus* Kamptner, *C. rugosus* Bukry and Bramlette, and *C. telesmus* Norris. The family first appears in the geologic record during the late Miocene, represented by the nonbirefringent-appearing species that constitute the genus *Amaurolithus*. A succession of these species persists into the early Pliocene. The distinctly birefringent forms assigned to the genus *Ceratolithus* first appear near the base of the Pliocene ; the succession has persisted to modern time.

The family Ceratolithaceae includes a group of late Neogene calcareous nannofossils found in association with coccoliths and discoasters in marine sediment. Ceratoliths generally have the shape of an asymmetrical horseshoe, unique among nannofossils, and hence are readily distinguishable in nannofossil assemblages. This distinctive appearance coupled with the relatively short geologic ranges of several species contributes to the usefulness of the family for biostratigraphy of late Miocene to Holocene marine sediments.

Bukry and Bramlette (1968) summarized the data on the stratigraphic significance of three distinctive species represented by ceratoliths, as then known from the pioneer studies of Bramlette on the biostratigraphy of coccolithophyceans: The still living *Ceratolithus cristatus* Kamptner (1950), a newly named *C. rugosus* of the Pliocene, and *C. tricorniculatus* of Gartner. Gartner's species was found to belong in a distinctive group that Bramlette had differentiated primarily on a very characteristic difference in showing little or none of the strong birefringence of calcite in the normal flat

orientation because the optic axis is vertical in that orientation, as with the discoasters. This marked difference from typical species of *Ceratolithus* is the basis for placing some species formerly assigned to *Ceratolithus* in a new genus—*Amaurolithus*.

Acknowledgment.—We are indebted to Dr. M. N. Bramlette, Scripps Institution of Oceanography, for many fruitful discussions on the nature of ceratoliths. We thank Dr. Bramlette and Dr. W. W. Hay, University of Miami, for reviewing the manuscript. Research was supported by National Science Foundation grant GA–35991.

THE LIVING ORGANISM

Although the common skeletal remains of ceratolith-secreting organisms have been known since 1950, Norris (1965) was the first to describe the living organisms from the Indian Ocean. The cells of these organisms, unlike other coccolithophyceans, do not have flagellae, but they bear two types of coccoliths. One type is the horseshoe-shaped ceratolith that occurs singly, surrounding the cell, and the other is a delicate circular to elliptical hoop of which numerous individual specimens are attached to the outer, spherical sheath. This second type is described in some detail by Norris (1971), although the tiny delicate coccoliths are not known to be preserved in sediments.

MORPHOLOGY AND ULTRASTRUCTURE OF CERATOLITHS

The unusual shape and ornamentation of ceratoliths makes it difficult to describe these forms accurately without a specialized terminology. The various terms applied to ceratolith morphology are shown in figure 1.

Ceratoliths were first named by Kamptner (1950), who later (1954) described the species *Ceratolithus cristatus* in detail from optical-microscope observations in ordinary transmitted light and in polarized light. *C. cristatus* has the shape of an asymmetrical horseshoe

---

[1] School of Marine and Atmospheric Sciences, University of Miami.

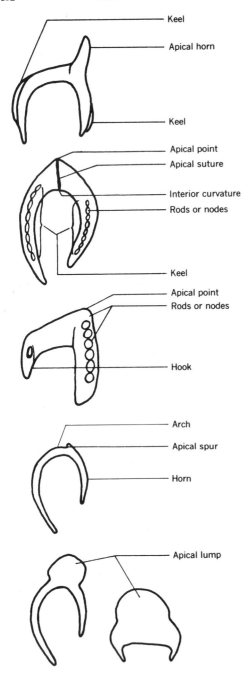

FIGURE 1.—Terms applied to ceratolith morphology.

because one side is shorter than the other (fig. 2). Kamptner describes the ultrastructure, here supplemented by some new observations using the electron microscope, as follows: With the ceratolith oriented so that the open end is upward and the shorter horn is to the left, a bladelike structure protrudes from either side perpendicular to the plane of the ceratolith horns. On the proximal surface (toward the viewer) this bladelike structure is composed of robust rods protruding along the median line of the ceratolith horns, each rod being joined to the adjacent rods along its edges. On the shorter horn the rods are perpendicular to the plane of the ceratolith, whereas on the longer horn the rods are inclined toward the point of the horn. The keel formed by these rods remains perpendicular to the plane of the ceratolith horns. Another keel protrudes in a similar fashion on the opposite side of the ceratolith (away from the viewer); this keel is featureless and very delicate. The inside of the horseshoe is smoothly but unevenly curved, the curvature being sharpest near the base of the shorter horn. The outside curvature is similar except for a blunt point that marks the near midpoint of the horseshoe. A thickening, commonly resembling a suture or ridge, extends from this blunt point to the inside of the horseshoe. The tips of the horns are regularly terminated, and the longer horn generally has a sharper point.

CRYSTALLOGRAPHY OF CERATOLITHS

Ceratoliths are constructed of calcite, and crystallographically every specimen behaves as a single calcite crystallite. But not all ceratoliths have the same crystallographic orientation relative to their general shape. On the basis of their appearance in cross-polarized light the various species can be divided into two groups, and this optical characteristic is used in this paper to assign ceratoliths to two genera. Species which

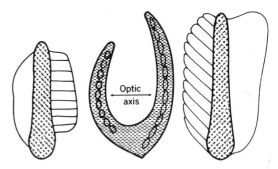

FIGURE 2.—Sketch of *Ceratolithus cristatus*, showing morphology and crystallography.

in preferred orientation show maximum birefringence are assigned to the genus *Ceratolithus* and are referred to as birefringent. Species which in preferred orientation show little or no birefringence are assigned to the genus *Amaurolithus* and for purposes of discussion are called nonbirefringent, although, strictly speaking, calcite is always birefringent, the apparent lack of birefringence being a result of orientation of the calcite optic axis.

The genus *Ceratolithus* includes *C. acutus, C. armatus, C. cristatus, C. rugosus,* and *C. telesmus.* The nonbirefringent species are *Amaurolithus amplificus, A. bizzarus, A delicatus, A. tricorniculatus,* and *A. primus.* Of these, *A. amplificus* and *A. tricorniculatus* may appear weakly birefringent, but this seems to be due to tilting of the specimens which causes the plane of the ceratolith to be inclined to the direction of illumination.

Kamptner (1954) determined that the crystallographic *c* axis (the optic axis of calcite) for *C. cristatus* is in the plane of the ceratolith and approximately at right angles to the long axis of the ceratolith (see fig. 2). All species assigned to *Ceratolithus* share these crystallographic and optical properties. For the nonbirefrigent forms, the crystallographic optic axis is perpendicular to the plane of the ceratolith, as with most discoasters. When the specimens are viewed in preferred orientation, which is with the plane of the ceratolith perpendicular to the direction of illumination, they remain dark in polarized light. As most such specimens do not extinguish completely (become completely black) in cross-polarized light but rather remain a dark gray (except when tilted), one may infer that a slight amount of random variation occurs in the position of the optic axis. Some specimens have either two arms of different thickness or overgrowths of calcite that can tilt the plane of the optic axis slightly, causing the effect in cross-polarized light.

The species of *Amaurolithus* that commonly show weak birefringence, *A. amplificus* and *A. tricorniculatus*, have accessory structures that cause specimens to come to rest on a flat surface in such a way that the plane of the ceratolith is somewhat tilted with respect to the surface of the slide, and hence the crystallographic optic axis is inclined slightly to the direction of illumination. On *A. amplificus* the tilting caused by accessory structures, probably nodes or rods homologous with the keel-forming rods of *C. cristatus*, is commonly in the direction in which the horns point, or at right angles to this direction. As a consequence, weakly birefringent specimens of this species generally are brightest with the horns pointing about 45° toward the direction of polarization. Rarely, specimens are

tilted at a random angle and in such orientations the position of brightest image is also random.

For the species *Amaurolithus tricorniculatus*, the tilt direction is generally determined by a spur below the base of the apical horn. Thus the crystallographic optic axis is usually inclined such that its projection in a surface perpendicular to the direction of illumination makes an angle about 45° with the direction in which the horns point. The brighter image for these specimens is therefore obtained when the horns are parallel to the direction of polarization.

Subtle differences in birefringence alone are not sufficient criteria for making species distinctions.

PHYLOGENY OF CERATOLITHACEAE

The earliest ceratolith-like nannofossils are of the rare species *Ceratolithina hamata* Martini (1967) from the Albian. This rather large species has a pronounced hook at the tip of one horn, is constructed of a single crystallographic unit, and shows birefringence in cross-polarized light. *Ceratolithoides kamptneri* Bramlette and Martini (1964) is a late Maestrichtian form on which the two horns are constructed of differently oriented calcite units. The late middle Eocene *Ceratolithina? vesca* Bukry and Percival (1971) also is constructed of a single crystallographic unit, but this species has relatively straight horns separated by about 75°. All three of the above species represent a discontinuous record of objects similar in shape but not lineally related to Neogene ceratoliths.

The origin of the family Ceratolithaceae remains unclear. The family first appears in the late Miocene, but among the diverse nannofossils of the middle and late Miocene there is no likely candidate for an ancestral ceratolith. Gartner (1967) suggested a possible relation between the genus *Ceratolithus* and *Triquetrorhabdulus rugosus* Bramlette and Wilcoxon (=*Ceratolithus farnsworthii* Gartner) but pointed out that the resemblance was closest to the Pleistocene and Holocene species *Ceratolithus cristatus* rather than the earlier representatives of the family, some of which coexisted with *T. rugosus.* Hence this relation must remain a remote possibility.

Norris (1965, 1971) has described delicate hoop-shaped coccoliths occurring on the surface of ceratolith-bearing cells, and this suggests another avenue by which ancestral ceratoliths may be sought. Unfortunately, these hoop-shaped coccoliths are so delicate that they are not likely to be preserved in the fossil record, and, indeed, none are known. Among the more robust types of coccoliths that might be considered in this context because of its hoop shape is the Oligocene to middle Miocene species *Coronocyclus nitescens;* there is, however, no structural evidence of any relation between delicate and robust types.

A probable scheme of evolutionary succession of the various species of the family Ceratolithaceae is suggested in figure 3. *Amaurolithus primus* is the earliest ceratolith, but *A. delicatus* appears about the same time. Within the resolution limits of pelagic-ooze sediments, it is not possible to determine which is ancestral to the other. It may be important that the heavily calcified forms of *A. primus* are most abundant in samples from tropical latitudes, whereas *A. delicatus* seems to have a wider latitudinal distribution. This would suggest that some ecological preference existed among these two species and that some differentiation of the parent organism occurred prior to the appearance of these species in the fossil record. A close relation between *A. delicatus* and *A. primus* is suggested by their common optical properties. In preferred orientation, with the plane of the ceratolith at right angles to the direction of illumination, both species are nonbirefringent; that is, they do not become alternately bright and dark when rotated in cross-polarized light but rather remain gray. Horns of these early ceratoliths are either featureless or bear only weak keels, nodes, or other structures. The most obvious secondary structures, when present, are confined to the apical region, where they most commonly take the form of apical spurs or lumps.

The next ceratolith to appear is *Amaurolithus amplificus*. Morphologically this species differs from the two earlier species in these respects: Specimens are generally robust and relatively large with the horns diverging from the apical region at about 80°. One of the horns bears a hook, the other one a row of median nodes similar to but shorter and more robust than the rods which form the keel of *Ceratolithus cristatus*. Commonly, these nodes are secondarily calcified and very irregular in size. Optically, *A. amplificus* is similar to *A. delicatus* and *A. primus* in birefringence. However, specimens of this species are commonly tilted, probably because of the nodes developed on one side, and consequently they show slight birefringence.

Two more nonbirefringent species appear later in the record: *Amaurolithus bizzarus* and *A. tricorniculatus*. *A. bizzarus* is rare but seemingly restricted to the early Pliocene, whereas *A. tricorniculatus* may occur in late Miocene but primarily in early Pliocene sediments. Clearly these two species are closely related, but whether they evolve from *A. amplificus* or from *A. delicatus* and *A. primus* is not clear. The delicate construction and overall shape of the horseshoe probably point to *A. delicatus* as being the most likely ancestor, although the apical structure of *A. amplificus* is more similar to the apical structures found in *A. bizzarus* and *A. tricorniculatus*.

All the remaining species show strong birefringence in cross-polarized light, hence an evolutionary jump has to be made. The earliest species having this property is *Ceratolithus acutus*, a species that morphologically most closely resembles *Amaurolithus amplificus*. During the early Pliocene a change in crystallite orientation came about that remains unexplained. From this time on, the evolution of the genus seems much more straightforward in that some of the remaining species grade into one another almost imperceptibly. *C. rugosus* is readily distinguishable from *C. acutus* and *C. armatus* but not from *C. cristatus*. In most well-preserved sediment, the two could not be distinguished readily, although the propensity of *C. rugosus* to calcify and to be transformed into heavy, rugose specimens, within the organism or after being shed, makes the identification of most specimens relatively easy. *C. telesmus*, the most recently developed species in the lineage, is, in its extreme development, also readily distinguished. But specimens transitional between it and its probable ancestor, *C. cristatus*, are not uncommon.

The earliest species of Ceratolithaceae appear in the fossil record with no likely ancestral forms. Although this can be explained by assuming that the ancestral form developed in a noncalcifying state, it is nevertheless not unusual for a genus to appear abruptly without its having an obvious predecessor in the fossil record of calcareous nannoplankton.

[*Editor's Note:* Systematic descriptions have been omitted at this point.]

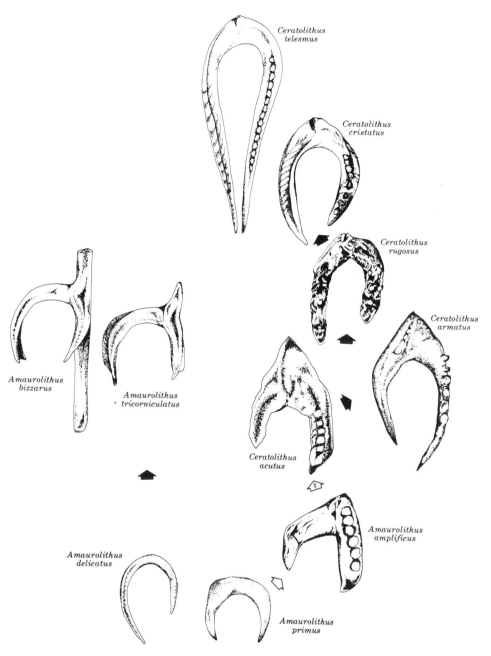

FIGURE 3.—Phylogenetic lineage of the family Ceratolithaceae. Only the end members, which constitute distinct species, are shown; for some branches, intermediate forms are recognizable. Solid arrows indicate a more direct relation between species than do open arrows.

REFERENCES CITED

Bramlette, M. N., and Martini, E., 1964, The great change in calcareous nannoplankton fossils between the Maestrichtian and Danian: Micropaleontology, v. 10, p. 291–322.

Bukry, David, 1971, Cenozoic calcareous nannofossils from the Pacific Ocean: San Diego Soc. Nat. History Trans., v. 16, p. 303–327.

—— 1973, Coccolith stratigraphy, eastern equatorial Pacific, Leg 16 Deep Sea Drilling Project: Deep Sea Drilling Proj. Initial Repts., v. 16, p. 653–711.

Bukry, David, and Bramlette, M. N., 1968, Stratigraphic significance of two genera of Tertiary calcareous nannofossils: Tulane Studies Geology, v. 6, p. 149–155.

—— 1969, Coccolith age determinations, Leg 1, Deep Sea Drilling Project: Deep Sea Drilling Proj. Initial Repts., v. 1, p. 369–387.

Bukry, David and Percival, S. F., Jr., 1971, New Tertiary calcareous nannofossils: Tulane Studies Geology and Paleontology, v. 8, p. 123–146.

Ellis, H. C., Lohman, W. H., and Wray, J. L., 1972, Upper Cenozoic calcareous nannofossils from the Gulf of Mexico (Deep Sea Drilling Project, Leg 1, Site 3): Colorado Mines Quart., v. 67, no. 3, 103 p.

Gartner, Stefan Jr., 1967, Calcareous nannofossils from Neogene of Trinidad, Jamaica, and Gulf of Mexico: Kansas Univ. Paleont. Contr., Paper 29, 7 p.

—— 1969, Correlation of Neogene planktonic foraminifer and calcareous nannofossil zones: Gulf Coast Assoc. Geol. Socs. Trans., v. 19, p. 585–599.

—— 1973, Absolute chronology of the late Neogene calcareous nannofossil succession in the equatorial Pacific: Geol. Soc. America Bull., v. 84, p. 2021–2034.

Gartner, Stefan, and Bukry, David, 1974, *Ceratolithus acutus* Gartner and Bukry n. sp. and *Ceratolithus amplificus* Bukry and Percival—nomenclatural clarification: Tulane Studies Geology and Paleontology, v. 11, p. 115–118.

Hekel, Heinz, 1973, Nannofossil biostratigraphy, Leg 20, Deep Sea Drilling Project: Deep Sea Drilling Proj. Initial Repts., v. 20, p. 221–247.

Kamptner, Erwin, 1950, Über den submikroskopischen Aufbau der Coccolithen: Österreichische Akad. Wiss., Math-Naturw. Kl., Anz., v. 87, p. 152–158.

—— 1954, Untersuchungen über den Feinbau der Coccolithen: Archiv Protistenkunde, v. 100, no. 1, 90 p.

Martini, E., 1967, *Ceratolithina hamata*, n.g., n. sp., aus dem Alb von N-Deutschland (Nannoplankton incertae sedis): Neues Jahrb. Geologie u. Paläontologie Abh., v. 128, p. 294–298.

Müller, Carla, 1974, Calcareous nannoplankton, Leg 25 (western Indian Ocean): Deep Sea Drilling Proj. Initial Repts., v. 25, p. 579–633.

Norris, R. E., 1965, Living cells of *Ceratolithus cristatus* (Coccolithophorineae): Archiv Protistenkunde, v. 108, p. 19–24.

—— 1971, Extant calcareous nannoplankton from the Indian Ocean: 2d Planktonic Conf., Roma 1971, Proc., p. 899–909.

Stradner, Herbert, 1973, Catalogue of calcareous nannoplankton from sediments of Neogene age in the eastern North Atlantic and the Mediterranean Sea: Deep Sea Drilling Proj. Initial Repts., v. 13, p. 1137–1199.

[Editor's Note: Figures 4–8 have been omitted.]

33

Transgressions, climatic change and the diversity of calcareous nannoplankton

BILAL UL HAQ

Woods Hole Oceanographic Institution, Woods Hole, Mass. (U.S.A.)

(Accepted for publication July 16, 1973)

ABSTRACT

Haq, B.U., 1973. Transgressions, climatic change and the diversity of calcareous nannoplankton. *Mar. Geol.*, 15: M25—M30.

Plankton-diversity models of various authors are examined in the light of available diversity data on calcareous nannoplankton. To account for the diversity peaks during times of widespread epicontinental seas in the Mesozoic and Tertiary, a new explanation is given.

INTRODUCTION

Most of the plankton-diversity models that have been proposed in recent years are aimed at explaining the events that culminated in the massive and relatively sudden extinction of oceanic faunas and floras at the end of the Cretaceous. Increased cosmic radiation (Uffen, 1965), eustatic changes in sea level (Newell, 1965), decrease in nutrient supply and subsequent failure of the productivity of the oceans (Bramlette, 1965), climatic deterioration (Hay, 1960; Cifelli, 1969), and the migration of the carbonate-compensation depth to the surface (Hay, 1960; Tappan, 1968; Worsley, 1971), are some of the well-known explanations that have been offered to account for this dramatic event (see Lipps, 1970, for a summary). The purpose of this letter is to examine the plankton-diversity models in the light of the known diversity data on calcareous nannoplankton. Together with other phytoplankton, these autotrophic marine algae comprise the basal link in the oceanic food chain and the more pronounced fluctuations in their abundances are bound to effect the abundances of other taxa.

PLANKTON DIVERSITY

Fig. 1 and 2 show the species-diversity data of calcareous nannoplankton from Jurassic to Pleistocene. In spite of the various factors that can bias such diversity data (see Haq, 1971, for a discussion), the present figures are acceptable in broad general outline, and especially in the case of sudden exponential increases and decreases in the taxonomic frequencies.

295

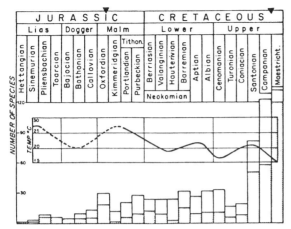

Fig. 1. Species diversity of calcareous nannoplankton through the Jurassic and Cretaceous. Shaded areas represent new species. The inset is the oxygen-isotope paleotemperature curve of Dorman (1968) for Europe, and triangles at the top mark periods of widespread sea-level maxima.

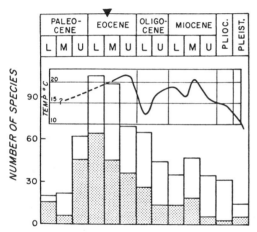

Fig. 2. Species diversity of calcareous nannoplankton through Cenozoic. Shaded areas represent new species. Paleotemperature curve (inset) is that of Devereux (1967) for New Zealand. Triangle at top marks the period of widespread Tertiary seas.

After their appearance on the phytoplankton scene in the lowermost Jurassic, the calcareous nannoplankton species increased insignificantly during the lower Jurassic. The first significant diversification occurred during the Oxfordian when a large number of new taxa originated. However, the diversity dropped back to the pre-Oxfordian level in the Kimmeridgian. A gradual increase in the number of species occurred in the remainder of the upper Jurassic and this level was maintained more or less up to the Cenomanian. An apparent decrease in the total species diversity is seen in the Turonian and Coniacian (Fig. 1); however, this is probably due to the lack of more reliable data for this period. The occurrence of numerous long-ranging species throughout the upper Cretaceous suggests a gradual increase in the diversity culminating in the highs of the Santonian and Campanian and an all-time abundance of calcareous nannofossils in the Maastrichtian (over 130 species).

The massive extinction of marine biota at the termination of the Maastrichtian depleted the world oceans of nannoplankton and only four species (Bramlette and Martini, 1964) managed to survive into the lower Danian and probably comprised the total genetic pool from which the Tertiary forms originated. The species density remains low, and at similar levels during the early and middle Paleocene. A marked radiation occurred during the late Paleocene and early Eocene; the latter being the period of highest diversity in the Cenozoic (over 120 species recorded). This exponential increase is also offset by the appearance of a new group of nannoplankton on the scene: the discoasters, which diversified rapidly during the early and middle Eocene adding substantially to the diversity figures of these periods. The total diversity declines gradually from the middle Eocene to early Oligocene and then sharply during the late Oligocene and early Miocene. A slight second bloom occurred during the middle Miocene and then the trend towards decreased diversity continued until the Pleistocene.

Fig. 3 and 4 show the rates of evolution calculated separately for the Cenozoic cocco-lithophores and discoasters (see Haq, 1971). The coccolithophores evolved rapidly from middle Paleocene to early Eocene and slowly during the remainder of the Tertiary, except during the middle Miocene when a slight change from negative to positive values occurs in the rate of change in total. The discoasters show similar trends but the middle-Miocene intensification in the rates is more pronounced and significant than that of the coccolitho-phores. When the taxonomic frequency data is compared with the paleotemperature curves — insets in Fig. 1 and 2 show oxygen-isotope temperature curves of Dorman (1968) for Europe, and Devereux (1967) for New Zealand, respectively — no close correlation can be observed between these two factors for the Mesozoic. In the Cenozoic there is a general agreement

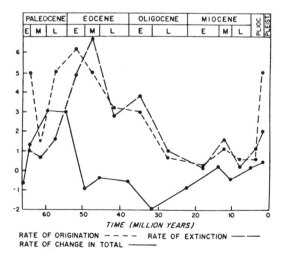

Fig. 3. Rates of evolution for Cenozoic coccolithophores. Vertical scale represents species/m.y. in case of rates of origination and extinction and a ratio in case of change in total. (From Haq, 1971).

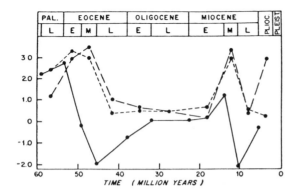

RATE OF ORIGINATION ‒‒‒‒‒   RATE OF EXTINCTION ‒‒ ‒‒ ‒‒
RATE OF CHANGE IN TOTAL ‒‒‒‒‒‒

Fig. 4. Rates of evolution for Cenozoic discoasters. Vertical scale represents species/m.y. in case of rates of origination and extinction and a ratio in case of change in total.(From Haq, 1971.)

between paleotemperatures and species diversity (e.g., highs in the early and middle Eocene and middle Miocene, and lows in the late Eocene, Oligocene and post-middle Miocene times) The discoasters, in particular, show a marked sensitivity to temperature when the rates of evolution of this group are compared with the paleotemperatures. Of special interest is the significant increase in the tempo of evolution of this group in the middle Miocene when the mean surface temperature in the higher latitudes increased from 5° to 8°C as compared to earlier values. The quantitative study of this group may thus prove valuable for paleo-climatology. However, a causal relationship between climatic fluctuations and the total species diversity of calcareous nannoplankton cannot be suggested from the available data.

PLANKTON-DIVERSITY MODELS

Two intriguing models of the evolution of planktonic biota have been offered recently. Lipps (1970) suggested that, during periods of cooler high-latitude seas, the vertical and hori-zontal thermal gradients are intensified and the resulting increase in ecologic barriers causes plankton speciation and provincialism. During times of warm high-latitude seas, on the other hand, these ecologic barriers are eliminated resulting in extinctions and a decrease in the plankton diversity. These conclusions, when tested against data presented here, fall short of a correlation between cooler high-latitude times and higher diversities, especially in the Cenozoic where the diversity data are relatively well documented. Both the coccolitho-phores and discoasters, in fact, show decreased diversities during periods of climatic deteri-oration (e.g., Oligocene, late Miocene and Pliocene). Similar relationships between species diversity and temperature have also been shown for the Cenozoic planktonic Foraminifera (Berggren, 1969).

A more promising model is that suggested by Tappan (1968), and Tappan and Loeblich (1972). They show evidence that continental physiography, sea-level changes, climatic fluctua-tions, nutrient supply, evolution of phytoplankton, and atmospheric-oxygen and CO_2 levels are all intimately related phenomena. On a broad scale, reduction of the continents to the base level and the concomitant spread of epicontinental seas is initially favorable to phytoplankton growth due to the proximity of nutrients, but eventually results in equable climates, decreased circulation and upwelling, lesser influx of land-derived detritus, and as Bramlette (1965) had hypothesized earlier, cause reduction in the nutrient levels inducing reduced phytoplankton diversification. This, in turn, leads to decreased productivity and depletion of atmospheric oxygen and increased pCO_2 levels. Continental rejuvenation, on the other hand, causes climatic changes, increased oceanic circulation and upwelling, renewed detritus supply from land causing nutrient replenishment, phytoplankton diversification and growth, and the expansion of dependent organisms.

DISCUSSION

The present data of calcareous nannoplankton, however, do not bear out the contention that widespread seas are followed by a reduction in diversity. In fact the times of rapid turnovers and high species diversities are associated with widespread epicontinental seas, and more specifically, the peaks of the diversity curves fall close to the large-scale sea-level maxima, e.g., Oxfordian, late Cretaceous (Maastrichtian) and early—middle Eocene (see Gignoux, 1955; Brinkmann, 1960; and Kummel, 1961). Widespread orogenic events and regressions are followed by reduction in the number of taxa, and the onset of large-scale transgressions signals the increase in taxonomic frequencies.

The available data on temperature tolerances of extant species show that the majority of them are adapted to relatively narrow ranges of maximum and minimum temperatures (Paasche, 1968). During times of extreme climates in the higher latitudes, diurnal and seasonal temperature changes would be relatively great and would have an adverse effect on the nannoplankton growth. During times of equable high-latitude climates (regardless of the mean surface temperatures) the maximum and minimum temperature ranges would become narrower, thereby increasing the potential ecospace for the growth and spread of organisms with narrow ranges of temperature tolerance.

In summary, a simple alternative explanation may be offered from the above experience: large-scale transgressions and spread of epicontinental seas cause equable "maritime" climates with lesser differences in maximum and minimum temperatures resulting in the extension of the potential ecospace for the growth and speciation of nannoplankton. Large-scale regressions and restriction of seas are followed by extreme "continental" climates with higher differences in the maximum and minimum temperatures, resulting in the narrowing of potential ecospace, reduced nannoplankton growth and speciation.

The model suggested here shows a closer agreement with the available data of calcareous nannoplankton and is offered as a plausible explanation only to account for the fluctuations in the diversity of these organisms. A more rigid test of this and the earlier plankton-diversity models will, however, have to wait until more detailed data can be obtained on a regional basis — a sedimentary basin with wide temporal and spatial (latitudinal) extend, e.g., North Atlantic Ocean.

ACKNOWLEDGEMENTS

The author thanks J.R. Heirtzler, W.A. Berggren and I. Premoli-Silva for their critique. This research was supported by NSF Grant GA-30723. This is Woods Hole Oceanographic Institution Contribution Number 3064.

REFERENCES

Breggren, W.A., 1969. Rates of evolution in some Cenozoic planktonic Foraminifera. *Micropaleontology*, 15: 351–365.

Bramlette, M.N., 1965. Massive extinction in biota at the end of Mesozoic time. *Science*, 148: 1696–1699.

Bramlette, M.N. and Martini, E., 1964. The great change in calcareous nannoplankton fossils between the Maastrichtian and Danian. *Micropaleontology*, 10: 291–322.

Brinkmann, R., 1960. *Geologic Evolution of Europe*. Ferdinand Enke, Stuttgart.

Cifelli, R., 1969. Radiation of the Cenozoic planktonic Foraminifera. *Syst. Zool.*, 18: 154–168.

Devereux, I., 1967. Oxygen isotope paleotemperature measurements on New Zealand Tertiary fossils. *N.Z. J. Sci.*, 10: 988–1011.

Dorman, F.H., 1968. Some Australian oxygen isotope temperatures and a theory for a 30-million-year world-temperature cycle. *J. Geol.*, 76: 297–313.

Gignoux, M., 1955. *Stratigraphic Geology*. Freeman, San Francisco, Calif.

Haq, B.U., 1971. Paleogene calcareous nannoflora, 4. Paleogene nannoplankton biostratigraphy and evolutionary rates in Cenozoic calcareous nannoplankton. *Stockholm Contrib. Geol.*, 25: 129–158.

Hay, W.W., 1960. The Cretaceous–Tertiary boundary in the Tampico embayment, Mexico. *Int. Geol. Congr., Norden*, 5: 70–77.

Kummel, G., 1961. *History of the Earth*. Freeman, San Francisco, Calif.

Lipps, J.H., 1970. Plankton evolution. *Evolution*, 24: 1–22.

Newell, N.D., 1962. Paleontological gaps and geochronology. *J. Paleontol.*, 36: 592–610.

Newell, N.D., 1965. Mass extinction at the end of the Cretaceous period. *Science*, 149: 922–924.

Paasche, E., 1968. Biology and physiology of coccolithophorids. *Ann. Rev. Microbiol.*, 22: 71–86.

Tappan, H., 1968. Primary production, isotopes, extinctions and the atmosphere. *Palaeogeogr. Palaeoclimatol. Palaeoecol.*, 4: 187–210.

Tappan, H. and Loeblich Jr., A.R., 1972. Fluctuating rates of protistan evolution, diversification and extinction. *Int. Geol. Congr., 24th, Montreal*, 7: 205–213.

Uffen, R.J., 1965. The evolution of the interior of the earth and its effects on biological evolution. In: C.H. Smith and T. Sorgenfrei (Editors), *The Upper Mantle Symposium (New Delhi, 1964)*. Int. Union Geol. Sci., Copenhagen, pp. 14–19.

Worsley, T., 1971. The terminal Cretaceous event. *Nature*, 230: 318–320.

Part V

NANNOPLANKTON AND THE
CRETACEOUS/TERTIARY BOUNDARY

Editor's Comments
on Papers 34 Through 37

34   BRAMLETTE
Massive Extinctions in Biota at the End of Mesozoic Time

35   WORSLEY
The Terminal Cretaceous Event

36   THIERSTEIN and OKADA
Abstract from *The Cretaceous/Tertiary Boundary Event in the North Atlantic*

37   ROMEIN
Calcareous Nannofossils from the Cretaceous/Tertiary Boundary Interval in the Barranco del Gredero (Caravaca, Prov. Murcia, S. E. Spain).I

Bramlette and Martini (1964) first demonstrated the massive reduction that occurred in the diversity of calcareous nannoplankton at the end of the Maastrichtian. Bramlette's article (Paper 34) ascribed this profound reduction in marine biota to a depletion of nutrients in the oceans received via detritus from land. Tappan (1968) has postulated that this drastic change was brought about as a consequence of a decrease in atmospheric oxygen levels due to reduced productivity at the end of the Maastrictian and an increase in the solubility of CO_2 in seawater, resulting in greater acidity and having a detrimental effect on calcareous plankton. Worsley (Paper 35) concurs with Tappan and maintains that this increased acidity may have led to the rise of calcite compensation depth to the ocean surface.

Papers by Thierstein and Okada (Paper 36) and Romein (Paper 37) are examples of detailed study of the nannofossil record, straddling the Cretaceous/Tertiary boundary in open ocean and in land-based marine sections, respectively. Both these studies and the study by Percival and Fischer (1977) record a dramatic increase in the dominance of thoracosphaerids and braarudosphaerids (both of which are stress-adapted forms) associated with the boundary event, and indicating a marked deterioration of the environment.

Recently there has been a revival of interest in the dramatic event at the Cretaceous/Tertiary transition. Both terrestrial and extra-terrestrial causes have been hypothesized. Among the terrestrial causes is the recent suggestion of the Arctic spill-over, or injection, of fresh to brackish water to the world ocean, made by Gartner and Keany (1978) and Thierstein and Berger (1978). This hypothesis requires the isolation of the Arctic at the end of Cretaceous, turning it into a fresh to brackish water basin, and requires its reconnection to the North Atlantic by rifting between Greenland and Norway, spilling over fresher waters over the world ocean in the form of a thin layer of cooler, low-salinity water, causing extinctions in plankton and the disruption of the food chain.

Gartner and Keany (1978) used the presence of thick nannofossil-bearing Maastrichtian sequences below and above a thin layer of Danian nannofossils in a North Sea well as their "crucial" evidence. They suggested that the "Danian" nannofossils actually developed in the isolated Arctic during the late Maastrichtian, and their occurrence within the Maastrichtian chalk represents the intrusion of Arctic waters into the North Sea. Perch-Nielsen et al. (1979) have, however, refuted this explanation and show reason for reinterpreting the Maastrichtian sequence above the Danian layer as a simple debris flow.

Thierstein and Berger's (1978) evidence for the injection hypothesis was that there is a marked decrease in the oxygen and carbon isotopic values in the carbonate fine-fraction across the Cretaceous/Tertiary boundary at a DSDP site in the South Atlantic. This shift has not been documented elsewhere and the authors now believe that the isotope measurements were done on poorly preserved, unsuitable material. In addition, there is considerable doubt about the possibility of the isolation of the Arctic in the late Maastrictian (see discussion in Haq, 1981).

The discovery of enrichment over the background levels of iridium and other noble elements in Cretaceous/Tertiary boundary layers from various sections of the world led Alvarez et al. (1980) to suggest a meteoritic impact as the probable cause for this event. They propose a post-impact distribution of pulverized rock in the atmosphere for several years, causing darkened skies and suppression of photosynthesis, leading to a collapse of the food chain. Discovery of iridium concentrations in the boundary layers of other sections, both on land and in the deep sea, is strengthening the argument for the extra-terrestrial cause for the radical change in life at the Cretaceous/Tertiary boundary (see Haq, 1981, for discussion).

REFERENCES

Alvarez, L. W., W. Alvarez, F. Asaro, and H. V. Michel, 1980, Extraterrestrial Cause for Cretaceous-Tertiary Extinction, *Science* **208:**1095-1108.

Bramlette, M. N., and E. Martini, 1964, The Great Change in Calcareous Nannoplankton Fossils between Maastrichtian and Danian, *Micropaleontology* **10:**291-322.

Gartner, S., and J. Keany, 1978, The Terminal Cretaceous Event: A Geologic Problem with an Oceanographic Solution, *Geology* **6:**708-712.

Haq, B. U., 1981, Paleogene Paleoceanography: Early Cenozoic Oceans Revisited, *Oceanol. Acta,* Special Issue: 71-82.

Perch-Nielsen, K., K. Ulleberg, and J. E. Evensen, 1979, Comments on the Terminal Cretaceous Event—Geologic Problem with an Oceanographic Solution" (Gartner and Keany, 1978), in *Cretaceous-Tertiary Boundary Events,* vol. II, W. K. Christiansen and T. Birkelund, eds., University of Copenhagen, pp. 106-111.

Percival, S. F., and A. G. Fischer, 1977, Changes in Calcareous Nannoplankton in the Cretaceous-Tertiary Biota Crisis at Zumaya, Spain, *Evol. Theory* **2:**1-35.

Tappan, H., 1968, Primary Production, Isotopes, Extinctions, and the Atmosphere, *Palaeogeography, Palaeoclimatology, Palaeoecology,* **4:**187-210.

Thierstein H., and W. H. Berger, 1978, Injection Events in Earth History, *Nature* **276:**461-464.

34

Copyright ©1965 by the American Association for the Advancement of Science

Reprinted from *Science* **148**:1696–1699 (1965)

MASSIVE EXTINCTIONS IN BIOTA AT THE END OF MESOZOIC TIME

M. N. Bramlette

The profound and geologically abrupt changes in the earth's biota, particularly those marking the end of Paleozoic and of Mesozoic time, have been the subject of much speculative discussion. A review by Newell (*1*) is one of the more recent and comprehensive on this subject, but does not include the possible explanation suggested here. These great changes primarily reflect a geologically sudden extinction of many important elements of the earth's population—the extinction of some large populations thriving toward the end of Mesozoic time being more demonstrably abrupt than that at the end of the Paleozoic. However, the causes of these major events were not necessarily similar, since the physical condition of the earth and the populations of organisms most affected were not close-

The author is professor emeritus of geology at the Scripps Institution of Oceanography, La Jolla, California.

ly similar at these two times. The extinction at the end of the Mesozoic seems, in the fossil record, to be most obviously and strikingly reflected by the planktonic life (plant and animal) of the oceans and by the larger forms dependent on plankton, such as ammonites and belemnites, whose extinction at that time has long been recognized. That so much marine life became extinct solely from a lack of adequate nutrition seems an oversimplification, but evidence that such a lack may have been the critical factor under the probable environmental conditions of that time deserves some consideration.

Among previous explanations, the one suggestion that the extinction occurred as a result of excessive radiation from an exceptional cosmic event might seem intriguing because such radiation could have had widespread and nearly instantaneous effects on life. Loeblich

and Tappan (*2*) suggest that such radiation might have induced mutations and more extinctions in the planktonic than in the more protected benthonic foraminifera at the end of the Mesozoic. However, even a thin layer of surface water would serve as an effective blanket, according to Urey (*3*), and, as mentioned by Newell (*1*), the radiation would have affected land plants much more than the record indicates. The suggestion of a climatic change with a reduction in temperature seems to have little support, and the effects of such a change should likewise be most apparent in the fossil land plants rather than in the marine life. Changes in sea level may have adversely affected nearshore marine life, as Newell and others have advocated, but such changes should not have affected the plankton populations to the unusual degree that is evident.

Life in the oceans may include an

even greater number of individuals (although not of taxa) than that on land because of the much greater surface area and depth of the marine part of the biosphere. Evidence is here summarized that this vast planktonic life was drastically affected at the end of Mesozoic time, and that this effect might be expected from certain physical conditions of the earth at that time which resulted in a reduced supply of detritus, with the required nutrients, to the ocean.

There appears to have been a proliferation of population sizes, as well as of the taxa, among some of the plankton groups during late Mesozoic time. However, that limitation of the population according to Malthusian principle would culminate in wholesale destruction seems improbable without some superimposed adversity, such as a decrease in the supply of nutrients to the ocean waters. The volume of nutrients in the depths of the vast oceanic reservoir might appear nearly inexhaustible to the biologist, but it appears that the supply of nutrients from the ultimate source on land decreased over some millions of years. This condition, however, should not have considerably affected the much greater amount of major inorganic constituents, or salinity, of the oceans. Whether or not the amount of nutrients in the oceans could have decreased to a level below the threshold for support of much of the phytoplankton, as in some laboratory experiments, and thus have resulted in mass extinctions in a geologically brief episode, is a serious and difficult question. This question is considered below after a review of the record of distribution of the fossils and of the environmental conditions which seem significant to any interpretation.

Changes in the plant and animal life on land were at a more nearly normal rate during late Cretaceous time. Even the conspicuous example of the extinction of the dinosaurs is not demonstrably such a sudden and wholesale destruction of thriving populations. Perhaps, too, a degree of circular reasoning is involved in the case of the dinosaurs, as certain strata, for whose ages adequate supplementary paleontologic evidence is lacking, have been reassigned from Cenozoic to late Mesozoic when some dinosaur remains have been discovered in them. The dinosaur remains are sparse, and synchroneity of the enclosing strata is commonly not well established from other evidence.

Record of the Fossil Plankton

The succession shown in Fig. 1 for the distribution of calcareous nannoplankton is evident from the direct superposition of strata in Alabama, and is similarly clear for the Danian resting directly on the earlier strata in Denmark and France. The correlation and age assignments agree with those from most recent studies of the foraminifera and other groups of fossils. Samples containing vast numbers of nannoplankton (dominantly protophyta) from the indicated taxa were taken from within a few meters below and above the top of the Mesozoic strata (upper Maestrichtian and the equivalent in Alabama). The skeletal remains of the marine calcareous plankton (nannoplankton and planktonic foraminifera) constitute about one-half the total in these chalk formations—countless millions of the "nanofossils" (averaging less than 10 microns) occurring in a few cubic centimeters of the chalk. The distribution of the identified taxa of nannoplankton shown in Fig. 1 is fairly representative of other known regions.

A surprisingly similar distribution, with a comparable number of extinctions at this same time, of the taxa of planktonic foraminifera is shown by unpublished results of several investigators of this group of fossils, and is indicated in part by Bolli, Loeblich, and Tappan (4) and by Berggren (5). The sparsity of preserved fossils from other groups of plankton, such as the radiolarians, diatoms, and dinoflagellates or "hystrichospherids," in strata of these ages precludes an equally clear comparison, although the meager evidence does suggest that marked extinctions occurred at the end of the Mesozoic. Among the larger forms of marine life which have preservable hard parts and are dependent on the smaller plankton, the complete extinction at that time of the belemnites and of the large, diversified, and long-existent group of ammonites is well known.

The data on calcareous nannoplankton and planktonic foraminifera are now adequate to indicate this worldwide extinction of most of the distinctive taxa, and to show that the extinction of these large populations was so abrupt that the stratal record of transition still remains obscure. A record, even though an abbreviated one, will doubtless be found which shows diminished numbers of Cretaceous taxa and individuals associated with progenitors of the few early Cenozoic

forms. Although some stratal discontinuity is commonly found at this horizon, much evidence indicates that the hiatus was not a long one in geological time, particularly because any large record of deposition missing in some areas should be represented by sedimentation elsewhere. The hiatus thus may involve many thousands of years but probably much less than a million; comparable changes in the fossil record normally require some millions of years. Such a long period of existence during Mesozoic time is indicated for many of the planktonic taxa which became extinct at the close of that era. It required several million years also for the meager assemblages of the nannoplankton and planktonic foraminifera surviving into the earliest Cenozoic to develop diversification comparable to that found in the late Mesozoic.

Significance of Land Conditions

There is no dispute with the principle of uniformitarianism in the view that the relative rate or intensity of the normal processes produced very different net results over much of the earth for long periods of time. In late Mesozoic time it seems that there were environmental factors which should have greatly reduced the supply of nutrients to the oceans, nearly all of which must be derived from the land surface. The character of the nearshore sedimentary deposits of the late Mesozoic indicates an almost senile earth for that time—or better termed a "hibernating" earth, because of the return to unusual vigor during the late Tertiary to Recent time. Some of the evidence is summarized below which suggests that the earth was not stirring with the usual amount of orogeny, uplifts, and erosion with the resulting supply of detritus to the oceans. No complete change in these conditions is probable; only an appreciable diminution from normal need be assumed, because there now exist large oceanic areas, for instance in the central north Pacific, where nutrients are inadequate for a prolific microplankton.

No large regions of unusual aridity appear to have caused a reduction in the supply of detritus in the late Mesozoic—apparently there was at that time even less aridity indicated by the strata than is normal for the earth. With normal rainfall and stream-flow from topographically reduced land surfaces,

Europe (Type Areas and SW-France)

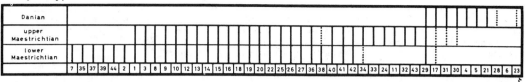

North America (Alabama)

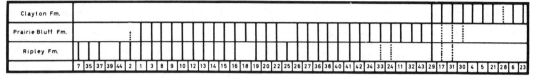

Fig. 1. Distribution of calcareous nannoplankton in stratigraphic arrangement, showing change at top of Maestrichtian, and equivalent in Alabama. Numerals correspond to names of taxa as given by Bramlette and Martini (9), where this illustration first appeared.

the amount of eroded detritus would be reduced, and in time the resulting old soils would have been depleted of nutrients, including the "soil extracts" containing such organic compounds as thiamine and vitamin B_{12}.

There is evidence over large parts of the earth that erosion and supply of detritus (presumably including the unrecognizable "soil extracts" which are so valuable in plankton cultures) were abnormally low in late Mesozoic time, although a stratal record of that particular time is lacking or is covered by later deposits in yet larger areas.

Most of northern Europe has marine chalk deposits of late Mesozoic age, indicating relatively little supply of land-derived detritus into these marginal sea deposits, and the same is true for much of southern Europe and northern Africa. Chalk accumulation around much of the Gulf of Mexico at this time likewise implies that little detritus was derived from most of the central part of North America, and a similar condition is indicated for western Australia.

Some regions with marine deposits of this age have extensive accumulations of richly glauconitic sediments rather than chalk, and these, too, indicate that the supply of detritus was reduced, so that the slowly accumulating glauconite formed a large proportion of the sediments. Although occurring in strata of many places and ages, such highly glauconitic sediments are conspicuous in most of the strata of latest Mesozoic age along the east coast of the United States, and even in considerable areas of "never quiet" California. In the Crimea, W. A. Berggren reports (6), "The contact is characterized

by abundant glauconite, immediately underlain by one to three meters of glauconitic sands with abundant oyster remains." In the major geosyncline of northern South America, Hedberg (7) reports glauconitic sediments only in the formation assigned to the latest Mesozoic and early Cenozoic, which suggests a reduced supply of sediment even in such a trough of generally rapid accumulation. Marine strata of this age in western Equatorial Africa are reported by Reyment (8) to be interbedded with extensive coal deposits, which may be evidence of some reduction of sedimentation.

Search of the literature for adequate descriptions of other areas in the latest Mesozoic and earliest Cenozoic remains to be done. Certainly there are some regions with deposits of this age, however, which indicate very active erosion and sedimentation, but, to repeat, only a subnormal supply from large areas of land should in time affect all the open oceans. One example of very active uplift and erosion during this period is known in the Rocky Mountain region, but most of the detritus accumulated in the same general region, much of it in non-marine basins, and presumably relatively little of that part supplied to the adjacent inland sea would have reached the open oceans.

Land conditions that evidently resulted in abnormally low supplies of detritus to many parts of the oceans should thus with time have affected the available nutrients of the entire volume of the ocean waters. The difficult problem is posed, however, as to whether a long period of decreasing supply could

culminate in a sudden extinction of much of the protophyta which formed the base of important food chains throughout the oceans. The relation of the time factor of geological events to that of laboratory experiments is too commonly an imponderable one.

Laboratory Experiments and Generalities

Some laboratory experiments show that cultures of phytoplankton may thrive in normal growth and rate of proliferation as they consume unreplenished nutrients in the medium until threshold conditions are reached that cause a sudden death of all. This likewise occurs in some cultures which include populations of more than one group, because the requirements, or limiting ones, are similar for many forms. Whether such results could be meaningful for the vast expanse of oceans would certainly seem very doubtful, except that some thousands of years for extension of the results upon plankton life would appear almost as brief in the geological record as the time involved in the laboratory experiments.

Much evidence indicates that the oceans were more uniform, at least with respect to surface water temperatures, in the late Cretaceous than at present, probably with an associated decrease in intensity of upwelling and other currents, and thus the conditions were then somewhat more comparable to those of a laboratory culture. If the indicated conditions could have resulted in widespread threshold effects on most of the phytoplankton, the disastrous consequences for many higher forms of life in the food chain of the oceans would surely have followed in a geological time so brief as to appear synchronous.

Consideration of this problem obviously should include data on other groups of fossils, but information on the significant aspects seems inadequate, or needs analysis by specialists on these fossils. The interesting histograms of Newell (*1*), showing the changes with time in the number of families within larger groups, suggest the magnitude of the event at the end of the Mesozoic. Possible causes should be reflected better, however, by an analysis of such changes shown in populous groups of similar habitat within these higher phylogenetic groupings. For example, Newell's histograms show the extinction of all the many families of ammonites at the end of the Mesozoic, a phenomenon which appears to be of particular causal significance because all of these ammonites seem to have belonged to the marine nekton. In contrast, his histogram on families of foraminifera shows little change at this time because the very populous planktonic taxa which became extinct are classed in only a few of the many families considered. The families of crinoids plotted likewise include family groups of planktonic, benthonic, and deep- and shallow-water habitats, and separate consideration of these groups should prove interesting.

Among the benthonic "shallow-water" forms, the *Rudistae* are conspicuous as a large group which became extinct at the end of the Mesozoic. Perhaps it is significant that the sessile rudistids seem most commonly to have existed on a calcareous substrate, which suggests a water environment with detritus and nutrient supply more nearly comparable with the open ocean than with that of those nearshore organisms which lived on a substrate of clastic detritus. The latter environment should have had a more nearly adequate food supply near shore even if less food reached the open oceans. One test of this possibility might be whether those taxa of corals commonly associated with rudistids were comparably affected.

Certain geochemical tests could, perhaps, prove significant for this or some other explanation—possibly by revealing differences between critical minor elements in clays or in the phosphatic skeletal remains of fish of the latest Maestrichtian and the clays or skeletal remains from earlier Maestrichtian strata of the same area—if diagenetic changes have not obscured any original differences in these constituents. In any case, some aspects of this discussion seem to bear on the ultimate solution of this intriguing and important problem in earth history.

References and Notes

1. N. D. Newell, *J. Paleontol.* **55**, 592 (1962).
2. A. R. Loeblich and H. Tappan, *Geol. Soc. Amer. Bull.* **75**, 386 (1964).
3. H. C. Urey, personal communication.
4. H. M. Bolli, A. R. Loeblich, Jr., H. Tappan, *U.S. Nat. Museum Bull.* **215**, 22 (1957).
5. W. A. Berggren, *Intern. Geol. Congr., 21st, Copenhagen, 1960, Rept. Session Norden*, pt. 5, 181 (1960). ·
6. W. A. Berggren, personal communication.
7. H. D. Hedberg, *Geol. Soc. Amer. Bull.* **48**, 1995 (1937).
8. R. A. Reyment, personal communication.
9. M. N. Bramlette and E. Martini, *Micropaleontology* **10**, 295 (1964).
10. I thank my colleagues, W. R. Riedel and F. L. Parker, for valued criticisms and suggestions.

Appendix

An explanation should be added on the usage of certain terms involved in the rules of stratigraphic nomenclature. The abrupt extinctions of a surprisingly large part of Mesozoic marine life took place prior to deposition of the strata of the Danian Stage and are commonly interpreted as occurring at the end of Mesozoic time. Others, however, place the Danian in the latest Mesozoic—in which case this marked change would have occurred within latest Mesozoic time. Paleontologic aspects of this problem have received extensive review recently by Berggren (*1*).

The Danian was originally assigned by Desor (1846) to the Mesozoic (*2*). Rules of priority in nomenclature are essential, and Desor's name for the Danian Stage should remain fixed. Such rules are not applicable, however, to Desor's placement of these Danian strata in the Cretaceous (upper Mesozoic). Precise limits at a type locality were not indicated by d'Halloy (1822) in his original designation of the Cretaceous (*3*), but the indicated Cretaceous strata in France have been generally accepted as including equivalents of the strata of the type Maestrichtian, which therefore was included as an upper stage of the Cretaceous. Priority would place the Danian Stage as an uppermost stage of the Cretaceous only if evidence indicated that the type Danian strata were equivalent in age to part of the chalk of the originally designated Cretaceous in France—and there seems little or no evidence for this. The Cretaceous System and other time-stratigraphic units would have little meaning as such if strata placed in them by correlations extended to the type locality from elsewhere were not subject to necessary adjustments, when and if justified by additional evidence (which, it is hoped, will eventually include consistent radiometric evidence).

References

1. W. A. Berggren, *Stockholm Contrib. Geol.* **11**, 5, 103 (1964).
2. E. Desor, *Bull. Soc. Geol. France*, ser. 2, **4**, 179 (1846).
3. J. J. O. d'Halloy, *Ann. Mines* **7**, 373 (1822).

35

Reprinted from *Nature* **230**:318–320 (1971)

THE TERMINAL CRETACEOUS EVENT

Thomas R. Worsley

Department of Oceanography, University of Washington,
Seattle, Washington

SEVERAL explanations have been offered (see refs. 1–3) for the abrupt faunal extinctions at the end of the Cretaceous and, with the advent of the JOIDES Deep Sea Drilling Project, it was hoped that the nature of the extinctions—at least for calcareous microfossils—would be found by coring a transitional sequence across the boundary in deep sea facies. Unfortunately, the results from the few JOIDES holes so far penetrating the Cretaceous–Tertiary boundary indicate that the unconformity is even greater in the deep ocean basins than on the continents and that a transitional sequence will probably never be found, especially in calcareous pelagic sediments. Although disappointing on first inspection, these data nevertheless provide the most tangible and significant clue yet as to the nature of the terminal Cretaceous event.

The late Maastrichtian was a time of general worldwide decrease of temperature with climatic belts becoming more sharply differentiated[4–9]. It was also a time of greatly reduced clastic influx into the oceans resulting from almost worldwide orogenic quiescence[1,2,10]. The only major orogeny was the Laramide Revolution in the western interior of North America, the sediments of which apparently never reached marine environments except in the restricted interior sea in which the Lance Group was deposited. It therefore seems likely that the Laramide Revolution, the sediments of which hold the key to terrestrial-marine correlations at the time of the Cretaceous–Tertiary event, was not a significant contributor of carbonate and nutrients to the world's oceans, to which the supply of nutrients and carbonate was then severely restricted[2,10].

These sediments do, however, preserve a remarkably complete record of megaphyta and dinosaurs that adds considerably to knowledge of the effects of the Cretaceous–Tertiary event on land. Although the megaphyta are thought to have been little affected by the event, Hall and Norton[6] described a palynologically significant change across the boundary in the North–Central United States (Larimide sediments) showing rapid replacement of thermophilic dicotyledons by temperate gymnosperms just above the highest dinosaur remains. There is no apparent sedimentary break at the locality. The area is especially significant because the lignite in which the Cretaceous–Tertiary boundary is preserved is laterally equivalent to the lignite just below the base of the fossiliferous marine Cannonball Formation of North Dakota, where there is a nearly continuous Cretaceous–Tertiary sequence[8,11]. Although calcareous nannofossils are virtually absent in the Cannonball, attesting to its restricted marine environment, rare planktonic foraminifera document its lowest Danian age[12].

In open marine environments, calcareous shelf sections in which rocks of latest Maastrichtian age are known everywhere present similar lithologies. The strata are always rich in glauconite and contain significant amounts of phosphate, both indications of slow deposition[2]. Most of these Upper Maastrichtian marl or chalk beds contain little detritus and many are rich in planktonic foraminifera and nannofossils, an indication that the water column above them was neither hypersaline nor brackish. There are usually hardgrounds or

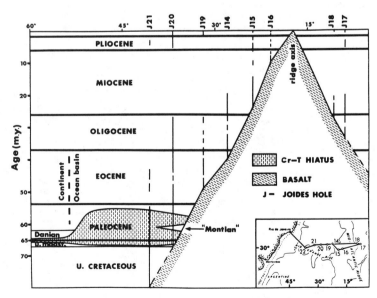

Fig. 1 Age relationships and present configuration of sediments draped across the mid-Atlantic Ridge at about latitude 30° S, showing the magnitude of the Cretaceous Tertiary Unconformity.

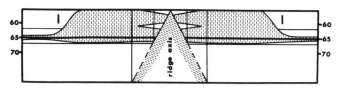

Fig. 2 Late Palaeocene palinspastic reconstruction of Fig. 1. Data are duplicated for the area east of the ridge (where no data exist) for symmetry.

phosphatic–glauconitic seams containing borings at the top of these strata[2].

There is good reason to believe that the unusual lithology at the boundary reflects submarine solution of carbonate at the end of the Cretaceous in all oceans; in other words, carbonate compensation depth (CCD) approached or reached the surface of the oceans at this time, as suggested by Tappan[2]. Recent deep sea drilling[13] substantiates this view by demonstrating a vertical migration of CCD in the Atlantic throughout the Tertiary of more than 1.5 km. Fig. 1 is a diagram intended to show the age relationships and present configuration of sediments draped over the southern part of the Mid-Atlantic Ridge at about latitude 30° S (ref. 14). Information about the Cretaceous–Tertiary boundary has been obtained only from sites J–20 and J–21, so that the Cretaceous–Tertiary unconformity plotted is a schematic lateral extrapolation of the drilling data. The data for the continent are also schematic in that they are not intended to represent South America alone but rather a composite of shelf sections from all continents. If the meagre data for the South Atlantic have been interpreted correctly, there would be a similar cross-section for all ocean basins where there has been seafloor spreading since the late Cretaceous.

Fig. 2 is an Upper Palaeocene palinspastic reconstruction of Fig. 1, which shows that the more continuous deep sea record of the Maastrichtian–Palaeocene interval occurs on the Maastrichtian–Danian Mid-Atlantic Ridge using calculated spreading rates[14]. This configuration is reasonable because the shallower ridge areas should have remained above CCD longer than the deep ocean basins but not as long as the shallower marine shelves. The available data suggest that the descent rate of points on the seafloor down the flanks of the Mid-Atlantic Ridge outstripped that of the downward migration of CCD after the Cretaceous–Tertiary event. Thus site J–20, which was almost at the crest of the Ridge in the late Maastrichtian, experienced a pause in calcareous sedimentation as the CCD rapidly migrated toward the surface during the Cretaceous–Tertiary event but received calcareous sediment again shortly after the event as the CCD migrated below the ridge crest. The time during which CCD was above the ridge top represents most of the Danian. Site J–21, which also contains Upper Maastrichtian calcareous sediments, was apparently lower on the ridge flank and never received calcareous sediment until the Upper Palaeocene, when the CCD finally descended low enough to permit calcareous sedimentation in the area of the Rio Grande Rise. The magnitude of the Cretaceous–Tertiary hiatus in the deep sea is therefore a function of palaeobathymetry with deeper water sections exhibiting a greater unconformity.

I will now outline a model of the terminal Cretaceous event in which most of the abrupt faunal and floral extinctions are direct consequences of the late Maastrichtian environmental conditions.

Before the Jurassic, $CaCO_3$ eroded from the continents was precipitated on the continental shelves by benthonic organisms and only a very small proportion was permanently lost to the deep ocean basins by the agencies of pelagic organisms with calcareous skeletons. With the advent of calcareous plankton in the Jurassic, a mechanism was provided for the removal of carbonate from the orogenic cycle, and the effects were first markedly apparent during the late Cretaceous, when there was no replacement from continents that were nearly base-levelled[1,10].

There followed during the late Maastrichtian a pronounced climatic deterioration, and among marine organisms, belemnites, several species of planktonic foraminifera and at least two species of calcareous nannofossils became climatically restricted[8,9]. Moreover, more rapidly evolving nannofossil taxa were selectively replaced by slowly evolving forms[15]. On the continents, cycads and other thermophilic floral elements were being replaced in middle latitudes by temperate conifers and hardwoods, with a concurrent decline of reptiles[5–7].

It is suggested that the cause of this deterioration was a decrease of CO_2 in the atmosphere brought about primarily by the late Cretaceous phytoplankton bloom which was responsible for the worldwide deposition of chalk[2]. Other mechanisms for converting CO_2 into O_2 involved the rise of the photosynthetically efficient angiosperms[16] and possibly the removal of carbon from the atmospheric cycle by lignite formation[17]. There is no method available for testing directly whether changes in CO_2 content in the atmosphere entail climatic changes, but Plass[18] has calculated that present temperatures would drop about 4° C if the CO_2 content of the atmosphere were reduced by 7%. The same forces would also reduce cloudiness and tend to accentuate the effects of climatic deterioration by making less effective the means of heat transfer between poles and equator, thus making all areas of the globe more dependent on direct insolation. Moreover, lower worldwide temperatures should decrease precipitation and terrestrial weathering and erosion rates, reducing still further the restricted supply of nutrients and carbonate available to marine phytoplankton.

With continuing climatic deterioration in the Maastrichtian, polar cooling of seawater would have increased the horizontal and vertical oceanic thermal gradients[3] and this would have increased the solubility of CO_2 in higher latitudes and deeper water. Ultimately, a CCD would have developed in the deep ocean basins and migrated to progressively shallower levels[2,13]. The absence of early Danian fossils in deep sea cores penetrating the Cretaceous–Tertiary boundary in calcareous sediments is evidence of the process[13,14,19,20]. Together with the CO_2 increase in the ocean, the lack of detritus reduced the effects of silicate buffering during this time[2], resulting in a slight lowering of *p*H. Evidence of this is found in numerous deep sea and shelf phosphate–glauconite layers and hardgrounds in this part of the column, which suggest submarine solution of carbonate.

The upward migration of the CCD through the late Maastrichtian suggests that the most continuous biological record across the Cretaceous–Tertiary boundary should indeed be found on the shallowest portions of shelves favourable to growth of calcareous nannoplankton. The CCD is supposed eventually to have approached the surface of the ocean and for the first time to have directly affected planktonic organisms[2,13], and this is the time of the massive extinctions that ended the Cretaceous. In the marine realm, the sharp phytoplankton reduction would have been associated with extinctions of dependent marine taxa throughout the pelagic food chain. Parenthetically, it is interesting that the sparse nannoflora surviving the terminal Cretaceous event were dominated by *Braarudosphaera* and *Thoracosphaera*, both of which are known to tolerate and even to prefer conditions adverse for the growth of other calcareous nannoplankton[21].

On land, the latitudinal thermal gradient together with the sharper seasonal and probably diurnal temperature differential readily explains the relatively abrupt extinction of the last surviving species of dinosaurs[5]. The megaphyta apparently suffered few extinctions across the Cretaceous–Tertiary boundary, but dicotyledons were abruptly shifted toward the Equator in response to fairly rapid chilling of middle latitudes[6,7].

The mechanism of recovery from this event is also a part of the model. The removal of a major portion of the phytoplankton would have severely curtailed the photosynthetic conversion of CO_2 to O_2 and of carbonate precipitation in the oceans, finally releasing CO_2 from the oceans into the atmosphere. This increase of atmospheric CO_2 would lead to amelioration of climate and the downward migration of the CCD. The oceans would, however, remain relatively sterile with a somewhat elevated CCD throughout the lower Palaeocene until the supply of nutrients and carbonate was again renewed by orogeny[2].

Received December 23, 1970.

[1] Newell, N., *J. Paleontol.*, **36**, 592 (1962).
[2] Tappan, H., *Palaeogeog., Palaeoclimatol., Palaeoecol.*, **4**, 187 (1968).
[3] Lipps, J., *Evolution*, **24**, 1 (1970).
[4] Lowenstam, H., in *Problems in Paleoclimatology* (edit. by Nairn, A. E. M.), 227 (1963).
[5] Axelrod, D., and Bailey, N., *Evolution*, **22**, 595 (1968).
[6] Hall, J., and Norton, J., *Palaeogeog., Palaeoclimatol., Palaeoecol.*, **3**, 121 (1967).
[7] Stanley, E., *Bull. Amer. Paleont.*, **49**, 179 (1965).
[8] Jeletsky, J., *Twenty-first Geol. Cong.*, part 5, proc. sect. 5, 25 (1960).
[9] Worsley, T., and Martini, E., *Nature*, **225**, 1242 (1970).
[10] Bramlette, M., *Science*, **148**, 1969 (1965).
[11] Fox, S., and Ross, jun., R., *J. Paleontol.*, **16**, 660 (1942).
[12] Fox, S., and Olsson, R., *Bull. Amer. Assoc. Petrol. Geol.*, **53**, 734 (1969).
[13] Hay, W., in Bader *et al.*, *Initial Reports of the Deep Sea Drilling Project*, **4**, 672 (1970).
[14] Maxwell, A., *et al.*, *Initial Reports of the Deep Sea Drilling Project*, **3** (1970).
[15] Worsley, T., thesis, Univ. Illinois (1970).
[16] Schindewolf, O., *Neues Jahrb. Geol. Paleontol., Monatsh.*, **10**, 457 (1954).
[17] Schwartzbach, M., *Climates of the Past*, 328 (1963).
[18] Plass, G., *Tellus*, **8** (1956).
[19] Ewing, M., *et al.*, *Initial Reports of the Deep Sea Drilling Project*, **1** (1969).
[20] Peterson, M., *et al.*, *Initial Reports of the Deep Sea Drilling Project*, **2** (1970).
[21] Hay, W., and Mohler, H., *J. Paleontol.*, **41**, 1505 (1967).

36

Reprinted from page 601 of *Initial Reports of the Deep Sea Drilling Project*
43:601–616 (1979)

THE CRETACEOUS/TERTIARY BOUNDARY
EVENT IN THE NORTH ATLANTIC

Hans R. Thierstein, Scripps Institution of Oceanography,
University of California, San Diego, La Jolla, California
and
Hisatake Okada,[1] Department of Geology,
Faculty of Science, Yamagata University, Yamagata, Japan

ABSTRACT

Paleomagnetic evidence suggests that the Cretaceous/Tertiary transition at DSDP Site 384 is continuous. Quantitative taxonomic analysis of the nannolith assemblages indicates that all taxa within the Cretaceous assemblage became extinct simultaneously. The evolutionary sequence of the Tertiary assemblages at Site 384 is characterized by an initial dominance of *Thoracosphaera, Zygodiscus sigmoides* and *Markalius astroporus*, a following *Braarudosphaera* bloom, and a subsequent gradual increase of *Cruciplacolithus primus* and *C. tenuis.* The preserved nannolith record can be explained by benthic mixing with an incompletely homogenized mixed layer of 21 cm in thickness. The carbonate record in the deep North Atlantic documents a major excursion (> 2 km) of the carbonate compensation surface to abyssal paleodepths from middle Maestrichtian to middle Danian time. Comparison with the scanty sedimentary evidence from the deep Pacific and Indian oceans suggests a late Mesozoic deep water fractionation between the Pacific and the Atlantic oceans with deep circulation reversals just prior and subsequent to the evolutionary Cretaceous/Tertiary boundary extinction event.

---

[1]Formerly at Lamont-Doherty Geological Observatory, Palisades, New York.

37

Copyright ©1977 by the Koninklijke Nederlandse Akademie van Wetenschappen

Reprinted from *Koninkl. Nederlandse Akad. Wetensch. Proc.* ser. B, **80**:256-279 (1977)

CALCAREOUS NANNOFOSSILS FROM THE CRETACEOUS/
TERTIARY BOUNDARY INTERVAL
IN THE BARRANCO DEL GREDERO
(CARAVACA, PROV. MURCIA, S.E. SPAIN).   I

BY

A. J. T. ROMEIN

ABSTRACT

The calcareous nannofossil assemblages are described from a sequence of closely spaced samples from the Gredero section, covering the Cretaceous/Tertiary boundary interval. An emended zonation for this interval is proposed. Additional evidence is given of provincialism and diachronous appearance levels of species in the Late Maastrichtian and Danian. Climatic deterioration plays an important role in the model which is presented to explain the terminal Cretaceous event. One new species is described, *Crepidolithus fossus*.

INTRODUCTION

Publications dealing with calcareous nannofossil assemblages in continuously exposed sections across the Cretaceous/Tertiary boundary are still very scarce. This is mainly due to the fact that most sections show considerable hiatuses in the boundary interval (Worsley, 1971, 1974). The most detailed information on the subject was given by Worsley (1974) in the report on the nearly complete Braggs section in Alabama.

One of the most complete sections across the boundary in Europe is to be found in the Barranco del Gredero near Caravaca (SE Spain), where a 225 meters thick sequence of Upper Maastrichtian to Lower Eocene sediments is continuously exposed. Von Hillebrandt (1975) studied the planktonic and larger foraminifera of this sequence and presented a zonation. The planktonic and benthonic foraminifera of the boundary interval were described by Abtahi (1975). He demonstrated the presence of the topmost Maastrichtian and the lowermost Danian planktonic foraminiferal zones, the *Abathomphalus mayaroensis* Zone and *Globigerina eugubina* Zone respectively.

The boundary interval was recently sampled in more detail (Smit, 1977) in order to obtain maximum information on the lithological, faunal and floral changes. The results of the study of the calcareous nannoplankton assemblages in these samples are given in the present paper. It is the first in a series on the distribution and evolution of calcareous nannoplankton in the Paleogene of S.E. Spain (in preparation).

For a detailed description of the locality and the lithology of the section the reader is referred to Smit (this volume).

ACKNOWLEDGMENTS

Thanks are due to J. Smit for providing the samples and for fruitful discussions. I am grateful to C. W. Drooger, R. R. Schmidt and J. W. Verbeek for their constructive criticism, to J. v. d. Linden and A. v. Doorn for drawing the figures, and to C. Bakker of the Laboratorium voor Elektronenmikroskopie of the University of Amsterdam for his assistance at the SEM.

METHOD OF STUDY; PRESERVATION

Altogether 36 of the samples collected by Smit were selected, covering an interval of 17 meters. A small amount of sediment ($\pm \frac{1}{2}$ cm^3) was chipped off or scratched off these samples and allowed to desintegrate in distilled water; then smear slides were prepared.

The following method was used to get an impression of the composition of the assemblages. On every slide the first 200 nannofossils were counted and the relative abundance of each species was recorded. It took about half an hour to count up to 200, but it turned out that in this way only the most common species were noted; therefore another half hour was spent on each sample to detect the rare species. Their frequencies were recorded on the range-chart as less than a half percent.

Broken specimens were included in the counts if preserved for more than 50%. Of some species (*Thoracosphaera operculata, Thoracosphaera saxea* and *Toweius petalosus*) only the spheres were counted; fragments of these species were indicated with P (present) or A (abundant).

Indeterminable specimens were not included in the final counts, as their percentage adds no information on the composition of the assemblages but only on the state of preservation. During the pre-investigation the percentage appeared to fluctuate around 20%.

All samples contained well preserved nannofossils which were easy to determine in the light microscope. SEM investigations of selected specimens (method described by Hansen *et al.*, 1975) revealed various degrees of secondary calcite overgrowth.

ZONATION

A good definition of the boundary between the uppermost Cretaceous and the Towermost Tertiary nannofossil zones is still lacking. The usual criterium to indicate this boundary is the "somewhat metaphysical" (Cepek and Hay, 1966) extinction level of Cretaceous species. The disadvantage of this definition is obvious: by their small size nannofossils are

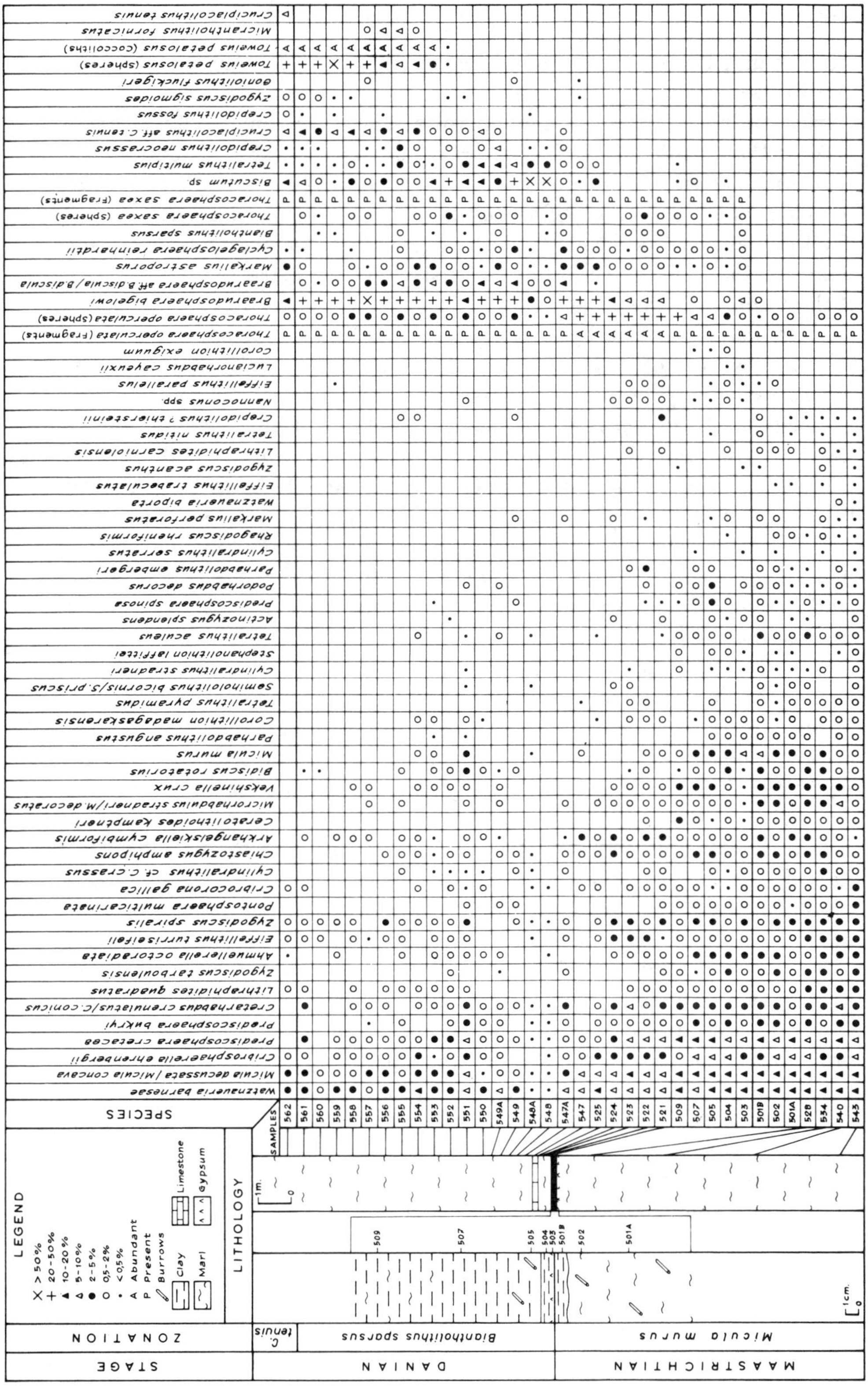

Fig. 1. Distribution of calcareous nannofossils in the Cretaceous/Tertiary boundary interval in the Gredero section.

315

easily exposed to reworking which makes this "level" of dubious correlation value.

In a study of calcareous nannofossil assemblages from Early Tertiary rocks at Pont Labau (S.W. France), Hay and Mohler (1967) introduced the *Markalius astroporus* Zone as the lowest zone in the Tertiary. The lower boundary of the zone was defined by the first occurrence of *M. astroporus*. The same authors assigned the Fish Clay and the immediately overlying chalks in the Holtug section at Stevns Klint, Denmark (cotype locality of the Danian) to this zone.

Later investigations by Perch-Nielsen (1969) of assemblages from Danish localities of Maastrichtian and Danian age indicated that *M. astroporus* already occurs in Maastrichtian sediments. She nevertheless retained the *M. astroporus* Zone as the lowermost Danian zone, but she based the lower boundary of the zone on the entry of *Biantholithus sparsus*.

The *M. astroporus* Zone was emended by Martini (1971) in his standard Tertiary and Quaternary calcareous nannoplankton zonation. He based the lower boundary of the *M. inversus (= M. astroporus)* Zone on the exit of *Arkhangelskiella cymbiformis* and other Cretaceous species. In the same year Perch-Nielsen introduced the *B. sparsus* Zone to replace the *M. inversus* Zone.

Edwards (1973) proposed the *Conococcolithus panis* Zone as the basal zone in the Danian in the report on the calcareous nannofossils recovered during Leg 21 (S.W. Pacific) of the Deep Sea Drilling Project. He based the lower boundary of this zone on the appearance of *C. panis*, and he recognized the zone also in the type Danian. *C. panis* might, however, be conspecific with the Cretaceous species *Biscutum constans* (Gorka) (Edwards, personal communication, 1976) and therefore the usefulness of the zone becomes questionable.

Bramlette and Martini (1964) already mentioned the abundance of long ranging species of the genera *Thoracosphaera* and *Braarudosphaera* in the Lower Danian of many regions. This abundance was confirmed in other investigations (Martini, 1964; Hay and Mohler, 1967; Perch-Nielsen, 1969; Edwards, 1973) and it was used by Haq and Lohmann (1976) to characterize Early Paleocene assemblages. It is equally apparent in the Gredero section and it is used in the zonation proposed below.

ZONATION

Micula murus Zone

Authors: Bukry and Bramlette, 1970, emend. this paper.

Diagnosis: Interval from the entry of *Micula murus* (Martini) to the entry of *Biantholithus sparsus*.

Distribution: Gredero section, samples 543–501b.

Assignment: Upper Maastrichtian.

Remarks: Only the upper part of this zone is present in the investigated sequence.

Biantholithus sparsus Zone

Author: Perch-Nielsen, 1971, emend. this paper.

Diagnosis: Interval from the appearance of *Biantholithus sparsus* to the entry of *Cruciplacolithus tenuis.*

Description: The zone is furthermore characterized by the high numbers of *Thoracosphaera operculata, Biscutum* sp., *Braarudosphaera* bigelowi and *Toweius petalosus,* accompanied by less frequent *Markalius astroporus.* Other typical species in this zone are *Thoracosphaera saxea, Braarudosphaera discula, Goniolithus fluckigeri, Crepidolithus neocrassus* and *Tetralithus multiplus.*

In addition to *B. sparsus* the following species appear in this zone *Braarudosphaera* aff. *B. discula, Cruciplacolithus* aff. *C. tenuis, Biscutum* sp., *Crepidolithus fossus* n.sp., *Neochiastozygus denticulatus, Zygodiscus sigmoides* and *Toweius petalosus.*

Reference locality: Barranco del Gredero, 3 km. S. of Caravaca, S.E Spain, samples 503–561. The zone has a thickness of 8 m. at this locality.

Assignment: Lower Danian.

The thickness of this zone in the Gredero section is rather large in comparison with the 4.5 m. reported from Alabama, along Dallas County Route 59 (Hay and Mohler, 1967).

Cruciplacolithus tenuis Zone

Authors: Mohler and Hay in: Hay et al., 1967.

Diagnosis: Interval from the first occurrence of *Cruciplacolithus tenuis* to the entry of *Fasciculithus tympaniformis.*

Only the uppermost sample of the investigated Gredero interval belongs to this zone.

Assignment: Lower Paleocene.

BIOSTRATIGRAPHY

The following concise account of the successive assemblages in the section is based on the range chart of figure 1.

Samples 543–501 b

These samples contain rich, highly diversified floras, typical for the Upper Maastrichtian *Micula murus* Zone. This zonal assignment is indicated by the presence of *M. murus, Lithraphidites quadratus* and *Ceratolithoides kamptneri.* Commen species in the assemblages are *Watznaueria barnesae, Micula decussata, Micula concava* and *Prediscosphaera cretacea. Kamptnerius magnificus* is absent in all samples, which might be due to the preference of this species for shallow water (Perch-Nielsen, 1972, p. 1023).

Sample 503

The assemblage in this sample differs strikingly from the previous ones by the low number of specimens per unit quantity of sediment. Yet the composition of the flora is still very well comparable with the foregoing ones. It differs only by the increased frequency of *B. bigelowi* and the presence of *B. sparsus*, *M. astroporus*, *T. saxea* and *Cyclagelosphaera reinhardtii* (Perch-Nielsen) nov. comb.

The sudden appearance of several species suggests a hiatus between this sample and sample 501b. Any statement on the extent of the hiatus is speculative, as the assumed gap is below the resolution level of any existing zonation. Yet, it is thought that the hiatus is small because of the considerable thickness of both the Upper Maastrichtian (at least 25 m) and the *Biantholithus sparsus* Zone (8 m).

The Cretaceous/Tertiary boundary is placed immediately below the level of sample 503.

Samples 504–509

The assemblages in this interval are again rich and well diversified. They are for more than 90% composed of species already present in the underlying Maastrichtian. The presence of *B. sparsus* could not be established, but *M. astroporus*, *C. reinhardtii* and *T. saxea* do occur.

Sample 521–561

The assemblages in this interval have low diversities. They show dominance waves of *T. operculata*, *B. bigelowi*, *Biscutum* sp. and *Toweius petalosus*. The percentage of typical Cretaceous species decreases rather sharply in the lower part of this interval. All samples are assigned to the *B. sparsus* Zone.

The increased frequencies of *W. barnesae*, *M. concava/M. decussata* and *P. cretacea* in sample 551 probably indicate a larger degree of reworking than in the other samples of this interval.

Sample 562

Toweius petalosus and *Cruciplacolithus tenuis* are the most important species in this sample which, based on the entry of the latter is assigned to the *Cruciplacolithus tenuis* Zone.

Final remarks

As may be seen from the range-chart, almost every Maastrichtian species continues into the *B. sparsus* Zone. This range extension is probably largely due to reworking. As already stated by Perch-Nielsen (1969) it is almost impossible to indicate which species are to be regarded as reworked and which are the ones that survived. Yet I am inclined to regard *C. reinhardtii*, *C. neocrassus* and *T. multiplus* as survivors, in addition to *T. operculata*, *B. bigelowi* and *M. astroporus*.

BOUNDARY PROBLEM

A problem arose with regard to the positioning of the Cretaceous/

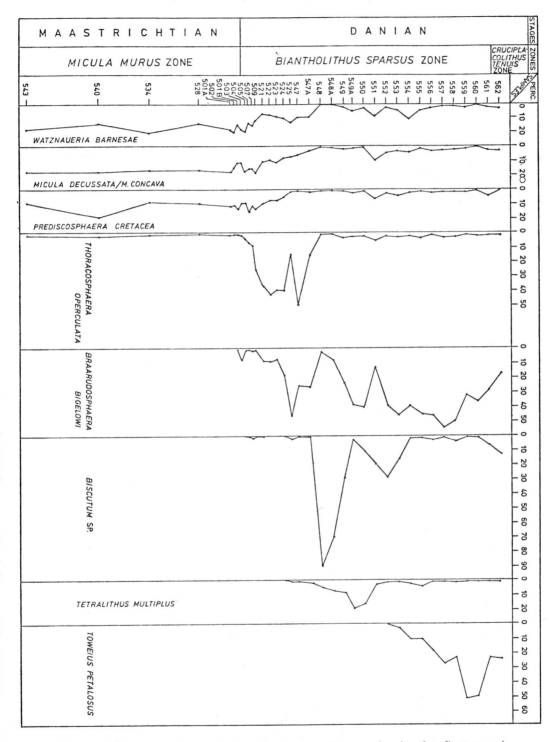

Fig. 2. Frequency curves of abundantly occurring species in the Cretaceous/ Tertiary boundary interval in the Gredero section.

Tertiary boundary. It was finally placed between sample 501b and 503 because of the first occurrence of *B. sparsus* in the latter sample. The assemblages in the interval just above this boundary (samples 504–509, of which 504–507 fall within Smit's Intermediate Zone) however, are nearly entirely composed of "typical" Cretaceous species. The next higher sample that could without any doubt be assigned to the *B. sparsus* Zone is sample 521 (indicated by the high frequencies of *T. operculata* and *B. bigelowi* and the presence of *B. sparsus* and *M. astroporus*). Thus, an interval is recognized which should be assigned to the Tertiary because of its stratigraphic position above the entry level of *B. sparsus*, but which belongs to the Cretaceous on the basis of the composition of the floras. In my opinion the assemblages in this interval are for the larger part (with the exception of *M. astroporus*, *T. saxea* and *C. reinhardtii*) the result of reworking. This supposition seems to be supported by the occurrences of *Nannoconus* spp. and *Lucianorhabdus cayeuxii* which species must have been derived from still older Cretaceous strata.

Acmes in the *Biantholithus sparsus* Zone

Several species show very high frequencies in the *B. sparsus* Zone (fig. 2). The first species showing an acme in this interval is *T. operculata*, which forms 42% of the assemblage in sample 522. An even higher percentage (49%) is reached in sample 547. Between these maximum values a decrease of the percentage occurs, caused by the first acme of *B. bigelowi* (46%). These peak occurrences are followed by the most striking acme in the interval, that of *Biscutum* sp. which forms 90% of the assemblage in sample 548. Another count of this sample in which *Biscutum* sp. was discarded revealed that *T. multiplus* has a maximum frequency of 48%. *B. bigelowi* has a second acme (54%) in the upper part of the zone (sample 557). It is followed by the last acme in the zone, that of *T. petalosus*, forming 51% of the assemblages in sample 559.

Provincialism

The composition of the assemblages in the Upper Maastrichtian of our section gives evidence supporting the assumption of the existence of nannofloral polar and equatorial provinces in the final part of the Cretaceous. *Micula murus*, *Ceratolithoides kamptneri*, *Cribrocorona gallica*, *Actinozygus splendens* and *Thoracosphaera operculata*, regarded as tropical species (Worsley and Martini, 1970; Worsley 1974) are indeed present while *Markalius astroporus* and *Nephrolithus frequens*, regarded as polar are absent.

Worsley (1974) also indicated that several Maastrichtian species *(M. murus; N. frequens)* show diachronous first occurrence surfaces and that these surfaces are youngest in lower latitudes. A comparison of the ranges of several species in the Upper Maastrichtian and Lower Danian in our section with those in other localities fits in with this pattern (fig. 3).

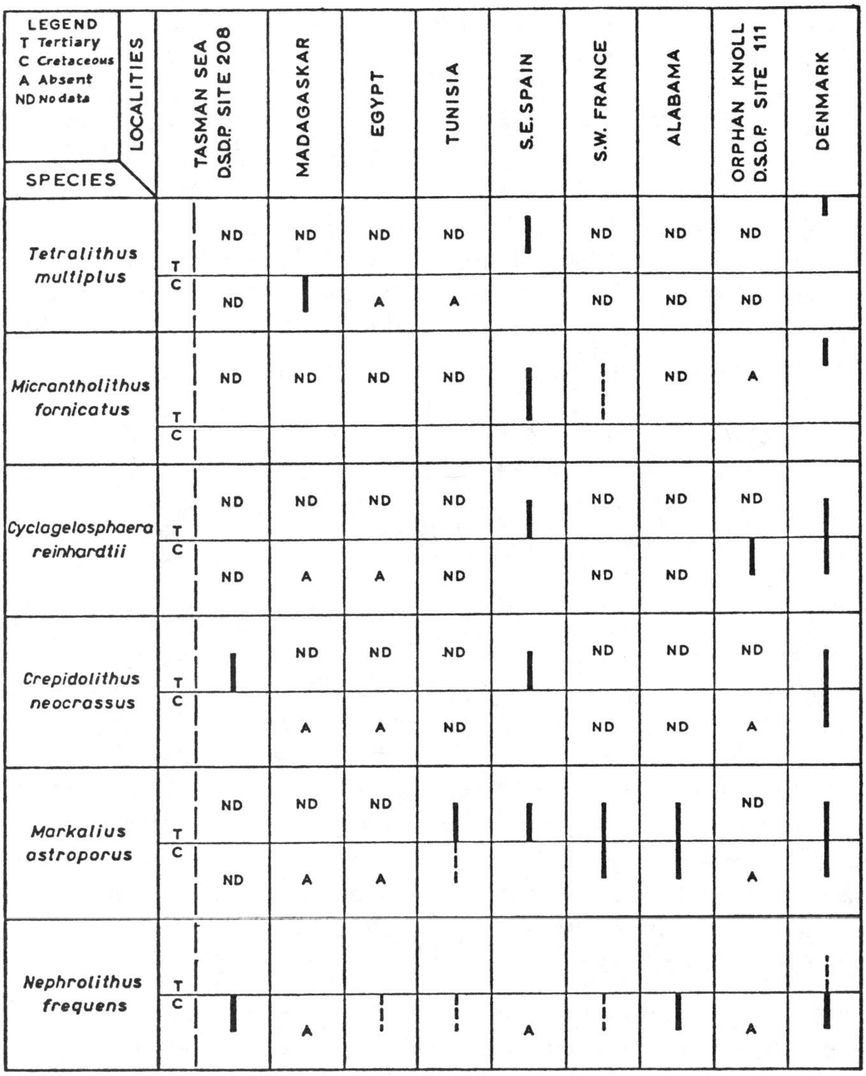

SPECIES	LEGEND T Tertiary C Cretaceous A Absent ND No data	TASMAN SEA D.S.D.P. SITE 208	MADAGASKAR	EGYPT	TUNISIA	S.E. SPAIN	S.W. FRANCE	ALABAMA	ORPHAN KNOLL D.S.D.P. SITE 111	DENMARK
Tetralithus multiplus	T	ND	ND	ND	ND	▮	ND	ND	ND	▮
	C	ND	▮	A	A		ND	ND	ND	
Micrantholithus fornicatus	T	ND	ND	ND	ND	▮	▮	ND	A	▮
	C									
Cyclagelosphaera reinhardtii	T	ND	ND	ND	ND	▮	ND	ND	ND	▮
	C	ND	A	A	ND		ND	ND	▮	▮
Crepidolithus neocrassus	T	▮	ND	ND	ND	▮	ND	ND	ND	▮
	C		A	A	ND		ND	ND	A	
Markalius astroporus	T	ND	ND	ND	▮	▮	▮	▮	ND	▮
	C	ND	A	A	▮				A	
Nephrolithus frequens	T	▮								▮
	C	▮	A	▮	▮	A	▮	▮	A	▮

Fig. 3. Comparison of distribution of species in the Cretaceous/Tertiary boundary interval at several localities. Based on data given by: Bramlette and Martini, 1964; Hay and Mohler, 1967; Edwards, 1972; Perch-Nielsen: 1968, 1969a, 1969b, 1972, 1973; Shafik and Stradner, 1971; Worsley, 1974.

Markalius astroporus was essentially restricted to higher latitudes in the Maastrichtian (Worsley, 1974) but it seems to have migrated to lower latitudes later on as it is present mainly in Lower Danian sediments in Tunisia and S.E. Spain.

Cyclagelosphaera reinhardtii and *Crepidolithus neocrassus* were probably also restricted to higher latitudes in the Maastrichtian. Thus far these species have only been reported from Maastrichtian sediments in Denmark (*C. reinhardtii* also from the Maastrichtian cored at Orphan Knoll) and they are absent in lower latitudes (Madagascar, Egypt, S.E. Spain). In our section these species appear in the Lower Danian and it is therefore assumed that they too, migrated to lower latitudes at the very end of the Cretaceous.

Tetralithus multiplus probably followed a different migration pattern The species is present in Maastrichtian sediments in Madagascar, absent in equivalents in Egypt, has its first occurrence in the Lower Danian *Biantholithus sparsus* Zone in S.E. Spain and in the Upper Danian *Cruciplacolithus tenuis* Zone in Denmark. Thus, it seems that the species was restricted to the southern hemisphere in the Maastrichtian, migrated northward to equatorial waters at the end of the Cretaceous and ultimately arrived in higher latitudes of the northern hemisphere in the Late Danian.

Migration to higher latitudes in the Danian is also shown by *Micrantholithus fornicatus*. In our section and in S.W. France this species has its first occurrence in the Lower Danian *(B. sparsus* Zone) but in Denmark it appears in the Late Danian *(C. tenuis* Zone).

Model to explain the terminal Cretaceous event

Calcareous nannoplankton species can tolerate fluctuations in temperature between certain limits and these limits differ from species to species (Paasche, 1968). As a consequence temperature is a primary factor governing the vertical and horizontal distribution of nannoplankton in the recent oceans (Mc Intyre and Bé, 1970; Honjo and Okada, 1974). Even short term fluctuations in temperature are reflected by the composition of living floras (Mc Intyre and Bé, 1967).

Hay (1960) and Worsley (1974) already suggested that temperature played an important role in the final Cretaceous event. This theory is adopted here and worked out in the following model:

1. Worldwide climatic deterioration towards the end of the Cretaceous caused the coming into existence of climatic belts in the former more uniform oceans. The existence of these belts is reflected by the latitudinally restricted occurrences of nannoplankton species in the Maastrichtian.

2. A further drop of temperature in the Late Maastrichtian caused an equatorially directed shift of nannoplankton provinces. This shift is indicated by the diachronous first occurrence surfaces of several species (Worsley and Martini, 1970; Worsley, 1974; this paper).

3. It is assumed here that the temperatures ultimately dropped to a level far below the lower temperature tolerance limit of most species, resulting in extinctions.

4. Climatic amelioration started in the Danian and opened large new ecospaces which became rapidly repopulated. Climatic amelioration is indicated by a poleward migration of several species (this paper).

Climatic deterioration will have caused not only diachronous entry levels but also diachronous exit levels. Thus, in theory, the "final Cretaceous extinction level" can be used to delineate the Cretaceous/Tertiary boundary only in tropical regions. In practice, most sections show hiatuses in the boundary interval, causing an apparent extinction level.

If temperatures dropped at a relatively slow rate one might expect to find a record in which the extinctions are a function of the individual lower temperature limits of species.

As a consequence of the model, those species which were restricted to higher latitudes in the Maastrichtian and of these only the ones with a low lower temperature limit had a good chance to survive the event. *M. astroporus, C. reinhardtii* and *C. neocrassus* survived and were indeed cold water species in the Maastrichtian.

That a large, empty ecospace existed in the Early Danian is indicated by the successive mass occurrences of several species in this interval. It is a well known principle in biology that repopulation of empty ecospaces follows a fixed pattern, characterized by a succession of "blooms" of immigrants and newly evolved species until finally an equilibrium situation is reached (a balanced ecosystem). The same pattern is observed in the Danian of our section.

Worsley (1974) presented a model in which final Cretaceous extinctions are explained by a raised C.C.D. level, caused by climatic deterioration. A raised C.C.D. level can indeed explain the lack of calcareous deposits and the occurrence of hardgrounds in the Upper Maastrichtian but it does not, in my opinion, explain the extinctions. Even if the C.C.D. level approached surface waters and affected the living nannoplankton directly, as suggested by Worsley, the only effect would have been a temporary loss of the ability to secrete skeletons. Investigations by Paasche (1967) showed that nannoplankton cells can survive a lowered pH as naked cells and start to secrete skeletons again when conditions turn back to normal.

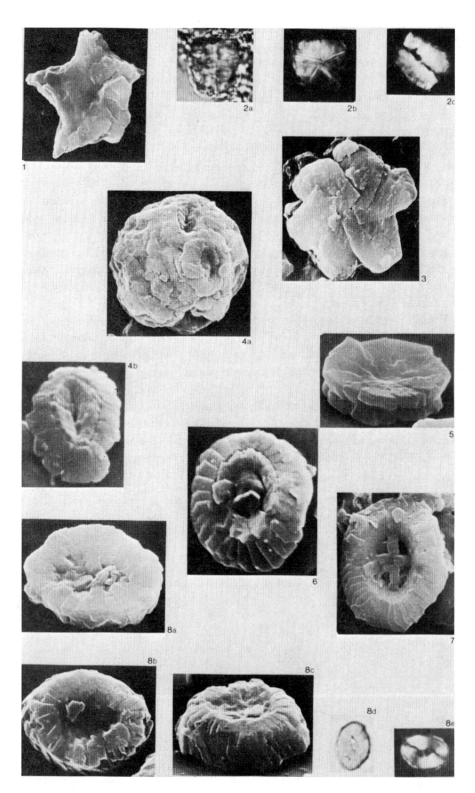

PLATE 1

Fig. 1. *Micrantholithus fornicatus* Martini, Sm 556, Utrecht slide CH 3707, location V +1, +2, 5000 ×.

Fig. 2. *Braarudosphaera* aff. *B. discula* Bramlette and Riedel, Sm 548, 2a, transmitted light, side view; 2b, cross-polarized light, distal view; 2c. cross-polarized light, side view; all 2500 ×.

Fig. 3. *Biantholithus sparsus* Bramlette and Martını, Sm 548, Utrecht slide CH 3708, location V −5, +2, distal view, 5000 ×.

Fig. 4. *Biscutum* sp., Sm 548, 4a, Coccosphere, Utrecht slide 3710, location V −6, −3, 12500 ×; 4b, Coccolith, Utrecht slide 3710, 16000 ×.

Fig. 5. *Biantholithus sparsus* Bramlette and Martini, Sm 562, Utrecht slide CH 3709, location V +4, +4, oblique proximal view, 5000 ×.

Fig. 6. *Cruciplacolithus* aff. *C. tenuis* (Stradner), Sm 562, Utrecht slide CH 3709, location V +3, −6, distal view of specimen with partly overgrown central area, 14000 ×.

Fig. 7. *Cruciplacolithus tenuis* (Stradner), Sm. 562, Utrecht slide CH 3709, distal view, 5000 ×.

Fig. 8. *Crepidolithus fossus* n.sp., 8a, paratype, Sm 562, Utrecht slide T 296, location V +4, −3, distal view, 6500 ×; 8b; holotype, Sm 562, Utrecht slide T 295, location V −4, +6, proximal view, 7000 ×; 8c, same specimen, tilted 50°, 7000 ×; 8d, paratype, Sm 562, Utrecht slide T 297, location V +5, +5, transmitted light, distal view, 2250 ×; 8e, same specimen, cross-polarized light.

PLATE 2

Fig. 1. *Markalius astroporus* (Stradner) Sm 560, Utrecht slide CH 3711, location V +1, −7, 5000 ×.

Fig. 2. *Cyclagelosphaera reinhardtii* (Perch-Nielsen), 2a, Sm 560, Utrecht slide CH 3711, location V +6, −2, distal view, 5000 ×; 2b, Sm 548a, cross-polarized light, proximal view, 2500 ×.

Fig. 3. *Crepidolithus neocrassus* Perch-Nielsen, 3a, Sm 562, Utrecht slide CH 3709, location V +4, −5, distal view, 6400 ×; 3b, Sm 562, Utrecht slide CH 3709, location V +1, +3, proximal view, 7000 ×; 3c, Sm 562, Utrecht slide CH 3709, location V −1, +1, oblique distal view, 7000 ×.

Fig. 4. *Prediscosphaera bukryi* Perch-Nielsen, Sm 543, Utrecht slide CH 3712, location V +1, +4, distal view, 14000 ×.

Fig. 5. *Toweius petalosus* Ellis and Lohman, 5a, Sm 562, Utrecht slide CH 3713,. location V +3, +6, interior of broken coccosphere, 5000 ×; 5b, Sm 561, coccosphere, cross-polarized light, 4500 ×; 5c, same specimen, transmitted light, 4500 ×.

Fig. 6. *Cylindralithus stradneri* (Perch-Nielsen), 6a, Sm 543, Utrecht slide CH 3714, location V −4, −6, side view, 5000 ×; 6b, Sm 543, Utrecht slide 3714, location V −6, +3, distal view, 5500 ×; 6c, Sm 543, cross-polarized light, distal view, 2500 ×.

Fig. 7. *Tetralithus multiplus* Perch-Nielsen, Sm 549a, cross-polarized light, 2000 ×.

Fig. 8. *Cylindralithus* cf. *C. crassus* Stover, Sm 543, Utrecht slide CH 3715, location V −1, −6, proximal view, 7000 ×.

The locations of the specimens are based on an X–Y coordinate system from the central V-marking of a 200 mesh copper grid (method described by Hansen et al., 1975).

[*Editor's Note:* Systematic descriptions from pages 269-277 have been omitted.]

REFERENCES

Abtahi, M. – Stratigraphische und mikropaläontologische Untersuchungen der Kreide/Alttertiär Grenze im Barranco del Gredero (Caravaca, Prov. Murcia SE Spanien). Technische Universität Berlin, Thesis (1975).

Bramlette, M. N. – Massive extinctions in biota at the end of Mesozoic time. Science, *148*, 1696–1699 (1965).

Bramlette, M. N. and E. Martini – The great change in calcareous nannoplankton fossils between the Maastrichtian and the Danian. Micropaleontology, *10*, 291–322 (1964).

Bramlette, M. N. and F. R. Sullivan – Coccolithophorids and related nannoplankton of the Early Tertiary in California. Micropaleontology *7*, 129–188 (1961).

Bukry, D. – Upper Cretaceous coccoliths from Texas and Europe. Univ. Kansas Paleont. Contrib., Protista, 2, *51*, 1–79 (1969).

Bukry, D. and M. N. Bramlette – Coccolith age determinations Leg 3, Deep Sea Drilling Project. *In* Maxwell, A. E. et al., Initial repts. Deep Sea Drilling Project, Washington (U.S. Government printing Office) *3*, 589–611 (1970).

Bukry, D., R. G. Douglas, et al. – Planktonic microfossil biostratigraphy of the Northwestern Pacific Ocean, Deep Sea Drilling Project, Leg 6. *In* Fischer, A. G., et al., Initial Repts. Deep Sea Drilling Project, Washington (U.S. Government Printing Office) *6*, 1253–1300 (1971).

Cepek, P. and W. W. Hay – Calcareous nannoplankton and biostratigraphic subdivision of the Upper Cretaceous. Trans. Gulf Coast Assoc. Geol. Soc., *19*, 323–336 (1969a).

Edwards, A. R. – Calcareous nannoplankton from the Uppermost Cretaceous and Lowermost Tertiary of the Mid-Waipara section, South Island, New Zealand. N.Z.J. Geol. Geophys., *9*, 481–490 (1966).

Edwards, A. R. – Key species of New Zealand calcareous nannofossils, N.Z.J. Geol. Geophys., *16*, 1, 68–89 (1973).

Edwards, A. R. – Calcareous nannofossils from the Southwest Pacific, Deep Sea Drilling Project, Leg 21. *In* Burns, R. E., Andrews, J. E., et al., Initial Repts. Deep Sea Drilling Project, Washington (U.S. Government Printing Office) *21*, 641–691 (1973).

Ellis, C. H. and W. H. Lohman – *Toweius petalosus* new species, a Paleocene calcareous nannofossil from Alabama. Tulane Studies Geol. Paleont., *10*, 2, 107–110 (1973).

Fütterer, D. – Kalkige Dinoflagellaten ("Calciodinelloideae") und die systematische stellung der Thoracosphaeroideae. N. Jb. Geol. Paläont. Abh. *151*, 2, 119–141 (1976).

Gartner, S. – Coccoliths and related calcareous nannofossils from Upper Cretaceous deposits of Texas and Arkansas. Univ. Kansas Paleont. Contribs., Protista, art. 1, 1–56 (1968).

Hansen, H. J., R. R. Schmidt and N. Mikkelsen – Convertible techniques for the study of the same nannoplankton specimen. Proc. Kon. Ned. Akad. Wetensch. ser. B, *78*, 3, 226–230 (1975).

Haq, B. U. and G. P. Lohmann – Early Cenozoic calcareous nannoplankton biogeography of the Atlantic Ocean. Marine Micropal. *1*, 2, 119–194 (1976).

Hay, W. W. – The Cretaceous-Tertiary boundary in the Tampico Embayment,
Mexico. 21st. Internat. Geol. Congr. Proc., part 5, 70–77 (1960).

Hay, W. W., H. P. Mohler, et al. – Calcareous nannoplankton zonation of the
Cenozoic of the Gulf Coast and Carribean – Antillean Area and Transoceanic
Correlation. Trans. Gulf Coast Assoc. Geol. Soc., *17*, 428–480 (1967).

Hay, W. W. and H. P. Mohler – Calcareous nannoplankton from Early Tertiary
rocks at Pont Labau, France, and Paleogene – Early Eocene correlations.
Journ. Paleontology, *41*, 1505–1541 (1967).

Hillebrandt, A. von – Bioestratigrafia del Paleogeno en el Sureste de España
(Provincia de Murcia y Alicante). Cuad. Geol. *5*, 135–153, (1974).

Honjo, S. and H. Okada – Community structure of Coccolithophores in the photic
layers of the mid-Pacific. Micropaleontology, *20*, 209–230, (1974).

Martini, E. – Nannoplankton aus dem Tertiär und der obersten Kreide von SW-
Frankreich. Senck. leth., *42*, 1, 1–41 (1961).

Martini, E. – Die Coccolithophoriden der Dan-Scholle von Katharinenhof (Fehmarn).
N. Jb. Geol. Paläont. Abh. *121*, 1, 47–54 (1964).

Martini, E. – Ein vollstandiges Gehäuse von *Goniolithus fluckigeri* Deflandre, N. Jb.
Geol. Paläont. Abh. *119*, 1, 19–21 (1964).

McIntyre, A. and A. W. H. Bé – Modern Coccolithophoridae of the Atlantic Ocean–
1. Placoliths and Cyrtoliths. Deep Sea Res. *14*, 561–597 (1967).

McIntyre, A., A. W. H. Bé and B. Roche – Modern Pacific Coccolithophoridae;
a paleontological thermometer. Trans. N.Y. Sci. series 2, *32*, 6, 720–731 (1970).

Moshkovitz, S. – First report on the occurrence of nannoplankton in Upper Cre-
taceous-Paleocene sediments of Israel. Jb. Geol. B. A. *110*, 135–168 (1967).

Noël, D. – Sur les Coccolithes du Jurassique Européen at d'Afrique du Nord.
Editions du C.N.R.S. (1965).

Noël, D. – Coccolithes Crétacés. La Craie Campanien du Bassin de Paris. Editions
du C.N.R.S., 1–129 (1970).

Paasche, E. – Biology and physiology of Coccolithophorids. Ann. Rev. Microbiol.
22, 71–86 (1968).

Perch-Nielsen, K. – Der Feinbau und die Klassifikation der Coccolithen aus dem
Maastrichtien von Dänemark. Biol. Skr. Kong. Danske Videnskab., *16*, 1,
1–96 (1968).

Perch-Nielsen, K. – Elektronmikroskopische Untersuchungen der Coccolithophoriden
der Dan-Scholle von Katharinenhof (Fehmarn). N. Jb. Geol. Paläont. Abh.
132, 317–332 (1969a).

Perch-Nielsen, K. – Die Coccolithen einiger Dänischer Maastrichtien – und Danien-
lokalitäten. Bull. Geol. Soc. Denmark, *19*, 1, 51–68 (1969b).

Perch-Nielsen, K. – Les nannofossiles calcaires de la limite Cretacé-Tertiaire. Mém.
B.R.G.M. *77*, 181–188 (1972).

Perch-Nielsen, K. – Remarks on Late Cretaceous to Pleistocene coccoliths from the
North Atlantic, Deep Sea Drilling Project, Leg 12. In Laughton, A. S., Berggren,
W. A., et al., Initial Repts. Deep Sea Drilling Project, Washington (U.S.
Government Printing Office) *12*, 1003–1069 (1972).

Perch-Nielsen, K. et M. C. Pomerol – Nannoplancton calcaire à la limite Crétacé-
Tertiaire dans le Basin de Majunga (Madagascar). C.R. Acad. Sc. Paris, *276*,
Série D, 2435–2438 (1973).

Perch-Nielsen, K. – Neue Coccolithen aus dem Maastrichtien von Dänemark,
Madagaskar und Ägypten. Bull. Geol. Soc. Denmark, *22*, 306–333 (1973).

Perch-Nielsen, K. – Neue Coccolithen aus dem Paleozän von Dänemark, der Bucht
von Biskaya und dem Eozän der Labrador See. Bull. Geol. Soc. Denmark,
21, 51–66 (1971).

Shafik, S. and H. Stradner – Nannofossils from the Eastern Desert, Egypt. Jb. Geol. B.A. Sonderband *17*, 69–104 (1971).

Smit, J. – Discovery of a planktonic foraminiferal association between the A. mayaroensis Zone and the G. eugubina Zone at the Cretaceous/Tertiary boundary in the Barranco del Gredero (Caravaca, S.E. Spain): a preliminary report. This volume, p. 280–301.

Stradner, H. – Vorkommen van Nannofossilien im Mesozoikum und Alttertiär. Erdoel Zeitschrift, *77*, 77–88 (1961).

Stradner, H. – In Gohrbandt, K., Zur Gliederung des Paläogen im Helvetikum nördlich Salzburg nach planktonischen Foraminiferen. Mitt. Geol. Ges. Wien, *56*, 72–81 (1963).

Worsley, T. R. and E. Martini – Late Maastrichtian nannoplankton provinces. Nature, *225*, 1242–1243 (1970).

Worsley, T. R. – Terminal Cretaceous events. Nature, *230*, 318–320 (1971).

Worsley, T. R. – The Cretaceous-Tertiary boundary event in the ocean. Soc. Econ. Pal. Min., Spec. Publ., *20*, 1, 94–125 (1974).

AUTHOR CITATION INDEX

Abtahi, M., 327
Adamiker, D., 9
Adams, C. G., 226
Adams, I., 8, 64, 79, 100
Adelseck, C. G., Jr., 56, 139, 153, 165
Aguirre, E., 227
Allen, M. B., 79
Alvarez, L. W., 304
Alvarez, W., 304
Anderson, T. F., 241
Andrews, J. E., 132
Arashkevich, Y. G., 139
Arrhenius, G., 132
Arthur, M. A., 226
Asaro, F., 304
Axelrod, D. I., 267, 311

Backman, J., 226
Bailey, H. P., 267
Bailey, N., 311
Barker, P., 178
Barnes, B., 63, 78, 89
Bartolini, C., 132
Bastin, E. E., 178
Bé, A. W. H., 57, 79, 89, 100, 101, 122, 132, 139, 154, 189, 192, 206, 227, 234, 238, 241, 267, 288, 328
Bender, M. L., 226
Benson, R. H., 226
Berge, G., 78, 234
Berger, W. H., 9, 57, 132, 139, 153, 154, 165, 166, 179, 226, 227, 238, 241, 304
Berggren, W. A., 121, 206, 226, 227, 300, 308
Bernard, F., 78, 153
Berner, E. K., 153
Berner, R. A., 153
Bernouilli, D., 227
Birkenes, E., 78
Black, M., 7, 56, 63, 78, 89, 238
Blackman, A., 192
Blackwelder, P. L., 57
Boersma, A., 226
Bolli, H. M., 178, 206, 308
Boney, A. D., 78

Borch, C. C. Von der, 178
Borg, S. G., 227
Bouché, P. M., 234
Boyce, R. E., 267
Boudreaux, J. E., 8, 101, 132, 267, 284, 286
Boyd, S. H., 154
Braarud, T., 7, 78, 89, 139, 234
Bradley, W. H., 132
Bradshaw, J. S., 153, 206
Bramlette, M. N., 7, 8, 89, 100, 128, 132, 139, 153, 206, 267, 284, 294, 300, 304, 308, 311, 327
Brinkmann, R., 300
Brinton, E., 121
Broecker, W. S., 153, 165, 192
Buchsbaum, R., 238
Buckley, H. A., 241
Bukry, D., 7, 100, 101, 153, 165, 178, 185, 226, 234, 267, 284, 294, 327
Burckle, L. H., 226, 227
Burrows, A., 78
Busson, G., 179
Bybell, L., 185

Carpenter, W. B., 7
Cepek, P., 327
Chadefaud, M., 78
Chen, C., 153
Chow, T. J., 79
Christensen, T., 78
Cifelli, R., 226, 300
Cita, M. B., 226, 227
Climap Project Members, 206, 226
Clocchiatta, M., 284
Cloud, P. E., Jr., 78
Cohen, C. L. D., 132
Cole, S. A., 241
Combaz, A., 179
Cooley, W. W., 226
Craig, H., 238, 241
Crenshaw, M. A., 78, 89
Crowley, T., 227
Culberson, C. H., 153
Curry, W. B., 227
Curtis, G. H., 227

Dacey, M. F., 153
Dalziel, I. W. D., 178
Dansgaard, W., 100
Deflandre, G., 7, 63, 78, 128
Degens, E. T., 234, 235
Desor, E., 308
Devereux, I., 206, 300
d'Halloy, J. J. O., 308
Dietrich, G., 121
Dinkelman, M. G., 178
DiNocera, S., 246
Dixon, P. S., 79
Dorf, E., 267
Dorman, F. H., 206, 300
Douglas, R. G., 153, 154, 166, 185, 206, 226, 227, 238, 267, 284, 327
Douglas, S. D., 78
Dow, D., 154
Dragesco, J., 78
Drake, R. E., 227
Dreyfus Rawson, M., 227
Droop, M. R., 78, 139
Dudley, W. C., 206, 241
Duplessy, J-C., 241

Edmond, J. M., 153
Edwards, A. R., 185, 238, 327
Edwards, G., 128
Ehrenberg, C. G., 7
Elliott, D. H., 178
Ellis, H. C., 294, 327
Emiliani, C., 128, 132, 189, 226
Eppley, R. W., 78, 153
Epstein, S., 238, 241
Erez, J., 165
Erickson, E., 227
Ericson, D. B., 100, 128, 189, 192
Esterly, C. O., 139, 153
Evensen, J. E., 304
Ewing, M., 100, 128, 189, 192, 311

Fabricius, F., 179
Fagerland, E., 78, 139
Fairbridge, R., 128
Feldmann, J., 78
Fert, C., 7, 63

Fischer, A. G., 179, 226, 267, 304
Fleming, R. H., 89
Flodén, T., 226
Foreman, J. B., 178
Fournier, R. O., 78, 153
Fox, S., 311
Franz, H. E., 247
Freeland, H., 154
Frost, B. W., 153
Früterer, D., 153
Fujita, Y., 122
Fütterer, D., 327

Gaarder, K. R., 64, 78, 89, 139, 234, 286
Gardner, J. V., 227
Garrison, R. E., 179, 227, 267
Gartner, S., 7, 185, 246, 267, 286, 294, 304, 327
Gavis, J., 154
Gealy, E. L., 267
Geehan, G. W., 56, 165
Geitzenauer, K. R., 57, 154, 267
Gentile, R., 246
Gignoux, M., 300
Gilmar, R. W., 153
Glass, B., 192
Glass, J. A., 192
Gombos, A. M., 178
Goodell, H. G., 192
Goodney, D. E., 206, 241
Gould, W., 153
Gran, H. H., 234
Green, J. C., 78
Guillard, R. R. L., 78, 89, 241
Gümbel, C. W., 246

Hall, J., 311
Halldal, P., 64, 78, 100
Hansen, H. J., 327
Haq, B. U., 7, 165, 166, 185, 206, 226, 286, 300, 304, 328
Harbison, G. R., 153
Harding, G. C. H., 139, 153
Hasle, G. R., 78, 121, 132, 139, 192, 234, 267
Hasumoto, H., 122
Hattin, D. E., 179
Hawley, J. E., 153
Hay, W. W., 7, 8, 101, 132, 166, 267, 284, 286, 300, 311, 327, 328
Hays, J. D., 192
Hedberg, H. D., 308
Heezen, B. C., 128, 132, 133, 166, 189, 192, 267
Heinrich, A. K., 139
Hekel, H., 294
Hellinger, S., 227
Hentschel, E., 78
Hill, M. E., III, 57
Hillebrandt, A. von, 328

Hjort, J., 139, 154
Hollister, C. D., 132, 226
Honjo, S., 57, 121, 122, 139, 153, 154, 165, 179, 206, 226, 227, 241, 328
Hopkins, D. M., 206
Hottman, W. E., 178
Hsü, K. J., 132, 166, 179, 226, 227
Hulbert, E. M., 78, 189, 234, 267
Hutner, S. H., 78
Hutson, W. H., 165
Huxley, T. H., 8

Imbrie, J., 206, 227
Ingle, J. C., Jr., 227
Ingle, S. E., 153
Isenberg, H. D., 78, 79

Jannasch, H. W., 153
Jantzen, R., 132
Jeffrey, S. W., 79
Jeletsky, J., 311
Jennings, D. H., 78
Johnsen, S. J., 100
Johnson, M. W., 121
Jöreskog, K. G., 227

Kagami, H., 178
Kamptner, E., 57, 64, 78, 79, 89, 128, 286, 294
Kaplan, I. R., 179
Keany, J., 7, 304
Keigwin, L. D., Jr., 226, 227
Keir, R. S., 153
Kellogg, T., 227
Kelts, K. R., 179
Kennett, J. P., 192, 206, 227, 238
Kidd, R. B., 226, 227
Killingley, J. S., 227, 241
Kipp, N. G., 206, 227
Kling, S. A., 153, 185, 267, 284
Klovan, J. E., 227
Knauss, J. A., 154
Kramer, D. D., 79
Krasheninnikov, V. A., 153, 185, 267, 284
Kroopnick, P. M., 206, 227, 238, 241
Ku, T. L., 192
Kuenen, P. H., 128
Kuenzler, E. J., 79
Kummel, G., 300
Kupferstein, D., 79

Lal, D., 153
Lalou, C., 241
Lancelot, Y., 154
Langway, C. C., Jr., 100
Lauer, G., 246
Lavine, L. S., 78, 79
Lear, P. E., 79
Lecal-Schlauder, J., 79

Leedale, G. F., 57, 79, 139
Lerman, A., 153
Letolle, R., 227, 241
Lewin, R. A., 79
Lidz, L., 206
Lipps, J. H., 286, 300, 311
Lisitzin, A. P., 267
Loeblich, A. R., Jr., 79, 165, 300, 308
Lohman, W. H., 294, 327
Lohmann, G. P., 165, 206, 226, 328
Lohmann, H., 64, 79, 122, 139, 153, 288
Lohmann, K. C., 227
Lohnes, D. R., 226
Lonardi, A., 178
Longoria, J. F., 178
Lowenstam, H. A., 128, 238, 311
Lyman, J., 89

McGill, S., 78
MacIlvaine, J., 235
McIntyre, A., 57, 79, 89, 100, 101, 122, 132, 139, 154, 166, 189, 192, 206, 227, 234, 238, 241, 267, 288, 328
McIntyre, R., 101, 132, 154
McKnight, B. K., 178
McLachlan, I. R., 179
McLachlan, J., 79
Magne, F., 79
Mahoney, M. E., 206, 241
Mandell, C., 79
Manton, I., 79, 139
Manton, J., 57
Margolis, S. V., 206, 227, 238, 241
Markali, J., 64, 78, 89, 100
Markali, T., 139
Marshall, H. G., 79, 122, 139, 153, 189
Marshall, S. M., 122, 139, 153, 154
Martini, E., 7, 8, 206, 238, 267, 288, 294, 300, 304, 308, 311, 327, 328, 329
Marumo, R., 122
Matthews, R. K., 227
Maxwell, A., 311
Mayeda, T., 241
Melguen, M., 165, 178, 179
Melieres, F., 227
Menzel, D. W., 79, 139, 154
Michel, H. V., 304
Mikkelsen, N., 327
Milliman, J. D., 132, 154, 165
Mjaaland, G., 79, 89, 234
Mohler, H. P., 7, 8, 101, 132, 267, 286, 311, 328
Møller, J., 100
Montadert, L., 227
Moore, T. C., 154

Moshkovitz, S., 328
Moss, M. L., 79
Müller, C., 227, 288, 294
Mullin, N. M., 139, 153, 154, 179
Munk, W. H., 154
Murray, J., 8, 13, 133, 139, 154

Nakai, T., 122
Nakamoto, N., 122
Natland, J., 178
Neugebauer, J., 165
Newell, N. D., 300, 308, 311
Noël, D., 8, 89, 165, 179, 246, 328
Nordli, E., 78, 89, 139
Norris, R. E., 79, 122, 294
Norton, J., 311

Okada, H., 57, 122, 139, 153, 154,
    166, 206, 226, 227, 241, 328
Olsson, R., 311
Opdyke, N. D., 192, 226, 238
Orr, A. P., 139, 153, 154
Osmond, J. K., 192
Outka, D. E., 139

Paasche, E., 79, 89, 154, 241, 300,
    328
Papp, A., 267
Parke, M., 8, 64, 79, 100
Parker, F. L., 133, 166, 206
Pasciak, W. J., 154
Peirson, J. F., 133
Perch-Nielsen, K., 8, 166, 206,
    238, 288, 304, 328
Percival, S. F., Jr., 235, 284, 294,
    304
Perras, J. P., 79
Peterson, M. N. A., 133, 154, 311
Phleger, F. B., 133, 206
Pierre, C., 227, 241
Pieterse, R. E., 179
Pimm, A. C., 267
Piper, D. J., 153
Pirini-Radrizzani, C., 246
Pitsyk, G. K., 234
Plafker, G., 178
Plass, G., 311
Pollard, L. D., 192
Pomerol, M. C., 328
Preikstas, R., 101, 238
Prell, W., 227
Premoli-Silva, I., 206, 226
Pringsheim, E. G., 79
Prins, B., 8, 246
Proto-Decima, F., 178
Provasoli, L., 78, 79
Pytkowicz, R. M., 153

Ramsay, A. T. S., 166
Ramsfjell, E., 64
Raup, D. M., 241
Rayment, R. A., 227

Rayns, D. G., 79
Remsen, C. C., 234
Renard, A. F., 8, 13
Rengers, E. H., 153
Revelle, R., 128
Reyment, R. A., 308
Richardson, P. L., 154
Riedel, W. R., 7, 128, 267
Riley, G. A., 154, 189
Riley, J. P., 79
Risatti, J. B., 166
Roche, M. B., 57, 122, 132, 154,
    227, 328
Rodman, J., 234
Romein, A. J. T., 247
Ross, D. A., 235
Ross, R., Jr., 311
Rossby, T., 154
Roth, P. H., 8, 56, 57, 101, 132,
    154, 165, 166, 179, 227, 238,
    247, 267, 286
Ruddiman, W. F., 132, 133, 166,
    227
Rust, C. W., 179
Ryan, W. B. F., 178, 226, 227
Ryther, J. H., 78, 79, 89, 154

Saito, T., 227, 235, 238
Sandberg, P. A., 8
Satake, K., 122
Savin, M. T., 235
Savin, S. M., 206, 226, 227, 238,
    241
Scandone, P., 246
Schiller, J., 8, 79, 139, 288
Schindewolf, O., 311
Schlanger, S. O., 154, 166
Schmidt, R. R., 8, 101, 132, 267,
    286, 328
Schmitz, W. J., 153
Schneidermann, N., 57, 154
Schnitker, D., 227
Schott, W., 133, 166, 206
Schrader, H.-J., 139, 154
Schussnig, B., 79
Schwartzbach, M., 311
Schwarz, E., 79
Sclater, J. G., 166, 227
Shackleton, N. J., 206, 226, 227,
    238, 241
Shaffer, B. L., 57
Shafik, S., 329
Sheridan, M. F., 227
Siesser, W. G., 57, 178
Simkiss, K., 79
Sissingh, W., 8
Sloan, P. R., 78, 79
Smayda, T. J., 79, 122, 139, 154,
    234, 235
Smirnova, L. T., 122
Smit, J., 329
Somayajulu, B. L. K., 192

Sommer, M. A., 226, 227
Spaeth, J. P., 79
Spicer, S. S., 78
Stanley, E., 311
Steemann Nielsen, E., 80
Stehli, F. G., 206, 227
Stosch, H. A. von, 80
Stradner, H., 8, 9, 267, 294, 329
Strickland, J. D. H., 79
Strong, A. E., 154
Stump, E., 227
Sullivan, F. R., 7, 267, 284, 327
Sweeney, R. E., 179
Swift, E., 5th, 80

Takahashi, M., 122
Takahashi, T., 154, 165
Talwani, M., 227
Tan, S. H., 267
Tappan, H., 57, 79, 165, 300, 304,
    308, 311
Tapscott, C., 227
Tarney, J., 178
Taylor, W. R., 80
Thiede, J., 166, 179
Thierstein, H. R., 9, 166, 185, 304
Thompson, P. R., 166
Thompson, R. W., 178, 179
Tjalsma, R. J., 178
Towe, K. M., 7
Tsujita, T., 122
Turekian, K., 154
Twenhofel, W. H., 128

Udintsev, G., 227
Uffen, R. J., 300
Ulleberg, K., 304
Ulrich, J., 79
Urey, H. C., 238, 308
Ushakova, M. G., 235

Valkanov, A., 80
Van Andel, T. H., 166, 179
Van Couvering, J. A., 226, 227
Van Donk, J., 238, 241
Venrick, E., 153
Vergnaud-Grazzini, C., 227, 241
Vincent, E., 226, 227, 241
Vinot, A. C., 241
Vinot-Bertouille, A-C., 241
von Rad, U., 132

Wada, K., 89
Wallich, G. C., 9, 122, 288
Watabe, N., 57, 80, 89, 100, 133,
    235, 241
Watkins, N. D., 192
Watson, S. W., 234
Weber, J. N., 241
Weiss, R. E., 57
Weissfellner, H., 78, 79
Wiebe, P. H., 154

Wilbur, K. M., 57, 80, 89, 100, 133, 235, 241
Wilcoxon, J. A., 7, 267, 284
Williams, D. C., 139
Wilson, T. R., 79
Wind, F. H., 166
Winget, C., 154
Wirsen, C. O., 153
Wise, S. W., Jr.,166, 178, 179, 206, 247

Wiseman, D. H., 241
Wolfe, J. A., 206
Wollin, G., 100, 189, 192
Wong, C. S., 227
Woodhead, P. M. J., 241
Woodruff, F., 226
Worsley, T. R., 166, 226, 300, 311, 329
Worthington, L. V., 154
Wray, J. L., 294

Wright, R. C., 226, 227
Wunsch, C., 153
Wyville Thompson, C., 7

Yesin, N. V., 235

Zenkevich, L. A., 235

SUBJECT INDEX

Atlantic Ocean
    migration of assemblages, 204–206, 220–221
    Miocene paleobiogeography, 213–220
    Paleogene paleobiogeography, 195, 196, 199–203
Atlantic paleoceanography
    Miocene, 207, 208, 222–225
    Pleistocene, 193, 194

Bathybius haekelii, 2, 3, 13, 37, 41, 42, 44
    designation, 36
Black shales, 168–178
    in Albian, 175–178
    in Aptian, 169–175
Braarudosphaera chalk, 161
    bloom in Danian, 312, 318, 319
Braarudosphaera rosa, bloom in Oligocene, 167
Bulldog, H.M.S., 24, 29

Calcite compensation depth (CCD), 129, 131, 136,
    161, 164, 302, 310, 311, 323
Calcite saturation depth, 140, 143, 144, 151
Carbon isotopes, in nannofossils, 236–238, 239
Catinaster mexicanus description, 265, 266
Catinaster ? umbrellus, description, 265, 266
Ceratolithus spp.
    crystallography, 290, 291
    evolutionary trends, 289–293
    morphology, 289, 290
    phylogeny, 291–293
Challenger, H.M.S., 3, 12, 42
Climates and nannoplankton diversity, 295–299
Coccolithophorids. *See also* Coccoliths;
    Coccospheres
    assemblages and relation to currents, 104–108
    biogeography in Pacific, 91–99, 111–118
    biology and physiology, 65–77
    diversity in Cenozoic, 297, 298
    feeding experiments, 134, 136, 147, 149
    growth and nutrition, 69–77, 83, 84
    life cycle, 67–69
    mineralogy, 58–63
    numbers in water column, 115
    photosynthesis and mineralization, 75–77
    significance in carbonate deposition, 124–128
    standing crop in Pacific, 108–111
    vertical distribution, 118, 119

Coccoliths. *See also* Coccolithophorids;
    Coccospheres
    calcification, 84–89
    designation, 14
    in Devonian, 244, 248–251
    diagenesis, 56
    discovery, 1
    dissolution indicators, 55, 129, 130
    diversity 296–298
    fine structure, 58–63
    formation of, 71–77
    lysocline, 55, 129–132
    mineralization of, 54, 71–73, 84–89
    mineralogy of, 59–63
    organic origin, 22–28
    in Pennsylvanian, 244
    in Permian, 244
    sedimentation of, 136, 139, 150–152
    in Silurian, 244, 248–251
    sinking of, 142, 143
    temperature effect on, 85, 86
    in Triassic, 245
Coccolithus huxleyi. See Emiliania huxleyi
Coccolithus pelagicus
    assemblage in Miocene Atlantic, 212, 214, 215
    in New Zealand climatology, 183
    in Paleogene Atlantic, 195, 199, 201, 203
    relationship to temperature, 99, 100
Coccosphaera, 47
    C. carterii, 47–49
    C. pelagica, 47–49
Coccospheres, 2, 15–17, 27, 34, 35, 39–46. *See
    also* Coccolithophorids; Coccoliths
    designation, 15
    sinking of, 56
Cretaceous
    diversity of coccolithophorids in, 296, 297
    mid Cretaceous assemblages, 228
Cretaceous-Tertiary boundary
    assemblages at, 302, 305–308, 312, 315, 318–322
    extraterrestrial causes of extinctions at, 303
    massive extinctions at, 305–308
    terrestrial causes of extinctions at, 303, 309–311,
    322, 323
Cyclococcolithus leptoporus
    biogeography in Pacific, 97, 98–105, 106

relationship to temperature, 99, 100
vertical distribution, 115–119
Cyclops, H.M.S., 14, 29

Danian
assemblages, 199, 307, 312, 315–322
blooms of *Braarudosphaera,* 312, 318, 319
Deep Sea Drilling Project (DSDP), 5, 56, 132, 167, 168, 169, 178, 197, 198, 206, 223, 236, 267, 309, 316
Devonian, occurrence of coccoliths in, 244, 248–251
Diagenesis, 164, 167
Discoasters
descriptions of
D. araneus, 260, 263, 264
D. berggreni, 260, 263, 264
D. braarudi, 260, 263, 264
D. calculosus, 261, 263, 264
D. signus, 263, 265, 266
D. variabilis decorus, 263, 265, 266
evolutionary rates of, 297–299
evolutionary trends of, 258, 259
Discosphaera tubifera
biogeography in Pacific, 94, 105, 106
relationship to temperature, 99, 100
vertical distribution, 115–119
Dissolution, 129–132, 140, 151, 152, 155–164, 167, 168
Diversity, 246
Cenozoic, 296
coccolithophores, 296, 297
Jurassic-Cretaceous, 296
relationship to climate and sea level, 295, 299

Emiliania huxleyi (= *Coccolithus huxleyi*)
biogeography in Pacific, 91, 105, 106
calcification in, 84–86
in culture, 81, 82
feeding experiment of, 134–136
relationship to temperature, 99, 100
vertical distribution, 115–119
Evolutionary rates, 297–299
Cenozoic coccolithophores, 297
discoasters, 245, 298
Evolutionary trends, 245
in arkhangelskiellids, 245
in ceratoliths, 289–293
in chiasmoliths, 245
in discoasters, 245, 298
in helicosphaerids, 268–271
in sphenoliths, 245

Fecal pellets, 134–139
role in coccolith sedimentation, 149–152
in sediment traps, 145–147, 148
Feeding experiments, 134, 136, 147, 149
Florisphaera profunda
description, 122, 123
vertical distribution, 115–119

Gephyrocapsa spp.
biogeography in Pacific, 94, 96, 105, 106
relationship to temperature, 99, 100
vertical distribution, 115–119
Glomar Challenger, R.V., 5, 90, 168

Helicopontosphaera. See Helicosphaera
Helicosphaera (=*Helicopontosphaera*)
description of spp., 272–287
evolutionary trends, 268–271
ranges of spp., 269
Heterococcoliths, 53
Holococcoliths, 53

Jurassic, diversity of coccolithophorids in, 295, 296

Lepispheres, 167
Lysocline, 55
coccolith lysocline, 129–132

Maastrichtian assemblages, 307, 312, 315–323
Massive extinctions
causes of extinctions, 303, 310, 311
at Cretaceous-Tertiary boundary, 305–308
Miocene assemblages, 207, 213–220

Nannofacies, 167–178
Albian black shale facies, 175–178
Aptian black shale facies, 169–175
Nannofloral migrations
in Miocene Atlantic, 220, 221
in Paleogene Atlantic, 195, 196, 204, 205
Nannofossil diagenesis, 167
Nannofossil dissolution
index of dissolution, 161–174
morphological effects of, 164
selective dissolution in late Cretaceous and early Tertiary spp., 155
susceptibility of taxa, 156, 161
Nannofossils
as paleoclimatic indicators, 186, 189, 190–194, 197–206, 220–225
as paleosalinity indicators, 229–234
stable isotopes in, 185, 236–241

Oligocene
biogeography in Atlantic, 195, 202, 203
blooms of *Braarudosphaera* in, 167
Oxygen isotopes, 184, 206, 222–224
in nannoplankton, 236–241

Pacific Ocean, biogeography, 91–99, 102–121
Paleobiogeography, 183–185
Cretaceous Atlantic, 184, 228
Tertiary Atlantic, 184, 195, 196, 207–220
Paleoclimatic indicators, 183, 184, 186–194, 196, 197, 204, 205, 220–225
Paleozoic, occurrence of coccoliths in, 244, 248–257

Pennsylvanian, occurrence of coccoliths in, 244
Permian, occurrence of coccoliths in, 244
Pleistocene, coccoliths as glaciation indicators,
    186–189, 190–192
Predation and preservation, 136, 139
Protococcus, 23, 29, 41, 44
Provincialism
    among spp. at Cretaceous-Tertiary boundary,
        320–322
    among Tertiary spp., 184

Quaternary, late paleoclimate, 190–192

Rhabdosphaera
    R. clavigera
        biogeography in Pacific, 105, 106
        vertical distribution, 115–119
    R. stylifera
        biogeography in Pacific, 97, 105, 106
        relationship to temperature, 99, 100
        vertical distribution, 115–119

Silurian, occurrence of coccoliths in, 244, 248–251
Standing crop, 108–111
Subantarctic Pacific, late Quaternary climate,
    190–192

Thoracosphaera flabellata, vertical distribution in
    Pacific, 115–119
Transgression and nannoplankton diversity, 295–299
Triassic, occurrence of coccoliths in, 245

Umbellosphaera
    U. irregularis
        biogeography in Pacific, 92, 94, 105, 106
        relationship to temperature, 99, 100
        vertical distribution, 115–119
    U. tenuis
        biogeography in Pacific, 93, 94, 105, 106
        relationship to temperature, 99, 100
        vertical distribution, 115–119
Urschleim, 2

Vital effect on stable isotopes, 240, 241

About the Editor

BILAL U. HAQ is a senior research specialist with the Exxon Production Research Company in Houston, Texas. From 1970 through 1982, he was on the scientific staff of the Woods Hole Oceanographic Institution. During 1980–81, he also served as Program Associate with the Division of Ocean Drilling Programs of the National Science Foundation in Washington, D.C., where his duties included the programmatic oversight of the Deep Sea Drilling Project. Dr. Haq has taken part in Legs 8 (Equatorial Pacific) and 35 (Antarctic) of the DSDP as micropaleontologist, and in Leg 63 (California Borderlands and East Pacific Rise) as co-chief scientist. His research interests include calcareous nannoplankton paleobiogeography, biostratigraphy and biochronology, paleoceanography, and marine geology.

Dr. Haq completed his undergraduate work at the University of Panjab in Lahore, Pakistan, where he also received his Masters in 1963. He began his nannofossil studies at the Geological Survey of Austria in Vienna in 1965, and continued his research at the University of Stockholm, Sweden, where he received the Ph.D. in 1967 and later the D.Sc. in 1972.

Dr. Haq is a fellow of the Geological Society of America and a member of the American Geophysical Union, the Geological Society of Sweden, the American Association of Petroleum Geologists, and the American Association for Advancement of Science. He has also served as a member of numerous working groups for the International Union of Geological Societies and has been on the National Research Council panel on pre-Pleistocene Climates. He is also an adjunct Docent with the University of Stockholm, Sweden.

Dr. Haq is the chief editor of the journal *Marine Micropaleontology*, and serves on the editorial board of *Micropaleontology*, published by the American Museum of Natural History. He has published extensively in the fields of marine micropaleontology and marine geology and has also coauthored the textbook *Introduction to Marine Micropaleontology*.